ENGINEERING QUANTUM MECHANICS

ENGINEERING QUANTUM MECHANICS

Doyeol Ahn
Seoung-Hwan Park

IEEE
IEEE PRESS

WILEY
A JOHN WILEY & SONS, INC., PUBLICATION

Published by John Wiley & Sons, Inc., Hoboken, New Jersey.
Published simultaneously in Canada.

Library of Congress Cataloging-in-Publication Data:

Ahn, Doyeol.
Engineering Quantum Mechanics/Doyeol Ahn, Seoung-Hwan Park.
p. cm.
Includes bibliographical references and index.
ISBN 978-0-470-10763-8
1. Quantum theory. 2. Stochastic processes. 3. Engineering mathematics.
4. Semiconductors–Electric properties–Mathematical models. I. Park, Seoung-Hwan.
II. Title.
QC174.12.A393 2011
620.001'53012–dc22

2010044304

oBook ISBN: 978-1-118-01782-1
ePDF ISBN: 978-1-118-01780-7
ePub ISBN: 978-1-118-01781-4

Printed in Singapore.

10 9 8 7 6 5 4 3 2 1

CONTENTS

Preface

Quantum mechanics is becoming more important in applied science and engineering, especially with the recent developments in quantum computing, as well as the rapid progress in optoelectronic devices. This textbook is intended for graduate students and advanced undergraduate students in electrical engineering, physics, and materials science and engineering. It also provides the necessary theoretical background for researchers in optoelectronics or semiconductor devices. In the task of providing advanced instruction for both students and researchers, quantum mechanics presents special difficulties because of its hierarchical structures. The more abstract formalisms and techniques are quite meaningless until one has mastered the earlier stages in classical physics, which most engineering students are lacking.

Quantum mechanics has become an essential tool for modern engineering. This book covers topics such as semiconductors and laser physics, which are traditionally quantum mechanical, as well as relatively new topics in the field, such as quantum computation and quantum information. These fields have seen an explosive growth during the past 10 years, as quantum computing or quantum information processing can have a significant impact on today's electronics and computations. The essence of quantum computing is the direct usage of the superposition and entanglement of quantum mechanics. The most challenging research topics include the generation and manipulation of quantum entangled systems, developing the fundamental theory of entanglement, decoherence control, and the demonstration of the scalability of quantum information processing.

In laser physics, there has been a growing interest in the model of semiconductor lasers with non-Markovian relaxation partially because of the dissatisfaction with the conventional model for optical gain in predicting the correct gain spectrum and the thermodynamic relations. This is mainly due to the poor convergence properties of the lineshape function, that is, the Lorentzian lineshape, used in the conventional model. In this book, a non-Markovian model for the optical gain of semiconductors is developed, taking into account the rigorous electronic band structure, many-body effects, and the non-Markovian relaxation using the quantum statistical reduced-density operator formalism for an arbitrary driven system coupled to a stochastic reservoir. Example programs based on Fortran 77 will also be provided for band structures of zinc blende quantum wells.

Many-body effects are taken into account within the time-dependent Hartree–Fock. Various semiconductor lasers including strained-layer quantum well lasers and wurtzite GaN blue-green quantum well lasers are discussed.

We thank Professor Shun-Lien Chuang, Doyeol Ahn's Ph.D. thesis adviser, for extensive enlightening and encouragement over many years. We are also grateful to many colleagues and friends, especially Frank Stern, B. D. Choe, Han Jo Lim, H. S. Min, M. S. Kim, Robert Mann, Tim Ralph, K. S. Seo, Y. S. Cho, and Chancellor Sam Bum Lee. The support of our research by the Korean Ministry of Education, Science and Technology is greatly appreciated. This book would not have been completed without the patience and continued encouragement of our editors at Wiley and above all the encouragement and understanding of Taeyeon Yim and Young-Mee An. Thanks for putting up with us.

Doyeol Ahn
Seoung-Hwan Park

PART I
Fundamentals

1 Basic Quantum Mechanics

1.1 MEASUREMENTS AND PROBABILITY

In the beginning of 20th century, it was discovered that the behavior of very small particles, such as electrons, the nuclei of atoms, and molecules, cannot be described by classical mechanics, which had been quite successful in explaining the macroscopic world until then. Nonetheless, it was soon discovered that the description of these phenomena on the atomic scale is possible by the set of laws described by quantum mechanics. Both classical mechanics and quantum mechanics are based on the description of measurements of observable quantities called dynamical variables, such as position, momentum, and energy. Consider an experiment in which we can make three measurements successive in time. Let's denote the first of observable quantities A, the second B, and the third C. We also denote a, b, and c as one of a number of possible results that could come from the measurement of A, B, and C, respectively. Let $P(b\,|\,a)$ be the conditional probability that if the measurement of A results in a, then the measurement of B will result in b. From the elementary probability theory, the conditional probability $P(b\,|\,a)$ can be written as follows:

$$P(b\,|\,a) = \frac{P(a,b)}{P(a)}, \tag{1.1}$$

where $P(a,b)$ is the joint probability that measurements of both A and B will give a and b, simultaneously, and $P(a)$ is the probability that the measurement of A will give the outcome a. For three successive measurements A, B, and C, the conditional probability $P(cb\,|\,a)$ that if the measurement of A results in a, then the measurement of B will result in b, then the measurement of C will result in c is given by:

$$P(cb\,|\,a) = P(c\,|\,b)P(b\,|\,a). \tag{1.2}$$

Engineering Quantum Mechanics, First Edition. Doyeol Ahn, Seoung-Hwan Park.

Moreover, if we sum Equation (1.2) over all the mutually exclusive alternatives for b, we obtain the conditional probability $P(c \mid a)$:

$$P(c \mid a) = \sum_b P(c \mid b) P(b \mid a). \tag{1.3}$$

In classical mechanics, the above relation described by Equation (1.3) is always true. However, it was found that the above relation sometimes fails on the atomic scale, and one needs to modify Equations (1.1) to (1.3) by introducing new complex quantities φ_{ba}, φ_{cb}, and φ_{ca}, called probability amplitudes, which are related to probabilities by [1,2]

$$P(b \mid a) = |\varphi_{ba}|^2, \tag{1.4}$$

and

$$P(c \mid a) = \left| \sum_b \varphi_{cb} \varphi_{ba} \right|^2. \tag{1.5}$$

Equations (1.4) and (1.5) describe the probability of measurement outcome in quantum mechanics. From the mathematical point of view, the probability amplitude is found to be the inner product of vectors in a special kind of vector space called the Hilbert space:

$$\varphi_{ba} = \langle b || a \rangle, \tag{1.6}$$

where $|a\rangle$ is the column vector, called the "ket vector," corresponding to the observable a, and $\langle b|$ is the row vector, called the "bra vector," corresponding to the observable b.

1.2 DIRAC FORMULATION

In quantum mechanics, a physical state corresponding to the observable quantity a is represented by the ket vector, $|a\rangle$, in a complex vector space H with dimension N. For example, when $N = 2$, the ket vector $|a\rangle$ is a column vector given by [1,3]

$$|a\rangle = \begin{pmatrix} a_1 \\ a_2 \end{pmatrix}, \tag{1.7}$$

where a_1 and a_2 are complex numbers. We can also consider the case where the dimension of the vector space is infinite. In this case, the ket vector $|a\rangle$ is represented by a column vector given by

$$|a\rangle = \begin{pmatrix} a_1 \\ a_2 \\ \cdot \\ \cdot \\ \cdot \\ a_\infty \end{pmatrix}, \tag{1.8}$$

with complex quantities $a_1, a_2, a_3, \ldots, a_\infty$. From now on, we will denote ket vectors simply as vectors.

If $|a\rangle$ and $|b\rangle$ are vectors in H, and α and β are complex numbers, then the linear superposition of these two vectors $\alpha|a\rangle + \beta|b\rangle$ is also a vector in H. For each vector $|a\rangle$ in H, we can relate a row vector $\langle a|$ called bra vector, which is given by

$$\langle a| = \begin{pmatrix} a_1^* & a_2^* \end{pmatrix}, \text{ for } N = 2, \tag{1.9}$$

and

$$\langle a| = \begin{pmatrix} a_1^* & a_2^* & a_3^* \ldots . a_\infty^* \end{pmatrix}, \text{ for } N = \infty, \tag{1.10}$$

where * is the complex conjugate. By comparing Equations (1.7) with (1.10), one can see that the ket vector $|a\rangle$ and bra vector $\langle b|$ are related by

$$\langle a| = (|a\rangle)^\dagger, \tag{1.11}$$

where $\dagger$ is the adjoint operation, which is the transposition, followed by the complex conjugation.

The inner product between two vectors $|a\rangle$ and $|b\rangle$ is a complex number $\langle b|a\rangle$, which is given by

$$\langle b|a\rangle = b_1^* a_1 + b_2^* a_2 + \ldots + b_\infty^* a_\infty = \langle a|b\rangle^*. \tag{1.12}$$

The vector space where one can define the inner product relation of Equation (1.12) is called the Hilbert space. The length of the vector $|a\rangle$ is defined by

$$\||a\rangle\| = \sqrt{\langle a|a\rangle}, \tag{1.13}$$

and when the vector has the unit length, we call it a normal vector.

The vectors in the Hilbert space H are mapped to different vectors in the Hilbert space H' by the linear transformation $\hat{A}$, which is often called the "linear operator" in quantum mechanics. Mathematically, the act of the linear operator $\hat{A}$ on the vectors of the Hilbert space H is expressed as $\hat{A}: H \to H'$. In most cases, the initial Hilbert space H and the final Hilbert space H' are the same, and unless otherwise specified explicitly, we will assume that is the case. Any two vectors $|a\rangle$ and $|b\rangle$ in H are transformed by $\hat{A}$ as

$$\hat{A}(\alpha|a\rangle + \beta|b\rangle) = \alpha\hat{A}|a\rangle + \beta\hat{A}|b\rangle, \tag{1.14}$$

and

$$\left(\hat{A}\alpha|a\rangle\right)^{\dagger} = \alpha^{*}\langle a|\hat{A}^{\dagger}, \tag{1.15}$$

where α and β are complex constants. If we set $|c\rangle = \hat{A}|a\rangle$ from Equation (1.11), we obtain $\langle c| = (|c\rangle)^{\dagger} = \langle a|\hat{A}^{\dagger}$, and then, using Equation (1.12), we get

$$\langle b|\hat{A}|a\rangle^{*} = \langle b|c\rangle^{*} = \langle c|b\rangle = \langle a|\hat{A}^{\dagger}|b\rangle. \tag{1.16}$$

For a given operator $\hat{A}$, there exists particular set of vectors $\{|a_1\rangle, |a_2\rangle, \ldots, |a_n\rangle, \ldots\}$ called eigenstates, which satisfy

$$\hat{A}|a_1\rangle = a_1|a_1\rangle, \hat{A}|a_2\rangle = a_2|a_2\rangle, \ldots, \hat{A}|a_n\rangle = a_n|a_n\rangle, \ldots \tag{1.17}$$

Here, the numbers $a_1, a_2, \ldots, a_n, \ldots$ are called eigenvalues. Of the many kinds of operators, one particular type of operator, called the Hermitian operator, plays an important role in quantum mechanics. If $\hat{A}$ is a Hermitian operator, then it satisfies the following properties:

(I) $\hat{A}^{\dagger} = \hat{A}$.
(II) Eigenvalues $a_1, a_2, \ldots, a_n, \ldots$ are real.
(III) Eigenvectors corresponding to different eigenvalues are orthogonal.
(IV) $\sum_n |a_n\rangle\langle a_n| = I$, where I is an identity matrix.

The proof is as follows: From $\hat{A}|a_i\rangle = a_i|a_i\rangle$ and $\hat{A}|a_j\rangle = a_j|a_j\rangle$, we have

$$\langle a_j|\hat{A} = \langle a_j|\hat{A}^\dagger = \left(\hat{A}|a_j\rangle\right)^\dagger = a_j^*\langle a_j|. \tag{1.18}$$

Then, we get

$$\langle a_j|\hat{A}|a_i\rangle = a_i\langle a_j|a_i\rangle, \tag{1.19}$$

and

$$\langle a_j|\hat{A}|a_i\rangle = a_j^*\langle a_j|a_i\rangle. \tag{1.20}$$

Subtracting Equation (1.20) from Equation (1.19), we obtain

$$(a_i - a_j^*)\langle a_j|a_i\rangle = 0. \tag{1.21}$$

When $i \neq j$, we get $\langle a_j|a_i\rangle = 0$, which proves the property (III). On the other hand, if $i = j$, then we have $a_i = a_i^*$, thus giving the property (II). The property (III) dictates that the set of eigenvectors $\{|a_i\rangle\}$ of a Hermitian operator $\hat{A}$ form an orthogonal basis if the state $|a_i\rangle$ is properly normalized—that is, $\langle a_i|a_i\rangle = 1$ and $\langle a_i|a_j\rangle = 0$ for $i \neq j$. Let the Hilbert space spanned by these basis vectors be denoted as H. Then any vector $|\psi\rangle$ that belongs to H can be expressed as

$$|\psi\rangle = \sum_m C_m|a_m\rangle, \tag{1.22}$$

where C_m is an expansion coefficient that can be calculated by taking an inner product between the state vectors $|\psi\rangle$ and $|a_m\rangle$:

$$C_m = \langle a_m|\psi\rangle. \tag{1.23}$$

By substituting Equation (1.23) into Equation (1.22), we obtain

$$\begin{aligned} |\psi\rangle &= \sum_m C_m|a_m\rangle \\ &= \sum_m |a_m\rangle\langle a_m|\psi\rangle \\ &= \left(\sum_m |a_m\rangle\langle a_m|\right)|\psi\rangle \\ &= I|\psi\rangle, \end{aligned} \tag{1.24}$$

which proves the property (IV). It would be handy to memorize that given a chain of vectors or operators, we can insert the identity operator defined by (IV) in any place at our convenience. For example, the Hermitian operator $\hat{A}$ can be written as

$$\begin{aligned}\hat{A} &= I\hat{A}I \\ &= \left(\sum_n |a_n\rangle\langle a_n|\right)\hat{A}\left(\sum_m |a_m\rangle\langle a_m|\right) \\ &= \sum_{n,m} |a_n\rangle\langle a_n|\hat{A}|a_m\rangle\langle a_m| \\ &= \sum_{n,m} |a_n\rangle(a_m\langle a_n|a_m\rangle)\langle a_m| \\ &= \sum_n a_n|a_n\rangle\langle a_n|,\end{aligned} \tag{1.25}$$

which is also called the spectral representation of an operator $\hat{A}$.

Let $\hat{A}$ and $\hat{B}$ be linear operators, and we define $\hat{D}$ as $\hat{D} = \hat{A}\hat{B}$. Moreover, we define $|b\rangle = \hat{B}|a\rangle$ and $|d\rangle = \hat{D}|a\rangle = \hat{A}|b\rangle$. Then, from Equations (1.11) and (1.15), we obtain

$$\begin{aligned}\langle d| &= \langle a|\hat{D}^\dagger \\ &= \langle b|\hat{A}^\dagger \\ &= \langle a|\hat{B}^\dagger\hat{A}^\dagger,\end{aligned} \tag{1.26}$$

or

$$\left(\hat{A}\hat{B}\right)^\dagger = \hat{B}^\dagger\hat{A}^\dagger. \tag{1.27}$$

1.3 BRIEF DETOUR TO CLASSICAL MECHANICS

Almost seven decades ago, Dirac made a connection between classical mechanics and quantum mechanics by assuming that the linear operators correspond to the dynamical variables at that time. By "dynamical variables," we mean quantities such as the coordinates and the components of velocity, momentum, and angular momentum of particles, and functions of these quantities, the variables in terms of which classical mechanics is built. Even now, 70 years later, his postulates of quantum mechanics are still valid and perhaps only plausible approaches. Dirac's postulates require that those dynamical variables shall also occur in quantum mechanics, but with the difference that they are now subject to an algebra in which the commutative axiom of multiplication does not hold. Nonetheless, the dynamical variables of quantum mechanics still have many properties in common with their classical counterparts, and it will still be possible to build up a theory of them closely analogous to the classical theory and form a generalization of it. In this spirit, the transition from classical mechanics to quantum mechanics can be made most conveniently and easily using the Hamiltonian formulation of classical mechanics [4,5].

Classical mechanics is based on the assumption that any physically interesting variable, that is, dynamical variable, can be measured with arbitrary precision and without mutual interference from any other such measurement. On the other hand, quantum mechanics is based on the realization that the measuring process may affect the physical system. The measurement of one variable affects other variables in such a way that it prevents us from knowing what their values might have been. The mathematical formulation of the law of physics that takes this basic idea into account is very different from the mathematical formulation of classical mechanics.

Hamilton's least action principle, which is equivalent to Newton's law, is formulated as follows.

> The laws of physics are such that the time integral over a certain function $L(q_i,\dot{q}_i,t)$, called Lagrangian of the physical system, assumes a minimum.

For mechanical systems, the variables q_i on which the Lagrangian depends on are the coordinates of all independent parts of the system. A system with f degrees of freedom has f coordinates $q_1,q_2,\cdots,q_f$ and the time integral J, which is defined by

$$J=\int_{t_1}^{t_2} L(q_i,\dot{q}_i,t)dt, \tag{1.28}$$

is minimum. In other words, when $\delta q_i=0$ at $t=t_1,t_2$, we get

$$\begin{aligned}\delta J &= \int_{t_1}^{t_2} \delta L(q_i,\dot{q}_i;t)dt\\ &= \int_{t_1}^{t_2}\left(\frac{\partial L}{\partial q_i}\delta q_i+\frac{\partial L}{\partial \dot{q}_i}\delta\dot{q}_i\right)dt\\ &= \int_{t_1}^{t_2}\left(\frac{\partial L}{\partial q_i}-\frac{d}{dt}\frac{\partial L}{\partial \dot{q}_i}\right)\delta q_i dt\\ &=0.\end{aligned} \tag{1.29}$$

Equation (1.29) dictates that the Lagrangian satisfies the following Euler equation:

$$\frac{\partial L}{\partial q_i} - \frac{d}{dt}\left(\frac{\partial L}{\partial \dot{q}_i}\right) = 0, \quad i = 1, 2, 3, \ldots, f. \tag{1.30}$$

The behavior of the physical system is thus completely specified by Euler's equation once the Lagrangian is known. The proper Lagrangian is the one that leads to a description of the physical system that is in agreement with experimental observations.

For example, if we choose $L = T - V = \frac{1}{2}m\dot{x}^2 - V(x)$ for a one-dimensional particle with mass m in the potential field $V(x)$, we obtain

$$\frac{\partial L}{\partial x} = -\frac{\partial V(x)}{\partial x} \text{ and } \frac{\partial L}{\partial \dot{x}} = m\dot{x}. \tag{1.31}$$

Then the Euler equation yields

$$m\ddot{x} = -\frac{\partial V(x)}{\partial x} = F(x), \tag{1.32}$$

which is the famous Newton's first law of mechanics.

For later purpose, we make the following canonical transformation:

$$p_i \equiv \frac{\partial L}{\partial \dot{q}_i}, \quad H \equiv \sum_i p_i \dot{q}_i - L, \tag{1.33}$$

where H is called the Hamiltonian of the system. Then we obtain the following, Hamilton's equation of motion:

$$\begin{aligned} &\frac{\partial H}{\partial p_i} = \dot{q}, \\ &\dot{p}_i = \frac{d}{dt}\left(\frac{\partial L}{\partial \dot{q}_i}\right) = \frac{\partial L}{\partial q_i} = \frac{\partial}{\partial q_i}\left(\sum_j p_j \dot{q}_j - H\right) = -\frac{\partial H}{\partial q_i}, \end{aligned} \tag{1.34}$$

or,

$$\dot{q}_i = \frac{\partial H}{\partial p_i}, \quad \dot{p}_i = -\frac{\partial H}{\partial q_i}. \tag{1.35}$$

Here, the Hamiltonian H represents the total energy of the system. For any dynamical variable F that depends on canonical variables (q_i, p_i), the time derivative of F is given by

$$\begin{aligned}\frac{dF}{dt} &= \sum_i \left\{ \frac{\partial F}{\partial q_i}\dot{q}_i + \frac{\partial F}{\partial p_i} \right\} + \frac{\partial F}{\partial t} \\ &= \sum_i \left\{ \frac{\partial F}{\partial q_i}\frac{\partial H}{\partial p_i} - \frac{\partial F}{\partial p_i}\frac{\partial H}{\partial q_i} \right\} + \frac{\partial F}{\partial t} \\ &= \{F, H\} + \frac{\partial F}{\partial t},\end{aligned} \tag{1.36}$$

where the Poisson bracket $\{A,B\}$ for dynamical variables A,B is defined by

$$\{A,B\} = \sum_i \left\{ \frac{\partial A}{\partial q_i}\frac{\partial B}{\partial p_i} - \frac{\partial A}{\partial p_i}\frac{\partial B}{\partial q_i} \right\} = -\{B,A\}. \tag{1.37}$$

Equations (1.36) and (1.37) imply that

$$\frac{dH}{dt} = \{H,H\} + \frac{\partial H}{\partial t} = \frac{\partial H}{\partial t}, \tag{1.38}$$

which says if H does not explicitly dependents on the time t, the Hamiltonian, the total energy of the system, is conserved. Among many interesting properties of the Poisson bracket, the following relation is especially useful for the latter purpose:

$$\{q_i, p_j\} = \delta_{ij}. \tag{1.39}$$

For a particle moving in a one-dimensional world specified by the coordinate x, the Lagrangian is given by $L = (1/2m)\dot{x}^2 - V(x)$, as before. Then the canonical transformation yields,

$$\begin{aligned} p &= \frac{\partial L}{\partial \dot{x}} = m\dot{x}, \\ H &= p\dot{x} - L = \frac{1}{2}m\dot{x}^2 + V(x) = \frac{p^2}{2m} + V(x). \end{aligned} \tag{1.40}$$

Equation (1.40) allows us to interpret the Hamiltonian H as the total energy of the system and p as the momentum. Hamilton's equation of motion, Equations (1.34) and (1.35), gives

$$\dot{x} = \frac{\partial H}{\partial p} = \frac{p}{m},$$

$$\dot{p} = -\frac{\partial H}{\partial x} = -\frac{\partial V(x)}{\partial x},$$

Newton's law.

If we have a charged particle in an electromagnetic field, the situation is a bit more complex. The electric field $\vec{E}$ and the magnetic flux density $\vec{B}$ can be expressed by a vector potential $\vec{A}$ and a scalar potential ϕ as

$$\begin{aligned} \vec{E} &= -\vec{\nabla}\phi - \frac{\partial \vec{A}}{\partial t}, \\ \vec{B} &= \vec{\nabla} \times \vec{A}. \end{aligned} \tag{1.41}$$

The divergence of a magnetic flux density is zero, so it can be written as the curl of the vector, the well-known vector potential $\vec{A}$. In static, the curl of the electric field is zero, and it can be written as the gradient of a scalar function. In the time-varying case, the electric field and the magnetic field are related by the following Maxwell's equations:

$$\begin{aligned} \vec{\nabla} \times \vec{E} &= -\frac{\partial \vec{B}}{\partial t}, \\ \vec{\nabla} \cdot \vec{B} &= 0, \\ \vec{\nabla} \cdot \vec{D} &= \rho, \\ \vec{\nabla} \times \vec{H} &= \vec{J} + \frac{\partial \vec{D}}{\partial t}, \end{aligned} \tag{1.42}$$

with $\vec{D} = \varepsilon\vec{E}$ and $\vec{B} = \mu\vec{H}$. Here, $\vec{D}$ is the electric fulx density, $\vec{H}$ is the magnetic field, ρ is the charge density, $\vec{J}$ is the current density, ε is the permittivity, and μ is the permeability. We have used the international system of units (SI units) to write Maxwell's equation. From substituting Equation (1.41) into (1.42), we find that

$$\vec{\nabla} \times \left(\vec{E} + \frac{\partial \vec{A}}{\partial t} \right) = 0.$$

Then it is obvious that we have

$$\vec{E} + \frac{\partial \vec{A}}{\partial t} = -\vec{\nabla}\phi. \tag{1.43}$$

Equation (1.30) implies that the generalized force Q_i is defined by

$$Q_i = -\frac{\partial V}{\partial q_i} + \frac{d}{dt}\left(\frac{\partial V}{\partial \dot{q}_i}\right). \tag{1.44}$$

The force acting on the charged particle, called the Lorentz force, is

$$\vec{F} = e\{\vec{E} + \dot{\vec{r}} \times \vec{B}\}. \tag{1.45}$$

Substituting Equation (1.41) into Equation (1.45), we obtain

$$\vec{F} = e\left\{-\vec{\nabla}(\phi - \dot{\vec{r}} \cdot \vec{A}) - \frac{d\vec{A}}{dt}\right\}, \tag{1.46}$$

where we have used the following relation:

$$\dot{\vec{r}} \times (\vec{\nabla} \times \vec{A}) = \vec{\nabla}(\dot{\vec{r}} \cdot \vec{A}) - (\dot{\vec{r}} \cdot \vec{\nabla})\vec{A}.$$

Equations (1.44) and (1.46) indicate that the generalized potential in the case of a charged particle moving in an electromagnetic field is given by

$$V = e(\phi - \dot{\vec{r}} \cdot \vec{A}), \tag{1.47}$$

and the corresponding Lagrangian is

$$L = \frac{1}{2} m\dot{\vec{r}}^2 + e\dot{\vec{r}} \cdot \vec{A} - e\phi. \tag{1.48}$$

The generalized canonical momentum is then

$$p_i = \frac{\partial L}{\partial \dot{q}_i} = m\dot{x}_i + eA_i, \quad q_i = x_i, \tag{1.49}$$

and the Hamiltonian is given by

$$\begin{aligned}
H &= \sum_i p_i \dot{q}_i - L \\
&= \sum_i (m\dot{x}_i + eA_i)\dot{x}_i - \frac{1}{2} m\dot{\vec{r}}^2 - e\dot{\vec{r}} \cdot \vec{A} + e\phi \\
&= \frac{1}{2} m\dot{\vec{r}}^2 + e\phi \\
&= \frac{(\vec{p} - e\vec{A})^2}{2m} + e\phi.
\end{aligned} \tag{1.50}$$

1.4 A ROAD TO QUANTUM MECHANICS

In classical mechanics, we dealt with functions of the coordinates q_i and momentum p_i such as energy; we describe these quantities collectively as "observable." The term "observable" describes any quantity accessible to the measurement processes. We assume that every physical observable is mathematically represented by a Hermitian operator, and every measurement of the physical observable will result in one of the eigenvalues of the corresponding Hermitian operator. The eigenvector is used to characterize the state of the physical system.

Quantum mechanics assumes that any arbitrary state of the physical system is characterized by a state vector that is not necessarily an eigenvector of any particular Hermitian operator. After a measurement has been performed, the state vector collapses to one of the eigenvectors with an eigenvalue E_n. If we describe the state of the system by a ket vector $|\Psi\rangle$, then we can represent this state vector by $|\Psi\rangle = \Sigma_n |E_n\rangle\langle E_n|\Psi\rangle$, where $|E_n\rangle$ is an eigenvector of a particular Hermitian operator corresponding to the measurement done on the system. How does $|\Psi\rangle$ relate to possible measurements when the result of a measurement must be one of the eigenvalues $\{E_n\}$? For this, we need the following postulate.

1.4.1 Postulate

The quantity $\langle\Psi|H|\Psi\rangle$ represents the average value of a series of measurements on an ensemble of systems that are all described by the state vector $|\Psi\rangle$ and is given by

$$\langle\Psi|H|\Psi\rangle = \sum_n |\langle E_n|\Psi\rangle|^2 E_n, \tag{1.51}$$

where $P_n = |\langle E_n|\Psi\rangle|^2$ is the probability of obtaining the value of E_n as a result of a measurement.

Previously, we described that in quantum mechanics the dynamical variables of classical mechanics are replaced by corresponding Hermitian operators. In classical mechanics, the dynamics of an observable are described by the Poisson bracket. We would like to extend the Poisson bracket to describe the dynamics of a quantum mechanical operator. The Poisson bracket for the classical observables A, B is defined in Equation (1.37) as

$$\{A,B\} = \sum_i \left\{ \frac{\partial A}{\partial q_i}\frac{\partial B}{\partial p_i} - \frac{\partial A}{\partial p_i}\frac{\partial B}{\partial q_i} \right\} = -\{B,A\}. \tag{1.37}$$

It is straightforward to show the following properties:

$$\{A+C,B\}=\{A,B\}+\{C,B\}, \tag{1.52a}$$

$$\{A,B+C\}=\{A,B\}+\{A,C\}, \tag{1.52b}$$

$$\{AB,C\}=\{A,C\}B+A\{B,C\}, \tag{1.53a}$$

$$\{A,BC\}=\{A,B\}C+B\{A,C\}. \tag{1.53b}$$

Let us assume that a quantum mechanical Poisson bracket is analogous to a classical one. So we assume that a quantum mechanical Poisson bracket satisfies all the conditions of Equations (1.37), (1.52), and (1.53). We shall denote the quantum mechanical Poisson bracket for operators $\hat{A}$ and $\hat{B}$ as $\left[\hat{A},\hat{B}\right]$. We use these conditions to determine the functional form of a quantum Poisson bracket by evaluating $\left[\hat{A}\hat{B},\hat{C}\hat{D}\right]$ in two different ways from Equation (1.53a,b):

$$\begin{aligned}\left[\hat{A}\hat{B},\hat{C}\hat{D}\right]&=\left[\hat{A},\hat{C}\hat{D}\right]\hat{B}+\hat{A}\left[\hat{B},\hat{C}\hat{D}\right]\\&=\left\{\left[\hat{A},\hat{C}\right]\hat{D}+\hat{C}\left[\hat{A},\hat{D}\right]\right\}\hat{B}+\hat{A}\left\{\left[\hat{B},\hat{C}\right]\hat{D}+\hat{C}\left[\hat{B},\hat{D}\right]\right\}\\&=\left[\hat{A},\hat{C}\right]\hat{D}\hat{B}+\hat{C}\left[\hat{A},\hat{D}\right]\hat{B}+\hat{A}\left[\hat{B},\hat{C}\right]\hat{D}+\hat{A}\hat{C}\left[\hat{B},\hat{D}\right],\end{aligned} \tag{1.54}$$

and

$$\begin{aligned}\left[\hat{A}\hat{B},\hat{C}\hat{D}\right]&=\left[\hat{A}\hat{B},\hat{C}\right]\hat{D}+\hat{C}\left[\hat{A}\hat{B},\hat{D}\right]\\&=\left\{\left[\hat{A},\hat{C}\right]\hat{B}+\hat{A}\left[\hat{B},\hat{C}\right]\right\}\hat{D}+\hat{C}\left\{\left[\hat{A},\hat{D}\right]\hat{B}+\hat{A}\left[\hat{B},\hat{D}\right]\right\}\\&=\left[\hat{A},\hat{C}\right]\hat{B}\hat{D}+\hat{A}\left[\hat{B},\hat{C}\right]\hat{D}+\hat{C}\left[\hat{A},\hat{D}\right]\hat{B}+\hat{C}\hat{A}\left[\hat{B},\hat{D}\right].\end{aligned} \tag{1.55}$$

Equating Equations (1.54) and (1.55), we obtain

$$-\left[\hat{A},\hat{C}\right]\left(\hat{B}\hat{D}-\hat{D}\hat{B}\right)+\left(\hat{A}\hat{C}-\hat{C}\hat{A}\right)\left[\hat{B},\hat{D}\right]=0. \tag{1.56}$$

Since the above condition holds for arbitrary Hermitian operators $\hat{A}$, $\hat{B}$, $\hat{C}$, and $\hat{D}$, we must have

$$\left[\hat{A},\hat{C}\right]=\left(\hat{A}\hat{C}-\hat{C}\hat{A}\right)\text{and}\left[\hat{B},\hat{D}\right]=\left(\hat{B}\hat{D}-\hat{D}\hat{B}\right). \tag{1.57}$$

From now on, we shall call a quantum Poisson bracket a commutator. We further assume that a commutator or a quantum Poisson bracket is proportional to the corresponding classical Poisson bracket:

$$\left[\hat{A},\hat{B}\right]=i\hbar\{A,B\}, \tag{1.58}$$

where $\hbar$ is Planck's constant divided by 2π. This suggests that quantum mechanics is based on the assumption that the Poisson bracket assumes the same physical meaning and the same numerical values as in classical mechanics. In particular,

$$[\hat{q}_i,\hat{p}_j]=i\hbar\{q_i,p_j\}=i\hbar\delta_{ij}, \tag{1.59}$$

which is the fundamental quantum condition.

The time evolution of a quantum mechanical operator can be obtained from the equation of motion for the classical observable:

$$\frac{dF}{dt}=\frac{\partial F}{\partial t}+\{F,H\}, \tag{1.36}$$

$$\frac{d\hat{F}}{dt}=\frac{\partial\hat{F}}{\partial t}+\frac{1}{i\hbar}\left[\hat{F},\hat{H}\right], \tag{1.60}$$

where $\hat{H}$ is the energy operator corresponding to the classical Hamiltonian. Equation (1.60) is called the equation of motion in the Heisenberg picture. This particular quantum mechanical representation assumes that the operators vary with time, while the state vector $|\Psi\rangle$ is time independent. This picture is formally analogous to classical mechanics, since the equations of motion for the operators closely resemble the corresponding classical equations.

Now, let us study a different picture, called the Schrödinger picture, where the operators are constant in time, while the time variation is expressed by the state vectors. In this picture, the explicit time dependence of an operator still remains. Different quantum mechanical pictures follow from each other by a unitary transformation. From now on, we denote a quantum mechanical operator $\hat{A}$ simply as A when we don't need to distinguish it from the corresponding classical observable. Let us denote the state vector and operator in the new picture by $|\Psi'\rangle$ and A', respectively. Then, they are related to $|\Psi\rangle$ and A of an old picture by

$$|\Psi'\rangle=U|\Psi\rangle,\ A'=UAU^{\dagger} \tag{1.61}$$

and

$$\langle\Psi'|A'|\Psi'\rangle=\langle\Psi|U^{\dagger}UAU^{\dagger}U|\Psi\rangle=\langle\Psi|A|\Psi\rangle, \tag{1.62}$$

since

$$U^{\dagger}U = UU^{\dagger} = I, \tag{1.63}$$

when U is a unitary operator. Equation (1.62) indicates that the unitary transformation does not change the physical content of the theory. Also, we have

$$\begin{aligned}\frac{dA'}{dt} &= \frac{dU}{dt}AU^{\dagger} + U\frac{dA}{dt}U^{\dagger} + UA\frac{dU^{\dagger}}{dt} \\ &= \frac{dU}{dt}AU^{\dagger} + U\frac{\partial A}{\partial t}U^{\dagger} + \frac{1}{i\hbar}U[A,H]U^{\dagger} + UA\frac{dU^{\dagger}}{dt}.\end{aligned} \tag{1.64}$$

If A is not explicitly dependent on time, $\partial A/\partial t = 0$, and the operator A in the Schrödinger picture is time independent except for an explicit time dependence. This implies that

$$\frac{dA'}{dt} - U\frac{\partial A}{\partial t}U^{\dagger} = 0. \tag{1.65}$$

From Equations (1.64) and (1.65), we get

$$\begin{aligned}&\frac{dU}{dt}AU^{\dagger} + \frac{1}{i\hbar}U[A,H]U^{\dagger} + UA\frac{dU^{\dagger}}{dt} \\ &= \frac{dU}{dt}AU^{\dagger} + UA\frac{dU^{\dagger}}{dt} + \frac{1}{i\hbar}U(AH - HA)U \\ &= \frac{dU}{dt}U^{\dagger}A' + A'U\frac{dU^{\dagger}}{dt} + \frac{1}{i\hbar}(A'H' - H'A') \\ &= \left(\frac{dU}{dt}U^{\dagger} - \frac{1}{i\hbar}H'\right)A' + A'\left(U\frac{dU^{\dagger}}{dt} + \frac{1}{i\hbar}H'\right) \\ &= 0.\end{aligned} \tag{1.66}$$

Equation (1.66) should hold for any quantum mechanical operators A' in the Schrödinger picture, and, as a consequence, we have

$$i\hbar\frac{dU}{dt} = H'U. \tag{1.67}$$

Since $|\Psi'\rangle = U|\Psi\rangle$, we obtain the following Schrödinger equation for the state vector

$$i\hbar\frac{d}{dt}|\Psi'\rangle = H'|\Psi'\rangle. \tag{1.68}$$

The Schrödinger picture is perhaps the most widely used because it leads directly to the formulation of wave mechanics. In this picture, eigenvectors of the position operator q are used to represent the state vector of the system. We consider the continuous set of eigenvalues q' of the position operator q given by

$$q|q'\rangle = q'|q\rangle. \tag{1.69}$$

We assume that corresponding eigenvectors $\{|q'\rangle\}$ form a complete set. For simplicity, we consider the one-dimensional case only, but the extension to the higher dimensional case is straightforward. The position representation of the state vector is defined by

$$\Psi(q') = \langle q'|\Psi\rangle. \tag{1.70}$$

From Equation (1.59), the canonical momentum operator p and the position operator q satisfy the following quantum Poisson's bracket or the commutator relation

$$[q, p] = (qp - pq) = i\hbar, \tag{1.71}$$

and

$$(qp - pq)|\Psi\rangle = i\hbar|\Psi\rangle. \tag{1.72}$$

If we take an inner product of Equation (1.72) by the bra vector $\langle q'|$, we obtain

$$\langle q'|(qp - pq)|\Psi\rangle = q'\langle q'|p|\Psi\rangle - \langle q'|pq|\Psi\rangle. \tag{1.73}$$

Also,

$$\langle q'|(qp - pq)|\Psi\rangle = i\hbar\langle q'|\Psi\rangle. \tag{1.74}$$

Equations (1.73) and (1.74) are consistent only if the following condition is satisfied

$$\langle q'|p|\Psi\rangle = -i\hbar\frac{\partial}{\partial q'}\langle q'|\Psi\rangle = -i\hbar\frac{\partial}{\partial q'}\Psi(q'), \tag{1.75}$$

since Equation (1.75) implies

$$\langle q'|pq|\Psi\rangle = -i\hbar\frac{\partial}{\partial q'}(q'\Psi(q')). \tag{1.76}$$

Moreover, we have the following relations:

$$\langle q'|p|q''\rangle = -i\hbar\frac{\partial}{\partial q'}\langle q'|q''\rangle = -i\hbar\frac{\partial}{\partial q'}\delta(q'-q''), \tag{1.77}$$

and

$$I = \int_{-\infty}^{\infty} dq|q\rangle\langle q|. \tag{1.78}$$

Then, we get

$$\begin{aligned}\langle\Phi|\Psi\rangle &= \int_{-\infty}^{\infty} dq\langle\Phi|q\rangle\langle q|\Psi\rangle \\ &= \int_{-\infty}^{\infty} dq\, \Phi^*(q)\Psi(q),\end{aligned} \tag{1.79}$$

$$\begin{aligned}\langle\Phi|p|\Psi\rangle &= \int_{-\infty}^{\infty} dq'\int_{-\infty}^{\infty} dq''\langle\Phi|q'\rangle\langle q'|p|q''\rangle\langle q''|\Psi\rangle \\ &= \int_{-\infty}^{\infty} dq'\int_{-\infty}^{\infty} dq''\Phi^*(q')\left(-i\hbar\frac{\partial}{\partial q'}\delta(q'-q'')\right)\Psi(q'') \\ &= \int_{-\infty}^{\infty} dq'\int_{-\infty}^{\infty} dq''\left(i\hbar\frac{\partial}{\partial q'}\Phi^*(q')\right)\delta(q'-q'')\Psi(q'') \\ &= \int_{-\infty}^{\infty} dq'\left(i\hbar\frac{\partial}{\partial q'}\Phi^*(q')\right)\Psi(q'') \\ &= \int_{-\infty}^{\infty} dq'\Phi^*(q')\left(i\hbar\frac{\partial}{\partial q'}\right)\Psi(q''),\end{aligned} \tag{1.80}$$

and

$$\langle q'|F(q,p)|\Psi\rangle = F\left(q',-i\hbar\frac{\partial}{\partial q'}\right)\Psi(q'). \tag{1.81}$$

As a special case, we consider $F = H$, where H is the Hamiltonian of the system. Then the Schrödinger equation (1.68) becomes

$$i\hbar\frac{\partial}{\partial t}\Psi(q',t) = H(q',-i\hbar\partial/\partial q')\Psi(q',t), \tag{1.82}$$

of the wave mechanics.

As an example, we consider a particle moving in a one-dimensional potential $V(x)$. If we set the position operator as $q' = x$, the classical Hamiltonian is given by Equation (1.40):

$$H = \frac{p^2}{2m} + V(x).$$

Then, the corresponding Schrödinger equation is given by

$$i\hbar\frac{\partial}{\partial t}\Psi(x,t) = \left(-\frac{\hbar^2}{2m}\frac{\partial^2}{\partial x^2} + V(x)\right)\Psi(x,t). \tag{1.83}$$

For three dimensions, we simply replace q' with $\vec{r}$ and $\partial/\partial q'$ with $\vec{\nabla}$. For example, the Schrödinger equation for a charged particle moving in an electromagnetic field is

$$i\hbar\frac{\partial}{\partial t}\Psi(\vec{r},t) = \frac{1}{2m}\left(-i\hbar\vec{\nabla} - e\vec{A}(\vec{r},t)\right)^2\Psi(\vec{r},t) + e\phi(\vec{r})\Psi(\vec{r},t). \tag{1.84}$$

When the Hamiltonian H is time independent, the time-dependent Schrödinger equation can be rewritten in time-independent form by trying the solution of the form

$$\Psi(\vec{r},t) = \exp(-iEt/\hbar)\psi_E(\vec{r}) \tag{1.85}$$

into Equation (1.83). Then, we obtain the following time-independent Schrödinger equation

$$-\frac{\hbar^2}{2m}\vec{\nabla}^2\psi_E(\vec{r}) + V(\vec{r})\psi_E(\vec{r}) = E\psi_E(\vec{r}), \tag{1.86}$$

where E is the energy of the particle.

One can also consider the eigenvectors of the momentum operator p, which are given by

$$p|p'\rangle = p'|p'\rangle. \tag{1.87}$$

From

$$\langle q'|p|p'\rangle = -i\hbar\frac{\partial}{\partial q'}\langle q'|p'\rangle = p'\langle q'|p'\rangle, \tag{1.88}$$

we get

$$\langle q'|p'\rangle = C\exp(iq'p'/\hbar), \tag{1.89}$$

where C is a normalization constant. If we assume that $\langle p'|p''\rangle = \delta(p'-p'')$, then it is straightforward to calculate the normalization constant, which is given by $1/\sqrt{2\pi\hbar}$. As a consequence, we have

$$\langle x|p\rangle = \frac{1}{\sqrt{2\pi\hbar}}\exp(ixp/\hbar), \tag{1.90}$$

and

$$\langle \bar{r}|\bar{p}\rangle = \left(\frac{1}{2\pi\hbar}\right)^{3/2}\exp(i\bar{r}\cdot\bar{p}/\hbar). \tag{1.91}$$

1.5 THE UNCERTAINTY PRINCIPLE

The most remarkable property of quantum mechanics is that the measurements of different dynamical variables are not necessarily independent. This property was first formulated by Heisenberg, and is known as the uncertainty principle. We start with a quantum Poisson bracket for the position and momentum operators q and p,

$$[q,p] = i\hbar.$$

Let $|\Psi\rangle$ be a state vector, and $\bar{q}$ and $\bar{p}$ be the average values of the measurements of q and p, respectively:

$$\bar{q} = \langle\Psi|q|\Psi\rangle, \bar{p} = \langle\Psi|p|\Psi\rangle.$$

We now calculate the mean-square deviations of measured values,

$$\begin{aligned}(\Delta q)^2 &= \langle\Psi|(q-\bar{q})^2|\Psi\rangle \\ &= \langle\Psi|q^2|\Psi\rangle - (\langle\Psi|q|\Psi\rangle)^2,\end{aligned} \tag{1.92}$$

and

$$\begin{aligned}(\Delta p)^2 &= \langle\Psi|(p-\bar{p})^2|\Psi\rangle \\ &= \langle\Psi|p^2|\Psi\rangle - (\langle\Psi|p|\Psi\rangle)^2.\end{aligned} \tag{1.93}$$

We also define $\alpha = q - \bar{q}$ and $\beta = p - \bar{p}$, and we get

$$[\alpha, \beta] = [q, p] = i\hbar.$$

Also,

$$(\Delta\alpha)^2 = (\Delta q)^2 \text{ and } (\Delta\beta)^2 = (\Delta p)^2. \tag{1.94}$$

From Equation (1.94), we get

$$\begin{aligned}(\Delta q)^2(\Delta p)^2 &= (\Delta\alpha)^2(\Delta\beta)^2 \\ &= \langle\Psi|\alpha^2|\Psi\rangle\langle\Psi|\beta^2|\Psi\rangle \\ &\geq |\langle\Psi|\alpha\beta|\Psi\rangle|^2,\end{aligned} \tag{1.95}$$

from the Schwarz inequality. Since $\alpha\beta = 1/2(\alpha\beta - \beta\alpha) + 1/2(\alpha\beta + \beta\alpha)$, we also obtain

$$\begin{aligned}|\langle\Psi|\alpha\beta|\Psi\rangle|^2 &= \frac{1}{4}|\langle\Psi|[\alpha,\beta]|\Psi\rangle + \langle\Psi|(\alpha\beta+\beta\alpha)|\Psi\rangle|^2 \\ &\geq \frac{1}{4}|\langle\Psi|[\alpha,\beta]|\Psi\rangle|^2 = \frac{\hbar^2}{4}.\end{aligned} \tag{1.96}$$

Equations (1.95) and (1.96) give Heisenberg's famous uncertainty relation:

$$\Delta q \Delta p \geq \frac{\hbar}{2}. \tag{1.97}$$

1.6 THE HARMONIC OSCILLATOR

The one-dimensional simple harmonic oscillator is perhaps the most important and useful problem in quantum mechanics. The Lagrangian for the harmonic oscillator is

$$L = \frac{1}{2}m\dot{x}^2 - \frac{1}{2}kx^2. \tag{1.98}$$

The canonical momentum is

$$p = \frac{\partial L}{\partial \dot{x}} = m\dot{x}, \tag{1.99}$$

and the Hamiltonian becomes

$$H = p\dot{x} - L = \frac{p^2}{2m} + \frac{1}{2}kx^2. \tag{1.100}$$

The time-independent Schrödinger equation for the harmonic oscillator is then given by

$$-\frac{\hbar^2}{2m}\frac{\partial^2 \psi}{\partial x^2} + \frac{1}{2}kx^2\psi = E\psi, \tag{1.101}$$

with the boundary condition $\psi = 0$ at $x = \pm\infty$. The solution of Equation (1.101) gives the energy eigenvalues of the harmonic oscillator and the eigenfunctions in terms of Hermite polynomials. In this section, we introduce an operator method that is based on the algebraic approach, and is much simpler than solving the Schrödinger directly. We introduce new operators

$$\begin{aligned} a &= \frac{1}{\sqrt{2m\hbar\omega}}(m\omega x + ip), \\ a^\dagger &= \frac{1}{\sqrt{2m\hbar\omega}}(m\omega x + ip), \\ \omega &= \sqrt{\frac{k}{m}}. \end{aligned} \tag{1.102}$$

From

$$\begin{aligned} aa^\dagger &= \frac{1}{2m\hbar\omega}(m\omega x + ip)(m\omega x - ip), \\ a^\dagger a &= \frac{1}{2m\hbar\omega}(m\omega x - ip)(m\omega x + ip), \end{aligned}$$

we obtain

$$aa^\dagger - a^\dagger a = \frac{i}{\hbar}(px - xp) = \frac{i}{\hbar}(-i\hbar) = 1. \tag{1.103}$$

We also have

$$
\begin{aligned}
x &= \sqrt{\frac{\hbar}{2m\omega}}\left(a + a^{\dagger}\right), \\
p &= i\sqrt{\frac{m\hbar\omega}{2}}\left(a^{\dagger} - a\right).
\end{aligned} \tag{1.104}
$$

Substituting Equation (1.104) into Equation (1.100), and using Equation (1.103), we get

$$
\begin{aligned}
H &= \frac{p^2}{2m} + \frac{1}{2}m\omega^2 x^2 \\
&= \frac{(-1)}{2m}\left(\frac{m\omega\hbar}{2}\right)\left(a^{\dagger} - a\right)\left(a^{\dagger} - a\right) + \frac{1}{2}m\omega^2\frac{\hbar}{2m\omega}\left(a + a^{\dagger}\right)\left(a + a^{\dagger}\right) \\
&= \frac{\hbar\omega}{2}\left(aa^{\dagger} + a^{\dagger}a\right) \\
&= \hbar\omega\left(a^{\dagger}a + \frac{1}{2}\right).
\end{aligned} \tag{1.105}
$$

Let's assume that the state is in a particular eigenstate $|E'\rangle$ with eigenvalue E',

$$
H|E'\rangle = E'|E'\rangle. \tag{1.106}
$$

Now,

$$
aH|E'\rangle = aE'|E'\rangle,
$$

with $aH = a\hbar\omega\left(a^{\dagger}a + \frac{1}{2}\right)$. From $[a, a^{\dagger}] = 1$, we get $aa^{\dagger}a = a^{\dagger}aa + a$, and

$$
aH|E'\rangle = Ha|E'\rangle + \hbar\omega a|E'\rangle = E'|E'\rangle. \tag{1.107}
$$

Equation (1.107) can be rewritten as

$$
Ha|E'\rangle = (E' - \hbar\omega)|E'\rangle. \tag{1.108}
$$

In other words, if $|E'\rangle$ is an eigenstate of a harmonic oscillator Hamiltonian H with an eigenvalue E', $a|E'\rangle$ is also an eigenstate of H but with different eigenvalue $E' - \hbar\omega$.

By the way,

$$\begin{aligned}\langle E|H|E\rangle &= \hbar\omega\langle E|a^{\dagger}a|E\rangle + \frac{1}{2}\hbar\omega\langle E|E\rangle \\ &= E\langle E|E\rangle.\end{aligned} \tag{1.109}$$

Since

$$\begin{aligned}\langle E|E\rangle &= \int dq\langle E|q\rangle\langle q|E\rangle \\ &= \int dq|\langle E|q\rangle|^2 \\ &\geq 0,\end{aligned}$$

we have

$$\langle E|a^{\dagger}a|E\rangle = \int dq|\langle q|a|E\rangle|^2 \geq 0. \tag{1.110}$$

Combining Equations (1.109) and (1.110), the energy eigenvalue E must be a nonnegative number. On the other hand, Equation (1.108) indicates that the operation of an operator a on the energy eigenstate $|E\rangle$ lowers the energy eigenvalue by an amount $\hbar\omega$. Nonetheless, the energy eigenvalue must be always nonnegative, so the reduction of the eigenvalue must be terminated at the ground state $|E_0\rangle$, that is, $a|E_0\rangle = 0$, and as a result, we obtain

$$E_0 = \frac{1}{2}\hbar\omega. \tag{1.111}$$

Likewise,

$$a^{\dagger}H|E'\rangle = E'a^{\dagger}|E'\rangle,$$

and after some mathematical manipulations, we get

$$Ha^{\dagger}|E'\rangle = (E' + \hbar\omega)a^{\dagger}|E'\rangle. \tag{1.112}$$

This indicates that $a^{\dagger}|E'\rangle$ is also an eigenstate of H, but with a new eigenvalue $E' + \hbar\omega$. We can see that the operator $a^{\dagger}$ raises the eigenvalue by the amount of $\hbar\omega$. So if $|E_n\rangle$ is the nth eigenstate, one can generate $|E_n\rangle$ from $|E_0\rangle$ by n successive applications of $a^{\dagger}$,

$$|E_n\rangle = A_n(a^\dagger)^n|E_0\rangle, \tag{1.113}$$

where A_n is the normalization factor. The normalization factor A_n is determined from the condition

$$\langle E_n|E_n\rangle = 1$$

with the relations

$$\begin{aligned} a(a^\dagger)^n &= aa^\dagger(a^\dagger)^{n-1} \\ &= (a^\dagger)^{n-1} + a^\dagger a(a^\dagger)^{n-1} \\ &= (a^\dagger)^{n-1} + a^\dagger\left[(a^\dagger)^{n-2} + a^\dagger a(a^\dagger)^{n-2}\right] \\ &= 2(a^\dagger)^{n-1} + (a^\dagger)^2 a(a^\dagger)^{n-2} \\ &= n(a^\dagger) + (a^\dagger)^n a \end{aligned} \tag{1.114}$$

and

$$a^n(a^\dagger)^n = na^{n-1}(a^\dagger)^{n-1} + a^{n-1}(a^\dagger)^n a.$$

Then we obtain

$$\begin{aligned} \langle E_n|E_n\rangle &= A_n^2\langle E_0|a^n(a^\dagger)^n|E_0\rangle \\ &= nA_n^2\langle E_0|a^{n-1}(a^\dagger)^{n-1}|E_0\rangle \\ &= n!\,A_n^2\langle E_0|E_0\rangle \\ &= n!\,A_n^2 \\ &= 1. \end{aligned}$$

The normalization factor is

$$A_n = \frac{1}{\sqrt{n!}}$$

and the normalized eigenstate is

$$|E_n\rangle = \frac{(a^\dagger)^n}{\sqrt{n!}}|E_0\rangle. \tag{1.115}$$

From Equations (1.114) and (1.115), we get

$$\begin{aligned} a|E_n\rangle &= \frac{a(a^\dagger)^n}{\sqrt{n!}}|E_0\rangle \\ &= \frac{n(a^\dagger)^{n-1}}{\sqrt{n!}}|E_0\rangle \\ &= \sqrt{n}|E_{n-1}\rangle \end{aligned} \tag{1.116}$$

and

$$\begin{aligned} a^\dagger|E_n\rangle &= \frac{(a^\dagger)^{n+1}}{\sqrt{n!}}|E_0\rangle \\ &= \sqrt{n+1}|E_{n+1}\rangle. \end{aligned} \tag{1.117}$$

We call the operator a an annihilation operator because it destroys a single quantum of energy. The operator $a^\dagger$ is called a creation operator since it raises (or creates) a quantum of energy. The operator $a^\dagger a$ is called the number operator because if operating on the state, one can see that its eigenvalue is the number of energy quantum n,

$$a^\dagger a|E_n\rangle = a^\dagger\sqrt{n}|E_{n-1}\rangle = n|E_n\rangle. \tag{1.118}$$

The operator method can also be used to determine the wave function of the one-dimensional harmonic oscillator in the coordinate space. Since $a|E_0\rangle = 0$, from Equation (1.102), we have

$$\begin{aligned} \langle x|a|E_0\rangle &= \langle x|\left(\frac{m\omega}{\sqrt{2m\hbar\omega}}x + i\frac{1}{\sqrt{2m\hbar\omega}}p\right)|E_0\rangle \\ &= \sqrt{\frac{m\omega}{2\hbar}}x\langle x|E_0\rangle + \sqrt{\frac{\hbar}{2m\omega}}\frac{\partial}{\partial x}\langle x|E_0\rangle \\ &= 0, \end{aligned} \tag{1.119}$$

where Equations (1.69) and (1.75) are used. If we denote $\psi_0(x) = \langle x|E_0\rangle$, then Equation (1.119) can be rewritten as

$$\frac{d}{dx}\psi_0(x) + \frac{m\omega x}{\hbar}\psi_0(x) = 0. \tag{1.120}$$

Solving the first-order differential equation is straightforward, and the solution is given by

$$\psi_0(x) = A\exp\left(-\frac{m\omega}{2\hbar}x^2\right), \tag{1.121}$$

where A is the normalization constant given by

$$A = \left(\frac{m\omega}{\hbar\pi}\right)^{1/4}. \tag{1.122}$$

In order to determine the normalization constant, we have used the condition

$$\int_{-\infty}^{\infty} |\psi_0(x)|^2\, dx = 1, \tag{1.123}$$

and the Gaussian integral

$$\int_{-\infty}^{\infty} \exp(-\alpha x^2)dx = \sqrt{\frac{\pi}{\alpha}}, \quad \alpha > 0. \tag{1.124}$$

The excited state wave functions can be obtained by successive application of a creation operator $a^\dagger$ on the state $|E_0\rangle$ and taking an inner product with the bra vector $\langle x|$. For example, the first excited state wave function is

$$\begin{aligned}
\psi_1(x) &= \langle x|a^\dagger|E_0\rangle \\
&= \langle x|\left(\sqrt{\frac{m\omega}{2\hbar}}x - i\frac{p}{\sqrt{2m\omega}}\right)|E_0\rangle \\
&= \sqrt{\frac{m\omega}{2\hbar}}x\psi_0(x) - \sqrt{\frac{\hbar}{2m\omega}}\frac{\partial}{\partial x}\psi_0(x) \\
&= \sqrt{\frac{2m\omega}{\hbar}}x\left(\frac{m\omega}{\pi\hbar}\right)^{1/4}\exp\left(-\frac{m\omega x^2}{2\hbar}\right),
\end{aligned} \tag{1.125}$$

and the nth state wave function is given by

$$\psi_n(x) = \frac{1}{\sqrt{2^n n!}}H_n\left(\sqrt{\frac{m\omega}{\hbar}}x\right)\left(\frac{m\omega}{\pi\hbar}\right)^{1/4}\exp\left(-\frac{m\omega x^2}{2\hbar}\right). \tag{1.126}$$

Here, $H_n(\zeta)$ is the Hermite polynomials, and the first few Hermite polynomials are

$$H_0(\zeta) = 1, H_1(\zeta) = 2\zeta, H_2(\zeta) = 4\zeta^2 - 2, \cdots \tag{1.127}$$

1.7 ANGULAR MOMENTUM EIGENSTATES

If the potential term $V(\vec{r})$ in the time-independent Schrödinger equation has a spherical symmetry, then the angular momentum $\vec{L}=\vec{r}\times\vec{p}$ is conserved. We need the following mathematical relations to prove the previous statement.

$$\begin{aligned}(\vec{r}\times\vec{\nabla})\cdot(\vec{r}\times\vec{\nabla}) &= \{(\vec{r}\times\vec{\nabla})\times\vec{r}\}\cdot\vec{\nabla}-\vec{r}\cdot\{(\vec{r}\times\vec{\nabla})\times\vec{\nabla}\}\\ &= \{-\vec{r}(\vec{\nabla}\cdot\vec{r})+\vec{\nabla}(\vec{r}\cdot\vec{r})\}\cdot\vec{\nabla}-\vec{r}\cdot\{-\vec{r}(\vec{\nabla}\cdot\vec{\nabla})+\vec{\nabla}(\vec{r}\cdot\vec{\nabla})\}\\ &= -3\vec{r}\cdot\vec{\nabla}+\vec{\nabla}(r^2)\cdot\nabla+r^2\nabla^2-(\vec{r}\cdot\vec{\nabla})(\vec{r}\cdot\vec{\nabla})\\ &= -r\frac{\partial}{\partial r}+r^2\nabla^2-r\frac{\partial}{\partial r}\left(r\frac{\partial}{\partial r}\right)\\ &= -r^2\frac{\partial^2}{\partial r^2}-2r\frac{\partial}{\partial r}+r^2\nabla^2.\end{aligned} \tag{1.128}$$

Here, we used the vector identity

$$(\vec{A}\times\vec{B})\times\vec{C}=-\vec{A}(\vec{B}\cdot\vec{C})+\vec{B}(\vec{C}\cdot\vec{A}) \tag{1.129}$$

and

$$\vec{r}\cdot\vec{\nabla}=r\frac{\partial}{\partial r}. \tag{1.130}$$

If we define the orbital angular momentum operator $\vec{L}$ as

$$\vec{L}=-i\hbar\vec{r}\times\vec{\nabla}, \tag{1.131}$$

then, from Equation (1.128), we obtain

$$\nabla^2=\frac{\partial^2}{\partial r^2}+\frac{2}{r}\frac{\partial}{\partial r^2}-\frac{\vec{L}\cdot\vec{L}}{\hbar^2 r^2}. \tag{1.132}$$

The time-independent Schrödinger equation (1.86) becomes

$$-\frac{\hbar^2}{2m}\left(\frac{\partial^2}{\partial r^2}+\frac{2}{r}\frac{\partial}{\partial r^2}-\frac{\vec{L}\cdot\vec{L}}{\hbar^2 r^2}\right)\psi_E(\vec{r})+V(r)\psi_E(\vec{r})=E\psi_E(\vec{r}), \tag{1.133}$$

and the Hamiltonian with spherically symmetric potential is given by

$$H = -\frac{\hbar^2}{2m}\left(\frac{\partial^2}{\partial r^2} + \frac{2}{r}\frac{\partial}{\partial r^2} - \frac{\vec{L}\cdot\vec{L}}{\hbar^2 r^2}\right) + V(r)\ldots \tag{1.134}$$

The angular momentum operator can be written, component-wise, as

$$\begin{aligned} L_x &= \left(-i\hbar\vec{r}\times\vec{\nabla}\right)_x = -i\hbar\left(y\frac{\partial}{\partial z} - z\frac{\partial}{\partial y}\right), \\ L_y &= \left(-i\hbar\vec{r}\times\vec{\nabla}\right)_y = -i\hbar\left(z\frac{\partial}{\partial x} - x\frac{\partial}{\partial z}\right), \\ L_z &= \left(-i\hbar\vec{r}\times\vec{\nabla}\right)_z = -i\hbar\left(x\frac{\partial}{\partial y} - y\frac{\partial}{\partial z}\right), \end{aligned} \tag{1.135}$$

and they satisfy the following commutation relations

$$\begin{aligned} [L_y, L_z] &= i\hbar L_x, \\ [L_z, L_x] &= i\hbar L_y, \\ [L_x, L_y] &= i\hbar L_z. \end{aligned} \tag{1.136}$$

Now consider the infinitesimal rotation about the z-axis by an angle θ. The rotated coordinates are given by

$$\begin{aligned} x' &= x\cos\theta + y\sin\theta \approx x + \theta y, \\ y' &= -x\sin\theta + y\cos\theta \approx -\theta x + y, \\ z' &= z. \end{aligned} \tag{1.137}$$

Since the Hamiltonian is invariant under the rotation, we have

$$H\psi_E(x,y,z) = E\psi_E(x,y,z), \tag{1.138}$$

and

$$H\psi_E(x',y',z') = E\psi_E(x',y',z'). \tag{1.139}$$

Also, we have

$$\begin{aligned} \psi_E(x+\theta y, -\theta x + y, z) &= \psi_E(x,y,z) + \theta\left(y\frac{\partial}{\partial x} - x\frac{\partial}{\partial y}\right)\psi_E(x,y,z) \\ &= \psi_E(x,y,z) - \frac{i\theta}{\hbar}L_z\psi_E(x,y,z). \end{aligned} \tag{1.140}$$

From Equations (1.138) to (1.140), we obtain

$$HL_z\psi_E = L_zH\psi_E,$$

or

$$[H, L_z] = 0. \tag{1.141}$$

In other words, the operators H and L_z share the same set of eigenfunctions. It is straightforward to show that

$$[L_z, L^2] = 0, \tag{1.142}$$

which implies that we can choose a common set of eigenfunctions for the operators H, L_z, L^2. Let's assume that the wave function $\psi_E(x, y, z)$ is given by

$$\psi_E(x, y, z) = R_E(r)Y_\lambda(\theta, \phi), \tag{1.143}$$

where (r, θ, ϕ) is a set of spherical coordinates, and $Y_\lambda(\theta, \phi)$ is a simultaneous eigenfunction of L_z and L^2 such that

$$\begin{aligned} L_zY_{lm}(\theta,\phi) &= m\hbar Y_{lm}(\theta,\phi), \\ L^2Y_{lm}(\theta,\phi) &= l(l+1)\hbar^2 Y_{lm}(\theta,\phi), \\ Y_{lm}(\theta,\phi) &= \langle\theta,\phi|lm\rangle. \end{aligned} \tag{1.144}$$

By using the definition of the differential operator $\vec{\nabla}$

$$\begin{aligned} \vec{\nabla} &= \hat{x}\frac{\partial}{\partial x} + \hat{y}\frac{\partial}{\partial y} + \hat{z}\frac{\partial}{\partial z} \\ &= \hat{r}\frac{\partial}{\partial r} + \hat{\theta}\frac{\partial}{r\partial\theta} + \hat{\phi}\frac{\partial}{r\sin\theta\partial\phi}, \end{aligned} \tag{1.145}$$

we get

$$\begin{aligned} \frac{\partial}{\partial x} &= \hat{x}\cdot\vec{\nabla} \\ &= \sin\theta\cos\phi\frac{\partial}{\partial r} + \cos\theta\cos\phi\frac{\partial}{r\partial\theta} - \frac{\sin\phi}{r\sin\theta}\frac{\partial}{\partial\phi}, \end{aligned} \tag{1.146}$$

$$\begin{aligned} \frac{\partial}{\partial y} &= \hat{y}\cdot\vec{\nabla} \\ &= \sin\theta\sin\phi\frac{\partial}{\partial r} + \cos\theta\sin\phi\frac{\partial}{r\partial\theta} + \frac{\cos\phi}{r\sin\theta}\frac{\partial}{\partial\phi}, \end{aligned} \tag{1.147}$$

and

$$\begin{aligned}\frac{\partial}{\partial z} &= \hat{z}\cdot\vec{\nabla} \\ &= \cos\theta\frac{\partial}{\partial r} - \sin\theta\frac{\partial}{r\partial\theta}.\end{aligned} \tag{1.148}$$

Then, by combining Equations (1.135) to (1.148), we obtain

$$\begin{aligned}L_x &= i\hbar\left[\cot\theta\cos\phi\frac{\partial}{\partial\phi} + \sin\phi\frac{\partial}{\partial\theta}\right], \\ L_y &= i\hbar\left[\cot\theta\sin\phi\frac{\partial}{\partial\phi} - \cos\phi\frac{\partial}{\partial\theta}\right], \\ L_z &= -i\hbar\frac{\partial}{\partial\phi}.\end{aligned} \tag{1.149}$$

We also define the ladder operators $L_\pm$ as

$$\begin{aligned}L_\pm &= L_x \pm L_y \\ &= \hbar\exp(\pm i\phi)\left(\pm\frac{\partial}{\partial\theta} + i\cot\theta\frac{\partial}{\partial\phi}\right).\end{aligned} \tag{1.150}$$

Then, we get the following relations among operators:

$$\begin{aligned}&[L_z, L_\pm] = \pm\hbar L_\pm, \\ &[L_+, L_-] = 2\hbar L_z, \\ &L_+L_- = L^2 - L_z^2 + \hbar L_z, \\ &L_-L_+ = L^2 - L_z^2 - \hbar L_z, \\ &(L_\pm)^\dagger = L_\mp.\end{aligned} \tag{1.151}$$

This implies that

$$\begin{aligned}\langle lm|L_+L_-|lm\rangle &= \langle lm|L^2 - L_z^2 + \hbar L_z|lm\rangle \\ &= \hbar^2\left(l(l+1) - m^2 + m\right) \geq 0,\end{aligned}$$

and as a result

$$-l \leq m \leq l+1. \tag{1.152}$$

Similarly, we also have

$$-l \le m \le l, \tag{1.153}$$

from

$$\langle lm| L_- L_+ |lm\rangle \ge 0. \tag{1.154}$$

Also, we have

$$\langle lm| L^2 |lm\rangle = \hbar^2 l(l+1) \ge 0. \tag{1.155}$$

Next, consider $L_\pm |lm\rangle$. From $[L_z, L_\pm] = \pm\hbar L_\pm$, we get

$$\begin{aligned} L_z L_\pm |lm\rangle &= ([L_z,\ L_\pm] + L_\pm L_z)|lm\rangle \\ &= (L_\pm L_z \pm \hbar L_\pm)|lm\rangle \\ &= \hbar(m \pm 1)|lm\rangle. \end{aligned} \tag{1.156}$$

Thus, $L_\pm |lm\rangle$ are eigenstates of L_z with eigenvalues $m \pm 1$:

$$L_\pm |lm\rangle = c_\pm |lm \pm 1\rangle. \tag{1.157}$$

For this reason, we call $L_\pm$ the ladder operators, in particular, the angular momentum-raising and -lowering operators. The normalization constants $c_\pm$ are evaluated with the aid of Equation (1.151):

$$\begin{aligned} |c_\pm|^2 &= \langle lm| L_\mp L_\pm |lm\rangle \\ &= \hbar^2 (l(l+1) - m(m \pm 1)). \end{aligned} \tag{1.158}$$

If we take $c_\pm$ as positive, then we have the following expressions:

$$c_\pm = \hbar\sqrt{l(l+1) - m(m \pm 1)}. \tag{1.159}$$

In order to find the eigenfunction of the orbital angular momentum, we start with

$$\langle \theta,\phi| L_z |lm\rangle = -i\hbar \frac{\partial}{\partial\phi} \langle \theta,\phi|lm\rangle = \hbar m \langle \theta,\phi|lm\rangle, \tag{1.160}$$

which gives

$$\langle \theta,\phi|lm\rangle = e^{im\phi} \langle \theta,0|lm\rangle. \tag{1.161}$$

By the way, Equations (1.157) and (1.159) imply that

$$L_+|ll\rangle = 0, \tag{1.162}$$

Or, from Equation (1.150),

$$\left(\frac{\partial}{\partial\theta} - l\cot\theta\right)\langle\theta,\phi|lm\rangle = 0. \tag{1.163}$$

Since the function $(\sin\theta)^l$ satisfies Equation (1.163), the orbital angular momentum eigenfunction has the following functional form:

$$\begin{aligned} Y_{ll}(\theta,\phi) &= \langle\theta,\phi|ll\rangle \\ &= \alpha e^{il\phi}(\sin\theta)^l. \end{aligned}$$

The normalization constant α is evaluated from the condition

$$\begin{aligned} 1 &= \int_0^{2\pi} d\phi \int_{-1}^{1} d(\cos\theta)|\alpha|^2(\sin\theta)^{2l} \\ &= 2\pi|\alpha|^2 \int_{-1}^{1} du(1-u^2)^l \\ &= 2\pi|\alpha|^2 \frac{2^{2l+1}(l!)^2}{(2l)!(2l+1)}, \end{aligned} \tag{1.164}$$

which gives

$$Y_{ll}(\theta,\phi) = \frac{(-1)^l}{2^l l!}\sqrt{\frac{(2l+1)(2l)!}{4\pi}}e^{il\phi}(\sin\theta)^l. \tag{1.165}$$

Here, $(-1)^l$ is the conventional phase factor. In order to get $Y_{lm}(\theta,\phi)$, we apply L_- to Equation (1.165) $(l-m)$ times and from Equation (1.157) we obtain

$$Y_{lm}(\theta,\phi) = \frac{(-1)^l}{2^l l!}\sqrt{\frac{(2l+1)(l+m)!}{4\pi(l-m)!}}\left(\frac{d}{d(\cos\theta)}\right)^{l-m}(\sin\theta)^l. \tag{1.166}$$

In mathematical physics, the function $Y_{lm}(\theta,\phi)$ is called as spherical harmonics.

1.8 QUANTIZATION OF ELECTROMAGNETIC FIELDS

There are cases where one needs to treat the electromagnetic field quantum mechanically, such as the spontaneous emission, the Lamb shift, the resonance florescence, and nonclassical light, such as squeezed states. Also, the fluctuation intensity of the laser near the threshold needs to be treated by the quantized electromagnetic fields. The classical electromagnetic fields are governed by Maxwell's equations (1.42). In most treatment of the quantized electromagnetic fields, unbound regions are considered, and vector potentials are used. In many applications, we are primarily considering the interaction between the laser radiation with electrons in atoms or quantum structures, such as the semiconductor quantum dot within the electric-dipole approximation. In such a case, it is convenient to develop the theory in a gauge-invariant form more appropriate for quantum electronics, emphasizing the electric and magnetic fields;

$$\begin{aligned} \vec{\nabla} \times \vec{H} &= \frac{\partial \vec{D}}{\partial t}, \\ \vec{\nabla} \times \vec{E} &= -\frac{\partial \vec{B}}{\partial t}, \\ \vec{\nabla} \cdot \vec{B} &= 0, \\ \vec{\nabla} \cdot \vec{E} &= 0, \end{aligned} \tag{1.167}$$

with $\vec{B} = \mu_0 \vec{H}$ and $\vec{D} = \varepsilon_0 \vec{E}$. We also take the electric field to have the spatial dependence appropriate for the cavity resonator with propagation axis in the z-direction such that [6]

$$E_x(z,t) = q(t)\left(\frac{2\Omega^2 M}{V\varepsilon_0}\right)^{1/2} \sin Kz \tag{1.168}$$

where V is the cavity volume, $K = \Omega/c$ is the wave number, M is the constant with the dimension of mass, and $q(t)$ is the quantum mechanical variable with the dimension of length. Then, from Maxwell's equations (1.167),

$$\begin{aligned} \vec{\nabla} \times \vec{H} &= -\hat{x}\frac{\partial H_y}{\partial z} \\ &= \hat{x}\varepsilon_0 \dot{q}(t)\left(\frac{2\Omega^2 M}{V\varepsilon_0}\right)^{1/2} \sin Kz, \end{aligned} \tag{1.169}$$

we obtain the magnetic field

$$H_y(z,t) = \dot{q}(t)\frac{\varepsilon_0}{K}\left(\frac{2\Omega^2 M}{V\varepsilon_0}\right)^{1/2} \cos Kz. \tag{1.170}$$

The total electromagnetic energy contained in the cavity is then

$$\begin{aligned} H &= \frac{1}{2}\int_V dv\{\varepsilon_0 E_x^2 + \mu_0 H_y^2\} \\ &= \frac{1}{2}\int_V dv\left\{q^2(t)\frac{2\Omega^2 M}{V}\sin^2 Kz + \dot{q}^2(t)\frac{2M}{V}\cos^2 Kz\right\} \\ &= \frac{p^2(t)}{2M} + \frac{1}{2}M\Omega^2 q^2(t), \end{aligned} \tag{1.171}$$

where we have defined the generalized momentum $p(t) = M\dot{q}(t)$. By comparing Equation (1.105), we see that the total electromagnetic energy inside the cavity is the same as that of the simple harmonic oscillator with mass M and angular frequency Ω. We can now define the creation and annihilation operator for a single mode of electromagnetic field quanta denoted as a photon by

$$\begin{aligned} a(t) &= \frac{1}{\sqrt{2\hbar M\Omega}}(M\Omega q(t) + ip(t)), \\ a^\dagger(t) &= \frac{1}{\sqrt{2\hbar M\Omega}}(M\Omega q(t) - ip(t)). \end{aligned} \tag{1.171}$$

The quantization rule for the photon is then

$$[q, p] = i\hbar, \quad [a, a^\dagger] = 1. \tag{1.172}$$

We also have

$$a(t) = a(0)\exp(-i\Omega t), \quad a^\dagger(t) = a(0)\exp(-i\Omega t). \tag{1.173}$$

Then, the quantized electric field can be written as

$$E_x(z,t) = (a + a^\dagger)E_o \sin Kz, \tag{1.174}$$

where E_o is the electric field per photon. The total electromagnetic energy contained in the cavity is

$$H = \hbar\Omega(a\ a + 1/2), \tag{1.175}$$

and the n-photon state in which n photons is defined as $|n\rangle$, where each photon has an energy $\hbar\Omega$. We interpret the vacuum state $|0\rangle$ as a state which contains no photon. Photons are quanta of a single mode of the electromagnetic fields, and are not localized at any particular position within the cavity. Rather, they are spread out over the entire cavity. The quantized electromagnetic fields seem to offer amazingly satisfying accounts of a very wide range of radiative phenomena, and there is no real need to have a corpuscular theory of photons.

When there is more than one mode of electromagnetic fields in the cavity, the electromagnetic fields can be expanded by multi-modes

$$\begin{aligned} \vec{E}(z,t) &= x\sum_n q_n(t)\left(\frac{2\Omega_n^2 M_n}{V\varepsilon_0}\right)^{1/2} \sin K_n z, \\ \vec{H}(z,t) &= y\sum_n \dot{q}_n(t)\frac{\varepsilon_0}{K_n}\left(\frac{2\Omega_n^2 M_n}{V\varepsilon_0}\right)^{1/2} \cos K_n z, \end{aligned} \tag{1.176}$$

with $\Omega_n = n\pi c/L$, $K_n = n\pi/L$, and L is the length of the cavity. The total electromagnetic energy contained in the cavity is

$$H = \frac{1}{2}\sum_n \left(M_n \Omega_n^2 q_n^2 + \frac{p_n^2}{M_n} \right), \tag{1.177}$$

and

$$[q_n, p_{n'}] = i\hbar\delta_{n,n'}, \quad [q_n, q_{n'}] = [p_n, p_{n'}] = 0. \tag{1.178}$$

The annihilation and creation operators for the mode n are

$$\begin{aligned} a_n(t) &= (2\hbar M_n\Omega_n)^{-1/2}(M_n\Omega_n q_n(t) + ip_n(t)), \\ a_n^\dagger(t) &= (2\hbar M_n\Omega_n)^{-1/2}(M_n\Omega_n q_n(t) - ip_n(t)), \end{aligned} \tag{1.179}$$

and a quantized electric field is given by

$$E_x(z,t) = \sum_n (a_n + a_n^\dagger) E_n \sin K_n z. \tag{1.180}$$

For a mode s, we have

$$a_s^\dagger a_s |n_s\rangle = \hbar\Omega_s |n_s\rangle.$$

The general eigenstate of H has n_1 photons in the first mode, n_2 in the second, n_s in the sth, and so forth, and can be written as

$$\begin{aligned} |n_1 n_2 \ldots n_s \ldots\rangle &\equiv |\{n_s\}\rangle \\ &= |n_1\rangle|n_2\rangle\ldots|n_s\rangle\ldots \end{aligned} \tag{1.181}$$

The annihilation operator a_s destroys a photon in the n_s mode alone:

$$a_s|n_1 n_2 \ldots n_s \ldots\rangle = \sqrt{n_s}\,|n_1 n_2 \ldots n_s - 1 \ldots\rangle. \tag{1.182}$$

The general state vector for the field is a superposition of these eigenstates:

$$|\psi\rangle = \sum_{n_1, n_2, \ldots n_s, \ldots} C_{n_1 n_2 \ldots n_s \ldots} |n_1 n_2 \ldots n_s \ldots\rangle. \tag{1.183}$$

This state includes the correlations between different field modes and is called an entangled state. The entangled state arises from perturbations such as the atom-field interaction coupled by an electric-dipole interaction.

1.9 PERTURBATION THEORY

In this section, we deal systematically with the technique for calculating the effect of small extra potential, or Hamiltonian H', acting on a system governed mainly by a free Hamiltonian H_0. Suppose for simplicity that our unperturbed Hamiltonian H_0 governs the noninteracting system with

$$H_0|\Psi_0\rangle = E_0|\Psi_0\rangle, \tag{1.184}$$

and the corresponding eigenstate is normalized. When the perturbation is applied, we shall need to solve the equation

$$(E - H_0)|\Psi\rangle = H'|\Psi\rangle. \tag{1.185}$$

Provided we make no condition on the normalization of the new eigenstate $|\Psi\rangle$, we may always split it into two terms [3]

$$|\Psi\rangle = |\Psi_0\rangle + |\Phi\rangle, \tag{1.186}$$

where the change due to the perturbation is to be orthogonal to the unperturbed state:

$$\langle \Psi_0 | \Phi \rangle = 0. \tag{1.187}$$

This condition can be expressed by the formal introduction of projection operator P:

$$P = |\Psi_0\rangle\langle\Psi_0|, \tag{1.188}$$

which acts on any quantum state $|\Phi\rangle$ as

$$P|\Phi\rangle = |\Psi_0\rangle\langle\Psi_0|\Phi\rangle.$$

The complementary projection operator is defined as

$$Q = 1 - P. \tag{1.189}$$

The complementary operator is idempotent, that is, $Q^2 = Q$, as in the case of P, and it projects any state upon the manifold in Hilbert space that is orthogonal to $|\Psi_0\rangle$. Then one can see that, from Equations (1.186) and (1.187), one can write

$$Q|\Psi\rangle = |\Phi\rangle. \tag{1.190}$$

Applying the unperturbed Hamiltonian operator to this function, we get

$$\begin{aligned} QH_0|\Phi\rangle &= H_0|\Phi\rangle - |\Psi_0\rangle\langle\Psi_0|H_0|\Phi\rangle \\ &= H_0|\Phi\rangle - E_0|\Psi_0\rangle\langle\Psi_0|\Phi\rangle \\ &= H_0|\Phi\rangle \\ &= H_0 Q|\Phi\rangle, \end{aligned} \tag{1.191}$$

showing that Q commutes with H_0. The one we write

$$(E - H_0)|\Psi\rangle = (E - E_0)|\Psi_0\rangle + (E - H_0)|\Phi\rangle, \tag{1.192}$$

$$Q(E - H_0)|\Psi\rangle = (E - H_0)Q|\Psi\rangle = (E - H_0)|\Phi\rangle, \tag{1.193}$$

and from Equations (1.185) and (1.193), we obtain

$$(E - H_0)|\Phi\rangle = QH'|\Psi\rangle. \tag{1.194}$$

Applying P to Equation (1.185), we get

$$E = E_0 + \langle\Psi_0|H'|\Psi\rangle. \tag{1.195}$$

Equation (1.194) can also be written as

$$|\Phi\rangle = (E - H_0)^{-1} QH'|\Psi\rangle, \tag{1.196}$$

which implies that there is an operator $(E - H_0)^{-1}$ which is the inverse of $(E - H_0)$. Then, Equations (1.185), (1.186), and (1.192) can be combined into

$$|\Psi\rangle = |\Psi_0\rangle + (E - H_0)^{-1} QH'|\Psi\rangle. \tag{1.197}$$

This is an exact equation, but can only be solved approximately by iteration. Thus

$$\begin{aligned} |\Psi\rangle &= |\Psi_0\rangle + (E - H_0)^{-1} QH'\big(|\Psi_0\rangle + (E - H_0)^{-1} QH'|\Psi\rangle\big) \\ &= |\Psi_0\rangle + (E - H_0)^{-1} QH'|\Psi_0\rangle \\ &\quad + (E - H_0)^{-1} QH'(E - H_0)^{-1} QH'|\Psi_0\rangle + \cdots. \end{aligned} \tag{1.198}$$

This series is known as the Brillouin-Wigner perturbation expansion. Let $\{|\Psi_n\rangle\}$ be the complete eigenset of the free Hamiltonian H_0, then we have

$$\sum_n |\Psi_n\rangle\langle\Psi_n| = I, \tag{1.199}$$

and

$$Q|\Psi_n\rangle = |\Psi_n\rangle, \quad n \neq 0. \tag{1.200}$$

Thus from Equations (1.195) to (1.200), we obtain the energy equation (1.195), and represent all states and operators in eigenstates of H_0. We get

$$\begin{aligned} E &= E_0 + \langle\Psi_0|H'|\Psi\rangle \\ &= E_0 + \langle\Psi_0|H'\left(|\Psi_0\rangle + \frac{1}{E - H_0} QH'|\Psi_0\rangle + \cdots\right) \\ &= E_0 + \langle\Psi_0|H'|\Psi_0\rangle + \sum_n \langle\Psi_0|(E - H_0)^{-1}|\Psi_n\rangle\langle\Psi_n|QH'|\Psi_0\rangle + \cdots \\ &= E_0 + \langle\Psi_0|H'|\Psi_0\rangle + \sum_{n\neq 0} \frac{|\langle\Psi_0|H'|\Psi_n\rangle|^2}{E - E_n} + \cdots. \end{aligned} \tag{1.201}$$

The projection operator Q automatically excludes certain matrix elements from the summations in each term of the series. It is an implicit equation to be solved for E. Consider the very simple case where H' couples just two states of nearly equal energy. If the expectation value of H' is zero in each of these states, we get

$$E = E_0 + \frac{|\langle \Psi_0 | H' | \Psi_1 \rangle|^2}{E - E_1} = E_0 + \frac{|H'_{01}|^2}{E - E_1}, \tag{1.202}$$

which has two roots to be obtained from the equation

$$E^2 - (E_0 + E_1)E + E_0 E_1 - |H'_{01}|^2 = 0. \tag{1.203}$$

Note that this is the same as the solution of the determinantal equation for the degenerate or nearly degenerate Rayleigh-Schrödinger perturbation theory, that is

$$\begin{vmatrix} E - E_0 & H'_{01} \\ H'_{10} & E - E_1 \end{vmatrix} = 0. \tag{1.204}$$

PROBLEMS

1. We consider the ket vector $|a\rangle$ represented by a column vector given by

$$|a\rangle = \begin{pmatrix} a_1 \\ a_2 \\ \cdot \\ \cdot \\ \cdot \\ a_\infty \end{pmatrix},$$

with complex quantities $a_1, a_2, a_3, \ldots, a_\infty$, and the bra vector given by $\langle b| = (b_1^* \quad b_2^* \quad b_3^* \ldots . b_\infty^*)$. The outer product of vectors $|a\rangle$ and $|b\rangle$ is defined by

$$|a\rangle\langle b| = \begin{pmatrix} a_1 b_1^* & a_1 b_2^* & \cdots & a_1 b_\infty^* \\ a_2 b_1^* & a_2 b_2^* & \cdots & a_2 b_\infty^* \\ \vdots & \vdots & \ddots & \vdots \\ a_\infty b_1^* & a_\infty b_2^* & \cdots & a_\infty b_\infty^* \end{pmatrix}.$$

In two-dimensional Hilbert space, the basis vectors are given by

$$|0\rangle = \begin{pmatrix} 1 \\ 0 \end{pmatrix} \quad \text{and} \quad |1\rangle = \begin{pmatrix} 0 \\ 1 \end{pmatrix}.$$

The Pauli operators are defined by

$$\sigma_x = \begin{pmatrix} 0 & 1 \\ 1 & 0 \end{pmatrix}, \quad \sigma_y = \begin{pmatrix} 0 & -i \\ i & 0 \end{pmatrix}, \quad \sigma_z = \begin{pmatrix} 1 & 0 \\ 0 & -1 \end{pmatrix}.$$

Obtain the spectral representation of Puali operators from Equation (1.25).

2. If A and B are two fixed noncommuting operators, and λ is a parameter, then show that

$$\exp(\lambda A)B\exp(-\lambda A) = B + \lambda[a,b] + \frac{\lambda^2}{2!}[A,[A,B]] + \frac{\lambda^3}{3!}[A,[A,[A,B]]] + \cdots.$$

When $[A,[A,B]] = [B,[A,B]] = 0$, show that

$$\exp(A+B) = \exp(A)\exp(B)\exp\left(-\frac{1}{2}[A,B]\right).$$

Hint: Consider the operator function $f(\lambda) = \exp(\lambda A)\exp(\lambda B)$, and differentiate $f(\lambda)$ with respect to λ to obtain $df(\lambda)/d\lambda = (A + \exp(\lambda A)B\exp(-\lambda A))f(\lambda)$, and solve the differential equation.

3. In one dimension, the Hamiltonian is given by $\hat{H} = \hat{p}^2/2m + V(\hat{x})$. Let $|E_n\rangle$ be the eigenvector of $\hat{H}$, and E_n be the eigenvalue. Then show that

$$\sum_{n'} |\langle E_{n''}|\hat{x}|E_{n'}\rangle|^2 (E_{n'} - E_{n''}) = \frac{\hbar^2}{2m}.$$

4. In two-dimensional Hilbert space, the basis vectors are given by

$$|0\rangle = \begin{pmatrix} 1 \\ 0 \end{pmatrix} \quad \text{and} \quad |1\rangle = \begin{pmatrix} 0 \\ 1 \end{pmatrix}.$$

When the Hamiltonian of this two-dimensional system is given by

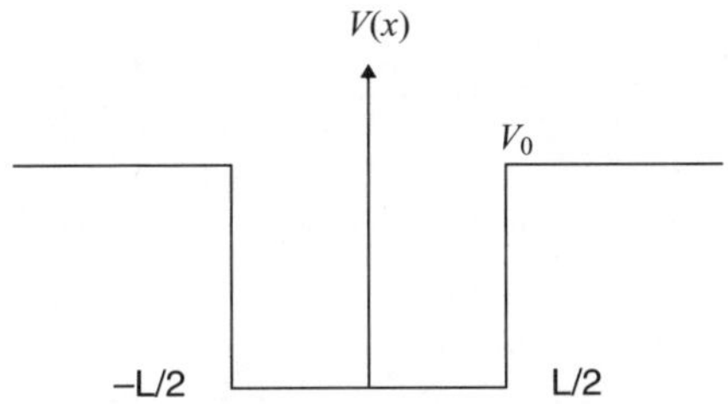

Figure 1.1. Finite potential well.

$$H = E_0|0\rangle\langle 0| + E_1|1\rangle\langle 1| + V(|1\rangle\langle 0| + |0\rangle\langle 1|),$$

find eigenvectors and eigenvalues of H.

5. A one-dimensional harmonic oscillator is under the electric field F, and the total Hamiltonian is given by $H = p^2/2m + \frac{1}{2}\omega^2 x^2 - eFx$. Using the perturbation theory equation (1.201), calculate the ground state energy.

6. Find the wave functions and the energy levels of a particle in the potential well given by $V(x) = 0$ if $|x| \leq L/2$ and $V(x) = V_0 > 0$ if $|x| > L/2$ (Fig. 1.1).

REFERENCES

[1] J. J. Sakurai, *Modern Quantum Mechanics*. New York: Addison Wesley, 1994.

[2] R. P. Feynman and A. R. Hibbs, *Quantum Mechanics and Path Integrals*. New York: McGraw-Hill, 1965.

[3] K. Gottfried, *Quantum Mechanics*. London: Benjamin Cummings, 1966.

[4] H. Goldstein, *Classical Mechanics*. Reading, MA: Addison Wesley, 1980.

[5] R. L. White, *Basic Quantum Mechanics*. New York: McGraw-Hill, 1966.

[6] M. Sargent, M. O. Scully, and W. E. Lamb, *Laser Physics*. London: Addison-Wesley, 1974.

2 Basic Quantum Statistical Mechanics

2.1 ELEMENTARY STATISTICAL MECHANICS

If a system in equilibrium can be in any of N states, then the probability of the system having energy E_n is given by $(1/Z)\exp(-E_n/k_BT)$, where the partition function Z is defined by [1]

$$Z = \sum_{n=1}^{N} \exp(-E_n/k_BT). \tag{2.1}$$

Here, k_B is a Boltzmann constant, and T is a temperature. For quantum mechanical system of states $\{|n\rangle\}$, where $|n\rangle$ denotes the state with energy E_n, the expected value of the quantum mechanical operator A is defined by

$$\langle A \rangle = \frac{1}{Z}\sum_{n} \langle n|A|n\rangle \exp(-E_n/k_BT). \tag{2.2}$$

This fundamental law is the essence of statistical mechanics, with the concept of thermal equilibrium and temperature. If a system is very weakly coupled to a heat bath at a given temperature, if the coupling is indefinite or not known precisely, if the coupling has been on for a long time, and if all the "fast" processes have happened and all the "slow" processes have not, the system is said to be in thermal equilibrium. Consider two different states of a system that has the same energy, $E_j = E_k$. The probabilities of the system being in either state j and k are then equal. Since the system remains in contact with the heat bath for a long time, one would expect states of equal energy to be equally likely. Also, states of different energies would be expected to have different probabilities.

Engineering Quantum Mechanics, First Edition. Doyeol Ahn, Seoung-Hwan Park.

Because two states of the same energy are equally probable, the probability of a state having energy E is a function only of energy; $P = P(E)$. Consider a system, S, in equilibrium with a large heat bath H. It is assumed that the heat bath is extremely large and its total energy E very great. Also, the possible energy levels of the heat bath may be assumed to be quasi-continuous. Let the energy levels of S be denoted by E_j. Then $H_i \gg E_j$ for all i, j. We now consider the bath plus the system as a new system, T_i, which is also in thermal equilibrium. T has a definite energy, but as that energy is not fixed exactly (the bath is in contact with the outside world), we can assume that the energy may be anywhere in the range $E_0 \pm \Delta$. If Δ is sufficiently small, we can assume that the states of the heat bath are equally likely in the range $H_i \pm \Delta$. Let $\eta(H_r)$ be the number of states per unit energy range in the heat bath H around energy H_r. The probability, $P(E_r)$, that S is in a state with energy E_r, is proportional to the number of ways S can have that energy. From $E_0 = E_r + H_i$, the number of states around E_r is the same as the number of states around H_i, and is given by $\eta(E_0 - E_r)\times 2\Delta$. Then the ratio of the probabilities $P(E_r)$ and $P(E_{r'})$ is

$$\begin{aligned}\frac{P(E_r)}{P(E_{r'})} &= \frac{\eta(E_0 - E_r)2\Delta}{\eta(E_0 - E_{r'})2\Delta} \\ &= \frac{\exp(\log\eta(E_0 - E_r))}{\exp(\log\eta(E_0 - E_{r'}))} \\ &= \exp(\log\eta(E_0 - E_r) - \log\eta(E_0 - E_{r'})).\end{aligned} \tag{2.3}$$

For $E_r \ll E_0$, we have the following approximation:

$$\log\eta(E_0 - E_r) \approx \log\eta(E_0) - E_r \frac{d}{dE}\log\eta(E)\bigg|_{E_0}. \tag{2.4}$$

Then we have

$$\frac{P(E_r)}{P(E_{r'})} = \exp(-\beta(E_r - E_{r'})), \tag{2.5}$$

or

$$P(E_r) \propto \exp(-\beta E_r), \tag{2.6}$$

where

$$\beta = \left.\frac{d\log\eta(E)}{dE}\right|_{E_0}. \tag{2.7}$$

From Equations (2.1), (2.6), and (2.7), we obtain

$$P(E_r) = \frac{\exp(-\beta E_r)}{\sum_r \exp(-\beta E_r)}, \tag{2.8}$$

with

$$\beta = k_B T. \tag{2.9}$$

We can regard Equations (2.7) and (2.9) as the definition of the temperature T.

We now consider a quantum system of N identical particles and assume that there is no interaction among them. Any two configurations of the system that differ only by an interchange of two or more identical particles are regarded as one and the same state. Thus, the state of the system is determined by giving the number of particles n_a with energy E_a. The problem is to calculate the partition function with the condition that $\sum_a n_a$ = constant. The values admitted for every possible n_a may be:

$$\begin{aligned} &(a)\ n_a = 0,\ 1,\ 2,\cdots: \quad \text{Bose–Einstein case,} \\ &(b)\ n_a = 0,\ 1: \quad \text{Fermi–Dirac case.} \end{aligned} \tag{2.10}$$

For Bose particles, any number of particles may occupy a given state. For Fermi particles, however, there can only be one particle at most in each state. This is due to the Pauli exclusion principle.

A state of the system is described by the set of numbers n_a, which can take any set of values allowed both by the statistics and by the condition

$$\sum_a n_a = N \tag{2.11}$$

and

$$Q = \sum_{n_1, n_2, n_3 \ldots} \exp\left(-\beta \sum_a n_a E_a\right). \tag{2.12}$$

We consider the system of particles to be a box connected to a large reservoir of particles, and by assuming that it is possible for particles to pass to and from the reservoir. We further assume that the statistical mechanics of the system act as if it took energy μ to move a particle from the box to the reservoir. For a given μ, the energy levels of a particle in the box becomes $E_a - \mu$ rather than E_a. The partition function also becomes then

$$\begin{aligned} Q^{(\mu)} &= \sum_{n_1, n_2, n_3 \dots} \exp\left(-\beta \sum_a n_a (E_a - \mu) \right) \\ &= \prod_a \left(\sum_{n_a} \exp(-\beta n_a (E_a - \mu)) \right) \end{aligned} \tag{2.13}$$

with $N = \sum_a n_a$. The second line of the Equation (2.13) can be easily derived:

$$\begin{aligned} &\prod_a \left(\sum_{n_a} \exp(-\beta n_a K_a) \right) \\ &= \left(\sum_{n_1} \exp(-\beta n_1 K_1) \right) \left(\sum_{n_2} \exp(-\beta n_2 K_2) \right) \cdots \left(\sum_{n_n} \exp(-\beta n_n K_n) \right) \\ &= \sum_{n_1, n_2 \dots} \exp\left(-\beta \sum_a n_a K_a \right), \end{aligned}$$

where $K_a \equiv E_a - \mu$. From simplified expression of the partition function

$$\begin{aligned} Q^{(\mu)} &= \sum_{n_1, n_2 \dots} \exp\left(-\beta \sum_a n_a (E_a - \mu) \right) \\ &\equiv \exp(-\beta g), \end{aligned} \tag{2.14}$$

we obtain

$$g = -\frac{1}{\beta} \ln Q^{(\mu)}. \tag{2.15}$$

Then we get the following relations:

$$
\begin{aligned}
\frac{\partial}{\partial\mu}Q^{(\mu)} &= \sum_{n_1,n_2\ldots}\beta\left(\sum_a n_a\right)\exp\left(-\beta\sum_a n_a(E_a-\mu)\right)\\
&= \sum_{n_1,n_2\ldots}\beta N\exp\left(-\beta\sum_a n_a(E_a-\mu)\right),
\end{aligned}
\tag{2.16}
$$

and

$$
\begin{aligned}
\langle N\rangle &= \frac{1}{Q^{(\mu)}}\sum_{n_1,n_2\ldots}N\exp\left(-\beta\sum_a n_a(E_a-\mu)\right)\\
&= \frac{1}{\beta Q^{(\mu)}}\frac{\partial}{\partial\mu}Q^{(\mu)}\\
&= -\frac{\partial}{\partial\mu}g.
\end{aligned}
\tag{2.17}
$$

For the purpose of comparing the probability of a state, we assumed that the system acts as if it had energy $E=\Sigma_a n_a(E_a-\mu)$. But μ was inserted only in order to provide the weighting factor that accounts for different probabilities of different number of particles in the system.

For the system of the ideal noninteracting Bose–Einstein particles, we have

$$
\begin{aligned}
\exp(-\beta g) &= Q^{(\mu)}\\
&= \sum_{n_1,n_2\ldots}\exp\left(-\beta\sum_a n_a(E_a-\mu)\right)\\
&= \left(\sum_{n_1}\exp(-\beta n_1(E_1-\mu))\right)\left(\sum_{n_2}\exp(-\beta n_2(E_2-\mu))\right)\cdots\\
&\quad \text{for } n_a=0,1,2,\cdots\\
&= \prod_a\left(\frac{1}{1-\exp(-\beta(E_a-\mu))}\right)
\end{aligned}
$$

or

$$
g=-\frac{1}{\beta}\sum_a\ln(1-\exp(-\beta(E_a-\mu))). \tag{2.18}
$$

From Equations (2.17) and (2.18), the average number of particles in the system is given by

$$\begin{aligned}\langle N\rangle &= -\frac{\partial}{\partial\mu}g \\ &= \sum_a \frac{1}{\exp(\beta(E_a-\mu))-1} \\ &= \sum_a \langle n_a\rangle,\end{aligned} \tag{2.19}$$

where $\langle n_a\rangle$ is the number of Bose–Einstein particles in the particular state with energy E_a, and is sometimes called the Bose–Einstein distribution. It is interesting to note that one can also write

$$\langle n_a\rangle = \frac{\partial}{\partial E_a}g. \tag{2.20}$$

For the system of the ideal noninteracting Fermi–Dirac particles, we have

$$\begin{aligned}\exp(-\beta g) &= Q^{(\mu)} \\ &= \sum_{n_1,n_2\ldots} \exp\left(-\beta\sum_a n_a(E_a-\mu)\right) \\ &= \left(\sum_{n_1}\exp(-\beta n_1(E_1-\mu))\right)\left(\sum_{n_2}\exp(-\beta n_2(E_2-\mu))\right)\cdots \\ &\quad \text{but } n_a = 0 \text{ or } 1 \\ &= \prod_a (1+\exp(-\beta(E_a-\mu))),\end{aligned}$$

or

$$g = -\frac{1}{\beta}\sum_a \ln(1+\exp(-\beta(E_a-\mu))), \tag{2.21}$$

and we obtain

$$\begin{aligned}\langle n_a\rangle &= \frac{\partial}{\partial E_a}g \\ &= \frac{1}{1+\exp(\beta(E_a-\mu))}.\end{aligned} \tag{2.22}$$

Below, we summarize the distribution functions for the Bose–Einstein and the Fermi–Dirac particles:

$$\langle n_a \rangle = \begin{cases} \dfrac{1}{\exp(\beta(E_a - \mu)) - 1}; \text{ Bose–Einstein particles} \\ \dfrac{1}{\exp(\beta(E_a - \mu)) + 1}; \text{ Fermi–Dirac particles} \end{cases} \tag{2.23}$$

2.2 SECOND QUANTIZATION

In the previous chapter, we considered the annihilation and creation operators for simple harmonic oscillators and quantized electromagnetic fields. These operators also provide a convenient representation of the many-particle states in the system of N identical particles. The quantum mechanic of a single particle is formulated in terms of the position operator $\hat{x}$ and the momentum operator $\hat{p}$, and the conventional representation, the coordinate representation, is defined in terms of the eigenfunctions of the position operator. For a system of many identical particles, however, it is convenient to define operators which create or annihilate a particle in specified states. Operators of physical interest can be expressed in terms of these creation and annihilation operators. This new presentation is called the second quantization formulation. Let's assume that the wave functions of N noninteracting identical particles $\varphi_1(x), \varphi_2(x), \cdots$ form the complete orthonormal set where x specifies the coordinates and spin variables that characterize the state of a particle. Instead of giving the complete wave functions of the system of particle, we can obviously describe the system by specifying the number of particles in a given state where the occupation numbers $n_1, n_2, \cdots$ are new variables. In this new representation, we define the new wave function or state vector as $\Phi_{n_1 n_2, \cdots}$, where n_1 particles are in the state 1, n_2 particles in the state 2.... Consider first the case of a system of Bose–Einstein particles. In this case, the complete wave function of a system of Bose–Einstein particles is symmetric with respect to permutations of the variables corresponding to different particles. Then the wave function corresponding to the occupation numbers $n_1, n_2, \cdots$ has the form [2]

$$\Phi_{n_1 n_2 \cdots} = \left(\frac{n_1! n_2! \cdots}{N!} \right)^{1/2} \sum_P \varphi_{p_1}(x_1) \varphi_{p_2}(x_2) \cdots \varphi_{p_N}(x_N) \tag{2.24}$$

where the p_i denotes the different states, and the sum is taken over all possible permutations of different numbers p_i. Now assume that $F^{(1)}$ is the operator that is symmetric in all particles and has the forms

$$F^{(1)} = \sum_n f_n^{(1)} \tag{2.25}$$

where $f_n^{(1)}$ is the operator acting on functions of x_n. When such an operator acts on the function $\Phi_{N_1N_2\cdots}$ it carries it either into the same function or into another function in which the state of one particle has been changed. As a result, the matrix elements of $F^{(1)}$ with respect to the wave function Equation (2.24) have the form

$$\begin{aligned} \bar{F}^{(1)} &= \sum_i f_{ii}^{(1)} n_i && \text{for diagonal elements,} \\ \left(F^{(1)}\right)_{N_i-1,N_k}^{N_i,N_k-1} &= f_{ik}^{(1)} \sqrt{n_i n_k} && \text{for nondiagonal elements,} \end{aligned} \tag{2.26}$$

where

$$f_{ik}^{(1)} = \int \varphi_i{}^*(x) f^{(1)} \varphi_k(x) dx. \tag{2.27}$$

The above relation Equation (2.26) can be conveniently expressed using the annihilation and creation operators and the number states introduced in Equations (1.181) and (1.182) of Chapter 1:

$$a_i \left| n_1 n_2 \cdots n_i \cdots \right\rangle = \sqrt{n_i} \left| n_1 n_2 \cdots n_i - 1 \cdots \right\rangle \tag{2.28}$$

and

$$a_i^\dagger \left| n_1 n_2 \cdots n_i \cdots \right\rangle = \sqrt{n_i + 1} \left| n_1 n_2 \cdots n_i + 1 \cdots \right\rangle$$

where $|n_1 n_2 \cdots\rangle$ denotes the abstract state vector where there are n_1 particles in state 1, n_2 particles in state 2, ... and so on. Using these new notations, the operator $F^{(1)}$ can be written in following second quantized form

$$F^{(1)} = \sum_{i,k} f_{ik}^{(1)} a_i^\dagger a_k. \tag{2.29}$$

The matrix elements of this operator coincide with those given by Equation (2.26) when it acts on the abstract state $|n_1 n_2 \cdots\rangle$. The creation operator $a_k^\dagger$ and the annihilation operator a_i satisfy the following commutation relations:

$$[a_i, a_k^\dagger] = \delta_{ik}, [a_i, a_k] = 0, [a_i^\dagger, a_k^\dagger] = 0. \tag{2.30}$$

A similar representation is possible for a symmetric operator of the form

$$F^{(2)} = \sum_{m,n} f_{mn}^{(2)} \tag{2.31}$$

where $f_{mn}^{(2)}$ acts on functions of x_m and x_n. In the second quantized representation, $F^{(2)}$ has the form

$$F^{(2)} = \sum_{i,k,l,m} f_{iklm}^{(2)} a_i^\dagger a_k^\dagger a_m a_l \tag{2.32}$$

where

$$f_{iklm}^{(2)} = \int \varphi_i * (x_a) \varphi_k * (x_b) f_{ab}^{(2)} \varphi_m(x_b) \varphi_l(x_a) dx_a dx_b. \tag{2.33}$$

In the case of Fermi–Dirac statistics, the complete wave function must be antisymmetric in all variables. This implies that in the case of noninteracting particles, the occupation numbers can only take the values 0 or 1, and the wave function has the form

$$\Phi_{n_1 n_2 \cdots} = \frac{1}{\sqrt{N!}} \sum_P (-1)^P \varphi_{p_1}(x_1) \varphi_{p2}(x_2) \cdots \varphi_{p_N}(x_N) \tag{2.34}$$

where all the numbers $p_1, p_2, \cdots, p_N$ are different, and the symbol $(-1)^P$ means that the odd number of permutations appear in the sum Equation (2.34) with a minus sign. For definiteness, the term in Equation (2.34) for which

$$p_1 < p_2 < \cdots < p_N \tag{2.35}$$

will be chosen with a plus sign. The operator $F^{(1)}$ of the type Equation (2.25) can also be written in the second quantized form

$$F^{(1)} = \sum_{i,k} f_{ik}^{(1)} c_i^\dagger c_k, \tag{2.36}$$

with $f_{ik}^{(1)} = \int \varphi_i * (x) f^{(1)} \varphi_k(x) dx$. However, in Equation (2.36), the fermion annihilation operator c_i and the creation operator $c_k^\dagger$ satisfy the following anticommutation relations:

$$\begin{aligned}\{c_i, c_k^\dagger\} &\equiv c_i c_k^\dagger + c_k^\dagger c_i = \delta_{ik}, \\ \{c_i, c_k\} &= \{c_i^\dagger, c_k^\dagger\} = 0.\end{aligned} \tag{2.37}$$

Similarly, the two-particle operator $F^{(2)}$ has the following second quantized representation:

$$F^{(2)} = \sum_{i,k,l,m} f_{iklm}^{(2)} c_i^\dagger c_k^\dagger c_m c_l, \tag{2.38}$$

where

$$f_{iklm}^{(2)} = \int \varphi_i^*(x_a)\varphi_k^*(x_b) f_{ab}^{(2)} \varphi_m(x_b)\varphi_l(x_a)\,dx_a dx_b.$$

2.3 DENSITY OPERATORS

When we solve complicated quantum mechanical problems, we usually divide the universe into two parts—the system in which we are interested and the rest of the universe. Then, normally, we assume that the system in which we are interested comprises the entire universe. Let us see what happens if we include something outside of the system. In this case, we assume $|\varphi_i\rangle$ to be a complete set of state vectors in the vector space describing the quantum system, and $|R_i\rangle$ to be a complete set for the rest of the universe or the reservoir. Then the most general wave function can be written as

$$|\Psi\rangle = \sum_{i,j} C_{ij} |\varphi_i\rangle \otimes |R_j\rangle. \tag{2.39}$$

Now let $\hat{A}$ be an operator that acts only on the system and does not act on the $|R_j\rangle$. In other words, when $\hat{A}$ acts on the tensor product states, we mean that

$$\hat{A}|\varphi\rangle \otimes |\psi\rangle = \left(\hat{A}|\varphi\rangle\right) \otimes |\psi\rangle. \tag{2.40}$$

Then

$$\begin{aligned}\langle A\rangle &= \langle \Psi|\hat{A}|\Psi\rangle \\ &= \sum_{i,j,k,l} C_{ij}{}^* C_{kl} \langle R_j|\left(\langle \varphi_i|\hat{A}|\varphi_k\rangle\right)|R_l\rangle \\ &= \sum_{i,j,k} C_{ij}{}^* C_{kj} \langle \varphi_i|\hat{A}|\varphi_k\rangle \\ &= \sum_{i,k} \langle \varphi_i|\hat{A}|\varphi_k\rangle \rho_{ki}\end{aligned} \tag{2.41}$$

$$\text{with } \rho_{ki} = \sum_j C_{ij}{}^* C_{kj}, \tag{2.42}$$

where ρ_{ki} is the matrix element of the density operator ρ, and is called the density matrix. The density operator is defined as

$$\rho_{ki} = \langle \varphi_k | \rho | \varphi_i \rangle. \tag{2.43}$$

Then, the expectation value of the operator is expressed as

$$\begin{aligned} \langle \hat{A} \rangle &= \sum_{i,k} \langle \varphi_i | \hat{A} | \varphi_k \rangle \langle \varphi_k | \rho | \varphi_i \rangle \\ &= \sum_i \langle \varphi_i | \hat{A}\rho | \varphi_i \rangle \\ &= Tr(\rho \hat{A}). \end{aligned} \tag{2.44}$$

From Equation (2.42), it is obvious that the density operator ρ is Hermitian. Therefore, it can be diagonalized with a complete orthonormal set of eigenvectors $|\psi_i\rangle$ and real eigenvalues w_i [1]

$$\rho = \sum_i w_i |\psi_i\rangle\langle\psi_i|, \tag{2.45}$$

and for $\hat{A} = \hat{I}$, we obtain

$$\langle \hat{I} \rangle = Tr\rho = \sum_i w_i = 1. \tag{2.46}$$

When $\hat{A} = |\psi_k\rangle\langle\psi_k|$, we have

$$\begin{aligned} w_k &= \langle \Psi | (|\psi_k\rangle\langle\psi_k|) | \Psi \rangle \\ &= \sum_j \langle \Psi | (|\psi_k\rangle | R_j \rangle \langle R_j | \langle \psi_k |) | \Psi \rangle \\ &= \sum_j |\langle R_j | \otimes \langle \psi_k || \Psi \rangle|^2 \\ &\geq 0. \end{aligned} \tag{2.47}$$

Therefore, we have

$$\sum_i w_i = 1 \text{ and } w_i \geq 0, \tag{2.48}$$

which means w_i can be interpreted as a probability of finding the system in the state $|\psi_i\rangle$. Above results suggest that we can reformulate quantum mechanics using density operator ρ, where ρ is of the form $\rho = \Sigma_i w_i|\psi_i\rangle\langle\psi_i|$, with $\Sigma_i w_i = 1$ and $w_i \geq 0$. Given an operator $\hat{A}$, the expectation value of $\hat{A}$ is given by $\langle\hat{A}\rangle = Tr\rho\hat{A}$. If all but one of the w_i are zero, we say that the system is in a pure state; otherwise, it is in a mixed state.

We now consider the operator $|\Psi\rangle\langle\Psi|$ and consider the partial trace of $|\Psi\rangle\langle\Psi|$ with respect to the environment state $|R_j\rangle$, then, from Equations (2.39), (2.42), and (2.43), we obtain

$$\begin{aligned} Tr_E(|\Psi\rangle\langle\Psi|) &= \sum_m \langle R_m|(|\Psi\rangle\langle\Psi|)|R_m\rangle \\ &= \sum_{i,k}\sum_j C_{kj}C_{ij}{}^*|\varphi_k\rangle\langle\varphi_i| \\ &= \rho. \end{aligned} \tag{2.49}$$

When there is no prospect of confusion, we will also denote the partial trace of the operator $|\Psi\rangle\langle\Psi|$ as $\overline{|\Psi\rangle\langle\Psi|}$. Then the density operator, more exactly the reduced density operator ρ, can be defined simply as

$$\rho = \overline{(|\Psi\langle\rangle\Psi|)} \tag{2.50}$$

Since the quantum state $|\Psi\rangle$ satisfies the Schrodinger equation

$$i\hbar\frac{\partial}{\partial t}|\Psi\rangle = H|\Psi\rangle,$$

the density operator satisfies the following equation of motion

$$\frac{\partial\rho}{\partial t} = \frac{1}{i\hbar}(H\rho - \rho H) = \frac{1}{i\hbar}[H,\rho]. \tag{2.51}$$

Now we would like to express the density operator in second quantization. Consider the system of electrons in solid with two energy bands. Details of the band structure will be discussed in the following chapter, and here we consider a model system.

It is often convenient to introduce the new operator, called field operator, which is defined as

$$\Psi(x) = \sum_k a_{ck}\varphi_{ck}(x) + \sum_k a_{vk}\varphi_{vk}(x), \tag{2.52}$$

where we assume the electrons to be a two-band semiconductor. Details of the semiconductor band structure are described in the next chapter, and we can treat it as a model in this section. Here c stands for the conduction band, v the valence band, and k the wave vector. We introduce the following simplified notations:

$$\begin{aligned} c &\equiv 1, \quad c_{1k} = a_{ck} \\ v &\equiv 2, \quad c_{2k} = a_{vk}, \end{aligned}$$

and the state vector

$$|\Psi\rangle = \sum_k (c_{1k}|1k\rangle + c_{2k}|2k\rangle), \quad \langle\Psi| = \sum_k \left(\langle 1k|c_{1k}^{\dagger} + \langle 2k|c_{2k}^{\dagger}\right). \tag{2.53}$$

The density operator is then defined as

$$\begin{aligned} \rho &= \overline{|\Psi\rangle\langle\Psi|} \\ &= \sum_{k.k'} \Big(\overline{c_{1k}c_{1k'}^{\dagger}}\,|1k\rangle\langle 1k'| + \overline{c_{1k}c_{2k'}^{\dagger}}\,|1k\rangle\langle 2k'| + \overline{c_{2k}c_{1k'}^{\dagger}}\,|2k\rangle \\ &\quad \langle 1k'| + \overline{c_{2k}c_{2k'}^{\dagger}}\,|2k\rangle\langle 2k'|\Big). \end{aligned} \tag{2.54}$$

Using random phase approximation, we obtain

$$\rho = \sum_{\alpha,\beta,k} \rho_{\alpha\beta}(k)|\alpha k\rangle\langle\beta k|, \quad \alpha,\beta = 1,2, \tag{2.55}$$

where

$$\rho_{\alpha\beta}(k) = \overline{c_{\beta k}^{\dagger}c_{\alpha k}}. \tag{2.56}$$

We assume that the Hamiltonian of the system is given by

$$H = \sum_{\alpha,\beta,k} E_{\alpha\beta}(k)c_{\alpha k}^{\dagger}c_{\beta k}. \tag{2.57}$$

Then, in the Heisenberg picture, we have

$$\begin{aligned} i\hbar\frac{\partial}{\partial t}c_{\beta k}^{\dagger}c_{\alpha k} &= [c_{\beta k}^{\dagger}c_{\alpha k}, H] \\ &= \sum_{\beta'} E_{\alpha\beta'}(k)c_{\beta k}^{\dagger}c_{\beta' k} - \sum_{\alpha'} E_{\alpha'\beta}(k)c_{\alpha' k}^{\dagger}c_{\alpha k}. \end{aligned} \tag{2.58}$$

From Equations (2.55) to (2.58), we obtain

$$\begin{aligned} i\hbar\frac{\partial}{\partial t}\rho &= \sum_{\beta'} E_{\alpha\beta'}(k)\rho_{\beta'\beta} - \sum_{\alpha'} \rho_{\alpha\alpha'}E_{\alpha'\beta}(k) \\ &= \left[\hat{E}(k),\rho\right] \end{aligned} \tag{2.59}$$

where $\hat{E}(k)$ is the matrix representation of the Hamiltonian Equation (2.57).

When two-particle interaction such as Coulomb interaction becomes important, the Hamiltonian Equation (2.57) need to be modified into

$$\begin{aligned} H &= \sum_{\alpha,\beta,k} E_{\alpha\beta}(k)c^{\dagger}_{\alpha k}c_{\beta k} \\ &+\frac{1}{2}\sum_{k,k',q\neq 0} V(q)\left[c^{\dagger}_{1k+q}c^{\dagger}_{1k'-q}c_{1k'}c_{2k} + c^{\dagger}_{2k+q}c^{\dagger}_{2k'-q}c_{2k'}c_{2k} + 2c^{\dagger}_{1k+q}c^{\dagger}_{2k'-q}c_{2k'}c_{1k}\right]. \end{aligned} \tag{2.60}$$

2.4 THE COHERENT STATE

In Chapter 1, we considered the quantum mechanical description of the radiation field through the use of a particular set of quantum states for the radiation field. These are the stationary states of the noninteracting field, which correspond to the presence of a precisely defined number of photons. At this stage, it is interesting to ask what wave function most closely describe the classical electromagnetic field given by $E_0\cos(\Omega t+\phi)$ and $H_0\sin(\Omega t+\phi)$. It is known that the wave function, which corresponds most closely to the classical field, turns out to be the "coherent" wave function, a displaced ground state of the simple harmonic oscillator. The corresponding state vector is the "coherent state," which is the eigenstate of the positive frequency part of the electric field operator. The coherent state $|\alpha\rangle$ for the single mode oscillator is defined as an eigenstate of the annihilation operator [3]:

$$a|\alpha\rangle = \alpha|\alpha\rangle. \tag{2.61}$$

We can expand the coherent state $|\alpha\rangle$ in terms of the number state $|n\rangle$ developed in Chapter 1 as

$$
\begin{aligned}
|\alpha\rangle &= \sum_n |n\rangle\langle n|\alpha\rangle \\
&= \sum_n |n\rangle\langle 0|\left(a^n/\sqrt{n!}\right)|\alpha\rangle \\
&= \langle 0|\alpha\rangle \sum_n \frac{\alpha^n}{\sqrt{n!}}|n\rangle,
\end{aligned}
\tag{2.62}
$$

where we have used Equation (1.115) of Chapter 1. Using the normalization condition $\langle\alpha|\alpha\rangle = 1$, we obtain that $\langle 0|\alpha\rangle = e^{-(1/2)|\alpha|^2}$. Then, we have

$$
|\alpha\rangle = \exp\left(-\frac{1}{2}|\alpha|^2\right)\sum_n \frac{\alpha^n}{\sqrt{n!}}|n\rangle. \tag{2.63}
$$

The photon number probability distribution $P(n)$ for the coherent state $|\alpha\rangle$ is then given by

$$
P(n) = |\langle n|\alpha\rangle|^2 = \exp\left(-|\alpha|^2\right)|\alpha|^{2n}/n! \tag{2.64}
$$

which turns out to be the Poisson distribution.

One can show that the coherent state $|\alpha\rangle$ is generated by the action on the vacuum state $|0\rangle$ of the Glauber displacement operator

$$
D(\alpha) = \exp\left(\alpha a^\dagger - \alpha^* a\right), \tag{2.65}
$$

with the Baker–Hausdorff relation for operators A and B

$$
\exp(A+B) = \exp(A)\exp(B)\exp(-[A,B]/2), \tag{2.66}
$$

such that

$$
|\alpha\rangle = D(\alpha)|0\rangle. \tag{2.67}
$$

The Glauber displacement operator satisfies the following properties:

$$
D(-\alpha)aD(\alpha) = a + \alpha, \tag{2.68}
$$

and

$$
D(-\alpha)a^\dagger D(\alpha) = a^\dagger + \alpha. \tag{2.69}
$$

One property of the coherent state $|\alpha\rangle$ is that, in contrast to the number states, the coherent states are not mutually orthogonal. The overlap between two coherent states $|\alpha\rangle$ and $|\beta\rangle$ is

$$\begin{aligned}\langle\alpha|\beta\rangle &= \exp\left(-|\alpha|^2/2-|\beta|^2/2\right)\sum_{n.m}\frac{(\alpha^*)^n(\beta)^m}{(n!m!)^{1/2}}\langle n|m\rangle \\ &= \exp\left(\alpha^*\beta-|\alpha|^2/2-|\beta|^2/2\right),\end{aligned} \tag{2.70}$$

and hence

$$|\langle\alpha|\beta\rangle|^2 = \exp\left(-|\alpha-\beta|^2\right). \tag{2.71}$$

Even though the coherent states lack mutual orthogonality, they form an overcomplete set such that the identity relation can be resolved in terms of the coherent states as

$$\frac{1}{\pi}\int d^2\alpha|\alpha\rangle\langle\alpha| = \frac{1}{\pi}\int d(\mathrm{Re}\,\alpha)d(\mathrm{Im}\,\alpha)|\alpha\rangle\langle\alpha| = I. \tag{2.72}$$

The overcomplete relation Equation (2.72) can be derived using

$$\begin{aligned}&\int d^2\alpha(\alpha^*)^n\alpha^m\exp(-|\alpha|^2) \\ &= \int|\alpha|^{n+m}\exp[i(m-n)\theta]\exp(-|\alpha|^2)|\alpha|d|\alpha|d\theta \\ &= \pi(n!)\delta_{n.m},\end{aligned} \tag{2.73}$$

and

$$\alpha = |\alpha|e^{i\theta}. \tag{2.74}$$

The use of coherent states leads to considerable simplification in the calculation of statistical averages related to the radiation field. The radiation field can be described by the density operator, which is defined in the previous section. The density operator is the outer product

$$\rho = |\psi\rangle\langle\psi| \tag{2.75}$$

when the complete state $|\psi\rangle$ is known. However, when one has only the statistical knowledge that the system has probability w_i of being the state $|\psi_i\rangle$, the density operator is given by Equation (2.45)

$$\rho = \sum_i w_i |\psi_i\rangle\langle\psi_i|$$

The density operator is often more easily applicable to the real problem than the state vector, and it can be expanded in terms of the coherent states. The values of the expansion coefficients express the degree to which the radiation field is actually coherent in the correct quantum mechanical sense. Any density operator ρ may be represented in a unique way by means of a function of two complex variables $R(\alpha^*,\beta)$, which is analytic throughout the finite α^* and β planes, and is defined by

$$R(\alpha^*,\beta) = \langle\alpha|\rho|\beta\rangle \exp\left(|\alpha|^2/2 + |\beta|^2/2\right). \tag{2.76}$$

We note that the coherent states form an overcomplete set, and then we can expand ρ as

$$\begin{aligned}\rho &= \frac{1}{\pi^2}\int d^2\alpha d^2\beta |\alpha\rangle\langle\alpha|\rho|\beta\rangle\langle\beta| \\ &= \frac{1}{\pi^2}\int d^2\alpha d^2\beta R(\alpha^*,\beta)\exp\left(-|\alpha|^2/2 - |\beta|^2/2\right)|\alpha\rangle\langle\beta|.\end{aligned} \tag{2.77}$$

By expanding the coherent states in terms of number states, we also get

$$R(\alpha^*,\beta) = \sum_{n,m} \frac{(\alpha^*)^n \beta^m}{(n!m!)^{1/2}} \langle n|\rho|m\rangle. \tag{2.78}$$

From the trace-preserving condition of the density operator, that is, $Tr\rho = 1$, we obtain

$$Tr(\rho) = \frac{1}{\pi}\int d^2\alpha R(\alpha^*.\alpha)\exp(-|\alpha|^2) = 1. \tag{2.79}$$

Although this expansion of the density operator is straightforward, it is quite complicated. We can often obtain a simpler expansion with real coefficient $P(\alpha)$:

$$\rho = \int d^2\alpha P(\alpha)|\alpha\rangle\langle\alpha|. \tag{2.80}$$

The expectation value of the operator A is given by $Tr(\rho A)$.

Then

$$\begin{aligned}\langle A\rangle &= Tr(\rho A)\\ &= \sum_n \langle n|\int d^2\alpha P(\alpha)|\alpha\rangle\langle\alpha|A|n\rangle\\ &= \int d^2\alpha P(\alpha)\sum_n \langle\alpha|A|n\rangle\langle n|\alpha\rangle\\ &= \int d^2\alpha P(\alpha)\langle\alpha|A|\alpha\rangle\\ &= \int d^2\alpha P(\alpha)A(\alpha),\end{aligned} \tag{2.81}$$

where

$$A(\alpha) = \langle\alpha|A|\alpha\rangle. \tag{2.82}$$

This leads to a simple calculation involving only *c* numbers, provided that the operator *A* is expressed in normal order, that is, so that creation operators stand to the left of the annihilation operators in the product appearing in *A*.

2.5 THE SQUEEZED STATE

Squeezed states arise in simple quantum models of a number of nonlinear optical processes including optical parametric oscillation and four-wave mixing [4–6]. The Heisenberg uncertainty principle $\Delta A\Delta B \geq \frac{1}{2}|\langle[A,B]\rangle|$ between the standard deviation of two arbitrary observables, $\Delta A = \sqrt{\langle(A-\langle A\rangle)^2\rangle}$, and similarly for ΔB, has a built-in degree of freedom: One can squeeze the standard deviation of one observable provided that one stretches that for the conjugate observable. All quantum mechanics requires is that the product be bounded from below. Given a field described by the annihilation operator *a*, we form two Hermitian conjugate operators giving its two quadratures as

$$x_\lambda = \frac{1}{2}(ae^{i\lambda} + a^\dagger e^{-i\lambda}), \quad p_\lambda = \frac{1}{2i}(ae^{i\lambda} - a^\dagger e^{-i\lambda}), \tag{2.83}$$

with $[x_\lambda, p_\lambda] = i/2$ so $\Delta x_\lambda \Delta p_\lambda \geq 1/4$. These two operators correspond to position and momentum for the case of a mechanical oscillator. The variance Δx_λ^2 is given by

$$\Delta x_\lambda^2 = \langle x_\lambda^2\rangle - \langle x_\lambda\rangle^2. \tag{2.84}$$

We now consider a quantum state such that the expectation values of the two Hermitian conjugate operators are zero, $\langle x_\lambda \rangle = \langle p_\lambda \rangle = \langle a \rangle = \langle a^\dagger \rangle = 0$. Then we have

$$\begin{aligned}\Delta x_\lambda{}^2 &= \frac{1}{4} + \frac{1}{2}\langle a^\dagger a \rangle + \frac{1}{4}(\langle aa \rangle e^{2i\lambda} + \langle a^\dagger a^\dagger \rangle e^{-2i\lambda}) \\ &= \frac{1}{4} + \frac{1}{2}\langle a^\dagger a \rangle + \frac{1}{2}\mathrm{Re}(\langle aa \rangle e^{2i\lambda}).\end{aligned} \tag{2.85}$$

The minimum variance is

$$\Delta x_\lambda^2 = \frac{1}{4} + \frac{1}{2}\langle a^\dagger a \rangle - \frac{1}{2}|\langle aa \rangle|. \tag{2.86}$$

Likewise, the variance of the conjugate operator is

$$\Delta p_\lambda^2 = \frac{1}{4} + \frac{1}{2}\langle a^\dagger a \rangle - \frac{1}{2}\mathrm{Re}(\langle aa \rangle e^{2i\lambda}), \tag{2.87}$$

so the minimum variance is given by

$$\Delta p_\lambda^2 = \frac{1}{4} + \frac{1}{2}\langle a^\dagger a \rangle + \frac{1}{2}|\langle aa \rangle|, \tag{2.88}$$

which is certainly greater than Equation (2.86). Equations (2.86) and (2.88) give the uncertainty principle

$$\Delta x_\lambda \Delta p_\lambda = \frac{1}{4}\sqrt{(1 + 2\langle a^\dagger a \rangle - 2|\langle aa \rangle|)(1 + 2\langle a^\dagger a \rangle + 2|\langle aa \rangle|)} \geq \frac{1}{4}. \tag{2.89}$$

Squeezing occurs for x_λ if Δx_λ^2 becomes smaller than $\frac{1}{4}$, that is, squeezed below the minimum uncertainty product value. This, however, does not violate the uncertainty principle, since p_λ has a correspondingly increased variance. In this case, it is the $|aa|$ term that leads to squeezing. The simplest single-mode squeezed states are generated by the action on the vacuum state $|0\rangle$ of the squeezing operator [4,5]

$$S(\zeta) = \exp\left(-\frac{\zeta}{2}a^\dagger a^\dagger + \frac{\zeta^*}{2}aa\right) \tag{2.90}$$

where $\zeta = r\exp(i\varphi)$ is the squeezing parameter with modulus r and argument φ. It is easy to show that the squeezing operator is unitary, that is, $S^\dagger(\zeta)=S(-\zeta)$. The single mode squeezed state $|\zeta\rangle$ is defined by

$$|\zeta\rangle = S(\zeta)|0\rangle, \tag{2.91}$$

and is given by

$$|\zeta\rangle = \frac{1}{\sqrt{\cosh r}}\sum_{n=0}^{\infty}\frac{\sqrt{(2n)!}}{n!}\left[-\frac{1}{2}\exp(i\varphi)\tanh r\right]^n |2n\rangle. \tag{2.92}$$

Here, we have used the definition of the exponential function of an operator A as

$$\exp(A) \equiv \sum_n \frac{1}{n!}A^n \tag{2.93}$$

with the fact that $(a^\dagger a + aa^\dagger)|0\rangle = |0\rangle$, and the ordered form of $S(\zeta)$ [4]:

$$\begin{aligned} S(\zeta) = \exp&\left[-\frac{1}{2}(a^\dagger)^2\exp(i\varphi)\tanh r\right]\exp\left[-\frac{1}{2}(aa^\dagger a^\dagger a)\ln(\cosh r)\right] \\ &\times \exp\left[\frac{1}{2}(a)^2\exp(-i\varphi)\tanh r\right]. \end{aligned} \tag{2.94}$$

The squeezed state $|\zeta\rangle$ is a right eigenstate of a ζ-dependent linear combination of a and $a^\dagger$, with eigenvalue 0:

$$[a\cosh r + a\ \exp(i\varphi)\sinh r]|\zeta\rangle = 0. \tag{2.95}$$

For this reason, $|\zeta\rangle$ is termed the "squeezed vacuum."

In the following, we derive the operator ordering theorem equation (2.98). For two operators A and B, which commute with their commutator $[A,B]$, we have

$$\exp[\theta(A+B)] = \exp(\theta A)\exp(\theta B)\exp\left(-\frac{1}{2}\theta^2[A,B]\right). \tag{2.96}$$

Generalized angular momentum operators $J_3(=J_z)$, J_+ and J_- satisfy the commutation relations $[J_3, J_\pm] = \pm J_\pm$ and $[J_+, J_-] = 2J_3$. We assume that $F(\theta) = \exp[i\theta(J_+ + J_-)]$ can be written as

$$F(\theta) = \exp[ip(\theta)J_+]\exp[iq(\theta)J_3]\exp[ir(\theta)J_-]. \tag{2.97}$$

Then,

$$\begin{aligned}
&\frac{d}{d\theta}F(\theta)\\
&\quad = i(J_+ + J_-)F(\theta)\\
&\quad = ip'(\theta)J_+ \exp[ip(\theta)J_+]\exp[iq(\theta)J_3]\exp[ir(\theta)J_-]\\
&\quad + \exp[ip(\theta)J_+]\{iq'(\theta)J_3\}\exp[iq(\theta)J_3]\exp[ir(\theta)J_-] \qquad (2.98)\\
&\quad + \exp[ip(\theta)J_+]\exp[iq(\theta)J_3]\{ir'(\theta)J_-\}\exp[ir(\theta)J_-]\\
&\quad = i[(p'(\theta) - iq'(\theta)p(\theta) + r'(\theta)\exp(-iq(\theta)p^2(\theta))J_+ + r'(\theta)\exp(-iq(\theta)J_-]\\
&\qquad F(\theta) + [(q'(\theta) + 2ir'(\theta)p(\theta)\exp(-iq(\theta))J_3]F(\theta),
\end{aligned}$$

where we have used

$$\exp(A)B\exp(-A) = B + [A,B] + \frac{1}{2!}[A.[A,B]] + \cdots \tag{2.99}$$

in the derivation of Equation (2.98), which in turn gives the following coupled differential equations:

$$\begin{aligned}
p' - ipq' + r'\exp(-iq)p^2 &= 1,\\
r'\exp(-iq) &= 1, \qquad (2.100)\\
q' + 2ir'p\exp(-iq) &= 0.
\end{aligned}$$

The initial conditions are: $p(0) = q(0) = r(0) = 0$. If we set

$$q(\theta) = i\log(\cos^2\theta) \tag{2.101}$$

then we also get

$$r(\theta) = \tan\theta, \tag{2.102}$$

and

$$p(\theta) = \tan\theta. \tag{2.103}$$

As a result, we obtain

$$\exp[i\theta(J_+ + J_-)] = \exp(iJ_+\tan\theta)\exp(-J_3\log[\cos^2\theta])\exp(iJ_-\tan\theta). \tag{2.104}$$

Likewise, for operators K_+, K_- and K_3 which satisfy the commutation relations

$$[K_3, K_\pm] = \pm K_\pm \tag{2.105}$$

and

$$[K_+, K_-] = -2K_3, \tag{2.106}$$

we obtain

$$\exp[i\theta(K_+ + K_-)] = \exp[iK_+ \tanh\theta]\exp[-K_3 \log(\cosh^2\theta)] \\ \exp[iK_- \tanh\theta], \tag{2.107}$$

which is generalized to

$$\begin{aligned} &\exp[\gamma_+ K_+ + \gamma_- K_- + \gamma_3 K_3] \\ &\quad = \exp(\Gamma_+ K_+)\exp(K_3 \log\Gamma_3)\exp(\Gamma_- K_-) \\ &\quad = \exp(\Gamma_- K_-)\exp(-K_3 \log\Gamma_3)\exp(\Gamma_- K_-), \end{aligned} \tag{2.108}$$

where

$$\Gamma_\pm = \frac{2\gamma_\pm \sinh\beta}{2\beta\cosh\beta - \gamma_3 \sinh\beta}, \tag{2.109}$$

$$\Gamma_3 = \left(\cosh\beta - \frac{\gamma_3}{2\beta}\sinh\beta\right)^2, \tag{2.110}$$

and

$$\beta = \left(\frac{1}{4}\gamma_3^2 - \gamma_+\gamma_-\right)^{1/2}. \tag{2.111}$$

If we define

$$K_3 = \frac{1}{4}\left(a^\dagger a + aa^\dagger\right), \tag{2.112}$$

and

$$K_+ = \frac{1}{2}(a^\dagger)^2 = (K_-)^\dagger, \tag{2.113}$$

one can easily show that they satisfy the commutation relations equations (2.105) and (2.106). By setting $\gamma_+ = -\zeta$, $\gamma_- = \zeta^*$ and $\gamma_3 = 0$, and using equations (2.108)–(2.111), we obtain the operator ordering relation equation (2.94).

For two modes A and B, with annihilation operators a and b, the two-mode squeezed state $|\zeta_{AB}\rangle$ is generated by the action on the two-mode vacuum state $|0_A;0_B\rangle$ of the two-mode squeezing operator [4,5]

$$S_{AB}(\zeta) = \exp(-\zeta a^\dagger b^\dagger + \zeta^* ba), \tag{2.114}$$

where $\zeta = r\exp(i\varphi)$ is any complex number. Both single-mode and two-mode squeezing arise in models of two-photon nonlinear optics being associated with degenerate and nondegenerate processes, respectively. It is also shown that the quantum black hole first suggested by Hawking can also be represented by a two-mode squeezed state. By setting

$$K_+ = a^\dagger b^\dagger = (K_-)^\dagger, \tag{2.115}$$

$$K_3 = \frac{1}{2}(a^\dagger a + b^\dagger b), \tag{2.116}$$

$\gamma_+ = -\zeta$, $\gamma_- = \zeta^*$, and $\gamma_3 = 0$, we obtain

$$\begin{aligned} S_{AB}(\zeta) = {} & \exp[-a^\dagger b^\dagger \exp(i\varphi)\tanh r] \\ & \times \exp[-(a^\dagger a + b^\dagger b)\log(\cosh r)]\exp[ba\exp(-i\varphi)\tanh r], \end{aligned} \tag{2.117}$$

from Equations (2.108) to (2.111).

The two-mode squeezed state $|\zeta_{AB}\rangle$ is given by

$$|\zeta_{AB}\rangle = S_{AB}(\zeta)|0_A;0_B\rangle, \tag{2.118}$$

and it then follows from Equation (2.117) that

$$|\zeta_{AB}\rangle = \frac{1}{\cosh r}\sum_{n=0}^{\infty}[-\exp(i\varphi)\tanh r]^n |n_A;n_B\rangle. \tag{2.119}$$

The state $|\zeta_{AB}\rangle$ is a superposition only of states in which the two modes contain the same number of photons and are normalized. Note that $|\zeta_{AB}\rangle$ is an entangled state in that it is not the product of a state for mode A and one for mode B. This leads to the correlations between the properties of the modes.

2.6 COHERENT INTERACTIONS BETWEEN ATOMS AND FIELDS

The simplest model of an atom includes only two discrete, orthonormal states $|1\rangle$ and $|2\rangle$, which, in the absence of coupling, are energy eigenstates with energies $\hbar\omega_1$ and $\hbar\omega_2 > \hbar\omega_1$, respectively. A general pure state $|\psi\rangle$ is given by the superposition $|\psi\rangle = c_1|1\rangle + c_2|2\rangle$, where $|c_1|^2 + |c_2|^2 = 1$. For this system, we define the four Pauli operators:

$$I = |2\rangle\langle 2| + |1\rangle\langle 2|, \tag{2.120}$$

$$\sigma_3 = |2\rangle\langle 2| - |1\rangle\langle 1|, \tag{2.121}$$

$$\sigma_+ = |2\rangle\langle 1|, \tag{2.122}$$

$$\sigma_- = |1\rangle\langle 2| = (\sigma_+)^\dagger. \tag{2.123}$$

We also define the Hermitian operators

$$\sigma_1 = \sigma_+ + \sigma_- \tag{2.124}$$

and

$$\sigma_2 = i(\sigma_- - \sigma_+). \tag{2.125}$$

The Hamiltonian describing the two-state atom interacting with a classical electric field $\vec{E}(t)$ within the electric dipole approximation is [4–6]

$$H = \frac{\hbar}{2}(\omega_1 + \omega_2)I + \frac{\hbar}{2}(\omega_2 - \omega_1)\sigma_3 - \vec{d}\cdot\vec{E}(t)(\sigma_+ + \sigma_-), \tag{2.126}$$

where $\vec{d}$ is the atomic dipole operator. The first two terms in this Hamiltonian are free Hamiltonian H_o for the atom having eigenstates $|1\rangle$ and $|2\rangle$:

$$\begin{aligned} H_o &= \frac{\hbar}{2}(\omega_2 + \omega_1)I + \frac{\hbar}{2}(\omega_2 - \omega_1)\sigma_3 \\ &= \begin{pmatrix} \hbar\omega_2 & 0 \\ 0 & \hbar\omega_2 \end{pmatrix}. \end{aligned} \tag{2.127}$$

In the quantum theory of light, the electric and magnetic fields are Hermitian operators, and the interaction between two-state atoms and the quantized electric field is described by the electric dipole Hamiltonian Equation (2.126), with $\vec{E}(t)$ replaced by the quantized electric field operator. For an atom placed at the origin $\vec{r} = 0$

interacting with a single plane wave mode of frequency ω, the Hamiltonian becomes

$$\begin{aligned} H_1 &= \frac{\hbar}{2}(\omega_2 + \omega_1)I + \frac{\hbar}{2}(\omega_2 - \omega_1)\sigma_3 \\ &\quad - i\hbar\lambda(\sigma_+ + \sigma_-)\left[a\exp(-i\omega t) + a^\dagger \exp(i\omega t)\right] \end{aligned} \tag{2.128}$$

where $\lambda = \vec{d}\cdot\hat{\varepsilon}(\omega/2\hbar\varepsilon_o V)^{1/2}$. An equivalent and time-independent form of Equation (2.128) is

$$\begin{aligned} H_2 &= \frac{\hbar}{2}(\omega_2 + \omega_1)I + \frac{\hbar}{2}(\omega_2 - \omega_1)\sigma_3 + \hbar\omega\left(a^\dagger a + \frac{1}{2}\right) \\ &\quad - i\hbar\lambda(\sigma_+ + \sigma_-)(a + a^\dagger). \end{aligned} \tag{2.129}$$

2.7 THE JAYNES–CUMMINGS MODEL

In quantum optics, we are often interested in the dynamics of atoms, molecules, or even nanostructures coupled to an electromagnetic field. The field may be described either classically or fully quantum mechanically, and may be monochromatic, have time-dependent amplitude or phase, or be broadband. In this section, we consider the Jaynes–Cummings model of the interaction between a two-state atom and a single-mode quantized electromagnetic field. The time-independent Hamiltonian is

$$\begin{aligned} H &= \frac{\hbar}{2}(\omega_2 + \omega_1)I + \frac{\hbar}{2}(\omega_2 - \omega_1)\sigma_3 + \hbar\omega\left(a^\dagger a + \frac{1}{2}\right) \\ &\quad - i\hbar\lambda(\sigma_+ + \sigma_-)(a + a^\dagger), \end{aligned} \tag{2.130}$$

as given in the previous section.

Quantum mechanics provides a number of equivalent descriptions or pictures of the dynamics of the system as discussed in the previous chapter. These are related by unitary transformations of both the state and the operators acting on it. In the Schrödinger picture, the operators are time independent, and all the dynamics are contained within the state. On the other hand, in the Heisenberg picture, the operators evolve while the state is time-independent. Intermediate between them are the interaction pictures in which part of the dynamics, usually associated with a free, uncoupled Hamiltonian, is contained within the operators, while that arising from the coupling appears in the state. Alternatively, the free evolution may be contained within the state, with the operators containing the effect of the coupling. These pictures are related to each other by unitary transformations. A measurable quantity is the expectation of the operator A in a state $|\psi\rangle$ given by $\langle\psi|A|\psi\rangle$, and must be independent of the picture in which it is evaluated. The interaction picture Hamiltonian H_I can be derived from the Schrödinger equation [5,6]

$$i\hbar\frac{\partial}{\partial t}|\psi\rangle = H|\psi\rangle, \tag{2.131}$$

together with the interaction picture state $|\psi_I\rangle = U|\psi\rangle$, described by the unitary operator U, and satisfies the transformed Schrödinger equation

$$i\hbar\frac{\partial}{\partial t}|\psi_I\rangle = H_I|\psi_I\rangle. \tag{2.132}$$

It is now straightforward to show that the interaction picture Hamiltonian is given by

$$H_I = i\hbar\dot{U}U^\dagger + UHU^\dagger. \tag{2.133}$$

We now consider the unitary transformation associated with the unitary operator

$$\begin{aligned} U &= \exp\left(\frac{i}{2}[(\omega_2+\omega_1)I + (\omega)\sigma_3]t\right) \\ &= \exp(i[\omega_2 - \omega - \Delta/2]t)|1\rangle\langle 1| + \exp(i[\omega_2 - \Delta/2])|2\rangle\langle 2|, \end{aligned} \tag{2.134}$$

where $\Delta = \omega_2 - \omega_1 - \omega$ is the detuning between the atomic transition energy and that of the driving field. The Hamiltonian in the interaction picture is then obtained by

$$\begin{aligned} H_I &= \frac{\hbar}{2}\Delta\sigma_3 + \hbar\omega\left(a^\dagger a + \frac{1}{2}\right) \\ &\quad - i\hbar\lambda[\sigma_+\exp(i\omega t) + \sigma_-\exp(-i\omega t)](a + a^\dagger). \end{aligned} \tag{2.135}$$

We apply the second unitary transformation $U_2 = \exp\left[i\sum_n (n+\frac{1}{2})\omega t|n\rangle\langle n|\right]$ on Equation (2.135), and get

$$H_I = \frac{\hbar}{2}\Delta\sigma_3 - i\hbar\lambda[\sigma_+\exp(i\omega t) + \sigma_-\exp(-i\omega t)][a\exp(-i\omega t) + a^\dagger\exp(i\omega t)]. \tag{2.136}$$

If we ignore terms with $\exp(\pm 2i\omega t)$ dependences, known as rotating wave approximation, we obtain the Jaynes–Cummings Hamiltonian

$$H_{JC} = \frac{\hbar}{2}\Delta\sigma_3 - i\hbar\lambda\left[\sigma_+ a + \sigma_- a^\dagger\right], \tag{2.137}$$

in which the explicit time dependence has been removed. Sometimes, the Jaynes–Cummings Hamiltonian is written as

$$\begin{aligned} H_{JC} &= \frac{1}{2}\hbar\omega_0 + \hbar\omega a^\dagger a + \hbar\eta\left(a^\dagger\sigma_+ + a\sigma_-\right) \\ &= \frac{1}{2}\hbar\omega_0 + \hbar\omega a^\dagger a + V, \end{aligned} \tag{2.138}$$

where $V = \hbar\eta\left(a^\dagger\sigma_+ + a\sigma_-\right)$. The unitary operator U is given by $U = \exp(iVt/\hbar)$, then the matrix representation of U is

$$U = \begin{pmatrix} \cos\left[\eta t\left(aa^\dagger\right)^{1/2}\right] & -i\dfrac{\sin\left[\eta t\left(aa^\dagger\right)^{1/2}\right]}{\left(aa^\dagger\right)^{1/2}}a \\ -i\dfrac{\sin\left[\eta t\left(aa^\dagger\right)^{1/2}\right]}{\left(aa^\dagger\right)^{1/2}}a^\dagger & \cos\left[\eta t\left(a^\dagger a\right)^{1/2}\right] \end{pmatrix}. \tag{2.139}$$

PROBLEMS

1. Assuming that a particle has discrete energy spectrum $E_0, E_1, \cdots$, and the probability of a particle with energy E_r is given by Equation (2.8). We define the partition function Z by $Z = \sum_n \exp(-\beta E_n)$. Then show that the average energy of a particle is given by $\langle E\rangle = -(\partial/\partial\beta)\log Z$.

2. Evaluate the following vacuum expectation value of $\langle 0|a_i a_j^\dagger a_k a_l^\dagger|0\rangle$, where the fermion annihilation operator a_m satisfies the relation $a_m|0\rangle = 0$.

3. For bosonic annihilation operator b, which satisfies the commutation relation $[b, b^\dagger] = 1$, show that $\left[b, (b^\dagger)^2\right] = 2b$ and $\left[b, (b^\dagger)^n\right] = n(b^\dagger)^{n-1}$.

4. Consider a two-level system with bases given by $\{|0\rangle, |1\rangle\}$. The Hamiltonian of the system H is given by the sum of an unperturbed

term H_0 and an interaction energy V, that is $H = H_0 + V$ with $H_0|0\rangle = E_0|0\rangle$ and $H_0|1\rangle = E_1|1\rangle$. The density operator of this two dimensional system is given by

$$\rho = \begin{pmatrix} \rho_{00} & \rho_{01} \\ \rho_{10} & \rho_{11} \end{pmatrix}.$$

Using Equation (2.51), show that

$$\frac{\partial \rho_{00}}{\partial t} = \frac{1}{i\hbar}(V_{01}\rho_{10} - \rho_{01}V_{10})$$

and

$$\frac{\partial \rho_{01}}{\partial t} = -\frac{1}{i\hbar}(E_1 - E_0) - \frac{1}{i\hbar}V_{01}(\rho_{00} - \rho_{11}).$$

Here, $V_{01} = \langle 0|V|1\rangle$ and diagonal elements of V are assumed to be zero.

5. Using Equations (2.62) to (2.66), derive Equation (2.67).

6. From Equations (2.96), (2.97) and (2.99), derive Equations (2.98) and (2.104).

REFERENCES

[1] R. P. Feynman, *Statistical Mechanics*. Reading, MA: W. A. Benjamins, 1972.

[2] A. A. Abrikosov, L. P. Gorkov, and I. E. Dzyaloshinski, *Methods of Quantum Field Theory in Statistical Physics*. New York: Dover, 1963.

[3] M. Sargent, M. O. Scully, and W. E. Lamb, *Laser Physics*. London: Addison-Wesley, 1974.

[4] S. M. Barnett and P. M. Radmore, *Methods in Theoretical Quantum Optics*. New York: Oxford University Press, 1997.

[5] W. H. Louisell, *Quantum Statistical Properties of Radiation*. New York: Wiley Interscience, 1990.

[6] P. Meystre and M. Sargent III, *Elements of Quantum Optics*. Berlin, Germany: Springer-Verlag, 1993.

3 Elementary Theory of Electronic Band Structure in Semiconductors

In this chapter, we derive the 8×8 Hamiltonian using the multiband effective theory. To understand the optical properties of semiconductors, we need to know the electronic band structure. In particular, the band structure calculations of the nanostructures, such as the quantum well, quantum wire, and quantum dot, are of great interest, because these structures have existing and potential applications for many important optoelectronic devices. First, we review several basic theories for band structure calculation, in particular, the effective mass theory. Next, we introduce the band structure calculation for nanostructures. Specially, we focus on the band structure of the quantum well and the crystal orientation effect on electronic properties.

3.1 BLOCH THEOREM AND EFFECTIVE MASS THEORY

3.1.1 Bloch Theorem

A fundamental theorem concerning electrons in a crystal was proved by Bloch in 1928. It states that the wave functions of the electrons in a crystal have the Bloch form

$$\Psi_{\mathbf{k}}(\mathbf{r}) = e^{i\mathbf{k}\cdot\mathbf{r}}u_{\mathbf{k}}(\mathbf{r}), \tag{3.1}$$

where $u_{\mathbf{k}}(\mathbf{r})$ is a function that has the periodicity of the lattice, that is, $u_{\mathbf{k}}(\mathbf{r}) = u_{\mathbf{k}}(\mathbf{r}+\mathbf{T}_m)$ where $\mathbf{T}_m$ is any primitive lattice translation. The Bloch function, in general, depends on the wave vector $\mathbf{k}$, and is just a free electron wave function, $\exp(i\mathbf{k}\cdot\mathbf{r})$, modulated by a function that has the periodicity of the lattice. That is, it is a modulated plane wave.

We present a proof of Bloch's theorem in the one-dimensional case. Consider a crystal of length $L = Na$, where N is the number of primitive unit cells, and a is the length of each cell. The wave equation is

Engineering Quantum Mechanics, First Edition. Doyeol Ahn, Seoung-Hwan Park.

$$-\frac{\hbar^2}{2m}\frac{d^2\Psi}{dx^2}+V(x)\Psi=E\Psi \tag{3.2}$$

with $V(x)=V(x+a)$. The periodic boundary condition requires that $\Psi(x)=\Psi(x+L)$, and the translational symmetry requires that the charge density is the same at x as at $x+ma$, where m is any integer, that is, $|\Psi(x+ma)|^2=|\Psi(x)|^2$. Therefore

$$\Psi(x+a)=A\Psi(x), \tag{3.3}$$

and

$$\Psi(x+ma)=A^m\Psi(x), \tag{3.4}$$

where A is a complex number such that $A^*A=1$ (the charge density is proportional to $\Psi^*\Psi$). By applying the translation N times, we have

$$\Psi(x+Na)=A^N\Psi(x)=\Psi(x) \tag{3.5}$$

$$A^N=1 \tag{3.6}$$

$$A=\exp(2\pi im/N), m=0,\pm1,\pm2,\ldots \tag{3.7}$$

Equation (3.5) shows that A is just one of the roots of unity, which is written explicitly in Equation (3.7). Thus, the wave functions must be of the form

$$\Psi(x+a)=\exp(2\pi im/N)\Psi(x) \tag{3.8}$$

which is the Bloch condition on the wave function. A function that satisfies this condition has the requirements demanded by translational symmetry.

This result suggests that in seeking an electron wave function (i.e., a function that satisfies Eq. (3.2)), we should try a modulated free electron function, $\Psi(x)=\exp(ikx)u(x)$, and determine the conditions that allow this function to satisfy the Bloch condition, Equation (3.8). Substituting this modulated plane wave in both sides of Equation (3.8), we see that the equality is kept if

$$k=2\pi m/Na \tag{3.9}$$

$$u(x+a)=u(x). \tag{3.10}$$

If $u(x)$ is a constant, then the free electron wave function with the same condition on k is recovered. Thus, a solution of the wave equation with the lattice periodicity can be written in one dimension in the form

$$\Psi(x) = \exp(ikx)u(x) \tag{3.11}$$

with $k = 2\pi m/Na$ and $u(x+a) = u(x)$. This proves Bloch's theorem in one dimension. The extension to three dimensions, Equation (3.1), follows straightforwardly.

3.1.2 The Schrödinger Equation and the Effective Mass Equation

The effective mass equations have been widely used to treat problems involving valence bands, such as quantum wells and superlattices. In this section, we will give a brief heuristic derivation of the main results. The reader should consult Reference [1] for additional details.

The one-electron Schrödinger equation describing the dynamics of an individual electron is given by

$$i\hbar\frac{\partial}{\partial t}\Psi_0(\mathbf{r},t) = -\frac{\hbar^2}{2m_0}\nabla^2\Psi_0(\mathbf{r},t) + U_T(\mathbf{r},t)\Psi_0(\mathbf{r},t), \tag{3.12}$$

where $U_T(\mathbf{r},t)$ is the total potential energy. Here, we used the subscript "0" for Ψ to distinguish it from the corresponding symbol in the effective mass equation, as discussed below. The potential energy $U_T(\mathbf{r},t)$ can be separated into macroscopic part and a microscopic part. The macroscopic potential $U_E(\mathbf{r},t)$ arises from macroscopic electric fields due to any externally applied voltage or macroscopic space charge. Also, an electron feels the microscopic electric fields due to the nearly periodic lattice of ions and the other valence electrons. We can separate it into two parts: A perfectly periodic potential $U_L(\mathbf{r},t)$ due to a perfect static lattice, plus a random fluctuating potential $U_S(\mathbf{r},t)$ representing the deviations. That is,

$$U_T(\mathbf{r},t) = U_L(\mathbf{r},t) + U_S(\mathbf{r},t) + U_E(\mathbf{r},t). \tag{3.13}$$

The randomly fluctuating potential $U_S(\mathbf{r},t)$ is called the scattering potential, which can arise from impurity scattering, phonon scattering, or electron–electron scattering. Then, Equation (3.12) is written as

$$i\hbar\frac{\partial}{\partial t}\Psi_0(\mathbf{r},t) = H_0\Psi_0(\mathbf{r},t) + U(\mathbf{r},t)\Psi_0(\mathbf{r},t), \tag{3.14}$$

where

$$\begin{aligned} U(\mathbf{r},t) &= U_E(\mathbf{r},t) + U_S(\mathbf{r},t), \\ H_0 &= -\frac{\hbar^2}{2m_0}\nabla^2 + U_L(\mathbf{r},t). \end{aligned} \tag{3.15}$$

If we consider only the periodic potential $U_L(\mathbf{r},t)$, the solutions to the Schrödinger equation can be written in the form of Bloch waves, as discussed in Section 3.1.1. The wave function of an electron in a band n with wave vector $\mathbf{k}$ is written as

$$\Psi_0(\mathbf{r},t) = u_{n,\mathbf{k}}(\mathbf{r})\exp(i\mathbf{k}\cdot\mathbf{r})\exp(-i\varepsilon_n(\mathbf{k})t/\hbar), \tag{3.16}$$

where $u_{n,\mathbf{k}}$ is a periodic function like $U_L(\mathbf{r})$ that is different for each n and each $\mathbf{k}$. The function $\varepsilon_n(\mathbf{k})$ is known as *energy band structures*. The calculated band structures can be used to write an effective mass equation in which $U_L(\mathbf{r})$ *does not appear explicitly*. That is, if we are interested in electrons in a band n, then the effective mass equation is written as

$$i\hbar\frac{\partial}{\partial t}\Psi(\mathbf{r},t) = \varepsilon_n(-i\nabla)\Psi(\mathbf{r},t) + U(\mathbf{r},t)\Psi(\mathbf{r},t). \tag{3.17}$$

The periodic lattice potential U_L does not appear directly in the effective mass equation. This makes Equation (3.17) much easier to solve than the original Schrödinger equation, Equation (3.14). Here, $\varepsilon_n(-i\nabla)$ is obtained by replacing $\mathbf{k}$ with $-i\nabla$ in the function $\varepsilon_n(\mathbf{k})$. That is, k_x with $-i\partial/\partial x$, k_y with $-i\partial/\partial y$, and k_z with $-i\partial/\partial z$. Equation (3.17) is known as the single-band effective mass equation, and can be used if the potential $U(\mathbf{r},t)$ introduces only negligible coupling to the other bands. As a result, it is inapplicable to the valence band, where multiple bands ("heavy" and "light" holes) overlap in energy, and even a weak static potential can induce interband transitions. In that case, we need multiband effective mass equation. We will consider the relation between the envelope function $\Psi(\mathbf{k},t)$ and the real wave function $\Psi_0(\mathbf{k},t)$ in Equation (3.12) for two cases, single and multiband effective mass equations, in the following two subsections, to understand the approximations involved in using Equation (3.17).

3.1.3 The Single-Band Effective Mass Equation

The eigenfunctions of the operator H_0 in Equation (3.15) can be written using two indices—the band index n and the wave vector $\mathbf{k}$.

$$H_0\psi_{n\mathbf{k}} = \varepsilon_n(\mathbf{k})\psi_{n\mathbf{k}}, \tag{3.18}$$

where

$$\psi_{n\mathbf{k}} = e^{i\mathbf{k}\cdot\mathbf{r}}u_{n\mathbf{k}}(\mathbf{r}) = \frac{e^{i\mathbf{k}\cdot\mathbf{r}}}{\sqrt{V}}\bar{u}_{n\mathbf{k}}(\mathbf{r}). \tag{3.19}$$

The functions $\varepsilon_n(\mathbf{k})$ are readily available for most semiconductors, though the periodic functions $u_{n\mathbf{k}}$ are not available. The functions $\psi_{n\mathbf{k}}$ in Equation (3.19), being eigenfunctions of the Hermitian operator H_0, are orthogonal, and are normalized to some arbitrary volume $V(\langle\psi_{n\mathbf{k}}|\psi_{n'\mathbf{k}'}\rangle = \delta_{nn'}\delta_{\mathbf{k}\mathbf{k}'})$. That is,

$$\langle\psi_{n\mathbf{k}}|\psi_{n'\mathbf{k}}\rangle = \delta_{nn'} = \int d^3r u_{n\mathbf{k}}^* u_{n\mathbf{k}}. \tag{3.20}$$

Hence,

$$\int d^3r u_{n\mathbf{k}}^* u_{n'\mathbf{k}} = \delta_{nn'}. \tag{3.21}$$

Since the functions $u_{n\mathbf{k}}$ are periodic with the same period as the lattice, the integral in Equation (3.21) yields the same value in each unit cell. If the volume V contains N unit cells, we can write

$$\int_{\text{unit cell}} d^3r u_{n\mathbf{k}}^* u_{n'\mathbf{k}} = 1/N\delta_{nn'} \tag{3.22}$$

As the periodic potential $U_L(\mathbf{r})$ goes to zero, the functions $\bar{u}_{n\mathbf{k}}$ in Equation (3.19) become constants equal to one. That is, Equation (3.19) then becomes just the plane wave states.

3.1.3.1 Actual Wave Function $\Psi_0(\mathrm{r},\mathbf{t})$ The wave functions $\Psi_0(\mathbf{r},t)$ in Equation (3.14) are expanded by using eigenfunctions $\psi_{n\mathbf{k}}$ as a basis set.

$$\Psi_0(\mathbf{r},t) = \sum_{n\mathbf{k}} \varphi_{n\mathbf{k}}(t)\psi_{n\mathbf{k}}. \tag{3.23}$$

Then, Equation (3.14) can be written in matrix form as follows:

$$i\hbar\frac{d}{dt}\varphi_{n\mathbf{k}}(t) = \sum_{n'\mathbf{k}'} \langle\psi_{n\mathbf{k}}|H_0 + U|\psi_{n\mathbf{k}'}\rangle\varphi_{n'\mathbf{k}'}(t). \tag{3.24}$$

Since the function $\psi_{n\mathbf{k}}$ are eigenfunctions of H_0 with eigenvalues $\varepsilon_n(\mathbf{k})$, we have

$$\langle\psi_{nk}|H_0|\psi_{n'\mathbf{k}'}\rangle = \varepsilon_{n'}(\mathbf{k}')\langle\psi_{n\mathbf{k}}|H_0|\psi_{n'\mathbf{k}'}\rangle = \varepsilon_n(\mathbf{k})\delta_{nn'}\delta_{\mathbf{kk}'}. \tag{3.25}$$

Hence, we can simplify Equation (3.24) to

$$i\hbar\frac{d}{dt}\varphi_{n\mathbf{k}}(t) = \varepsilon_n(\mathbf{k})\varphi_{n\mathbf{k}}(t) + \sum_{n'\mathbf{k}'} U_{n\mathbf{k};n'\mathbf{k}'}\varphi_{n'\mathbf{k}'}(t), \tag{3.26}$$

where

$$U_{n\mathbf{k};n'\mathbf{k}'} = \langle\psi_{n\mathbf{k}}|U|\psi_{n'\mathbf{k}'}\rangle \equiv \int d^3r u^*_{n\mathbf{k}}(\mathbf{r})u_{n'\mathbf{k}'}(\mathbf{r})e^{i\mathbf{k}\cdot\mathbf{r}}e^{i(\mathbf{k}'-\mathbf{k})\cdot\mathbf{r}}U(\mathbf{r},t). \tag{3.27}$$

3.1.3.2 Envelope Function $\Psi(t,\mathbf{t})$ Next, we use the plane waves $\psi_{\mathbf{k}} = e^{i\mathbf{k}\cdot\mathbf{r}}/\sqrt{V}$ to expand the envelope function $\Psi(\mathbf{k},t)$.

$$\Psi(\mathbf{r},t) = \sum_{\mathbf{k}} \varphi_{\mathbf{k}}(t)\psi_{\mathbf{k}}. \tag{3.28}$$

The effective mass equation can then be written in matrix form as follows:

$$i\hbar\frac{d}{dt}\varphi_{\mathbf{k}}(t) = \sum_{\mathbf{k}'}\langle\psi_{\mathbf{k}}|E_n(-i\nabla)+U|\psi_{\mathbf{k}'}\rangle\varphi_{\mathbf{k}'}(t). \tag{3.29}$$

Using the relation

$$\left(-i\frac{\partial}{\partial x}\right)^p\left(-i\frac{\partial}{\partial y}\right)^q\left(-i\frac{\partial}{\partial z}\right)^r\psi_{\mathbf{k}'} = (k_{x'})^p(k_{y'})^q(k_{z'})^r\psi_{\mathbf{k}'}, \tag{3.30}$$

we obtain

$$\varepsilon_n(-i\nabla)\psi_{\mathbf{k}'} = \varepsilon_n(\mathbf{k})\psi_{\mathbf{k}'}. \tag{3.31}$$

Thus, using the orthogonality relation $\langle\psi_{\mathbf{k}}|\psi_{\mathbf{k}'}\rangle = \delta_{\mathbf{kk}'}$, we obtain

$$\langle\psi_{\mathbf{k}}|\varepsilon_n(-i\nabla)|\psi_{\mathbf{k}'}\rangle = \varepsilon_n(\mathbf{k})\delta_{\mathbf{kk}'}. \tag{3.32}$$

Using Equation (3.32), we can write Equation (3.29) as

$$i\hbar\frac{d}{dt}\varphi_{\mathbf{k}}(t) = \varepsilon_n(k)\varphi_{n,\mathbf{k}}(t) + \sum_{\mathbf{k}'} U_{\mathbf{k};\mathbf{k}'}\varphi_{\mathbf{k}'}(t). \tag{3.33}$$

where

$$U_{\mathbf{kk}'} = \langle \psi_{\mathbf{k}} | U | \psi_{\mathbf{k}'} \rangle \equiv \int \frac{d^3 r}{V} e^{i(\mathbf{k}-\mathbf{k})\cdot\mathbf{r}} U(\mathbf{r},t). \tag{3.34}$$

3.1.3.3 The Relation betwen $\Psi_0(\mathrm{r},\mathbf{t})$ ***and*** $\Psi(t,\mathbf{t})$ Now let us compare Equations (3.26) and (3.33). Equation (3.26) is the matrix representation of the Schrödinger equation (Eq. (3.15)) using the Bloch function $\psi_{n\mathbf{k}}$ as the basis, while Equation (3.33) is the matrix representation of the effective mass equation (Eq. (3.17)) for a particular band using the plane waves $\psi_{\mathbf{k}}$ as the basis. The two equations would be identical if the following were true:

$$\langle \psi_{n\mathbf{k}} | U | \psi_{n'\mathbf{k}'} \rangle = \langle \psi_{\mathbf{k}} | U | \psi_{\mathbf{k}'} \rangle \delta_{nn'}. \tag{3.35}$$

That is, the matrix elements of the external potential between states from different bands should be negligible. This is the basic approximation on which the accuracy of the effective mass equation rests. It is fairly accurate for potentials U that are slowly varying on an atomic scale. It is easy to see that for a constant potential $U(\mathbf{r},t) = U_o$, Equation (3.35) is exact because of the orthogonality of the eigenfunctions.

$$\langle \psi_{n\mathbf{k}} | U | \psi_{n'\mathbf{k}'} \rangle = U_o \langle \psi_{n\mathbf{k}} | \psi_{n'\mathbf{k}'} \rangle = \delta_{nn'} U_o \delta_{\mathbf{kk}'}. \tag{3.36}$$

$$\langle \psi_{\mathbf{k}} | U_o | \psi_{\mathbf{k}'} \rangle = U_o \delta_{\mathbf{kk}'}. \tag{3.37}$$

In the case of a spatially varying potential U, we start from Equation (3.27) to prove Equation (3.35). The right-hand side requires us to integrate over the entire sample volume V. But first, let us consider the effect of integrating over a single unit cell, say the nth one.

$$\begin{aligned} &\int_{n\text{th unit cell}} d^3 r u_{n\mathbf{k}}^*(\mathbf{r}) u_{n'\mathbf{k}'}(\mathbf{r}) e^{i(\mathbf{k}'-\mathbf{k})\cdot\mathbf{r}} U(\mathbf{r},t) \\ &\simeq e^{i(\mathbf{k}'-\mathbf{k})\cdot\mathbf{r}_n} U(\mathbf{r}_n,t) \int_{n\text{th unit cell}} d^3 r u_{n\mathbf{k}}^*(\mathbf{r}) u_{n'\mathbf{k}'}(\mathbf{r}) \\ &\simeq e^{i(\mathbf{k}'-\mathbf{k})\cdot\mathbf{r}_n} U(\mathbf{r}_n,t) \delta_{nn'}(1/N), \end{aligned} \tag{3.38}$$

where $\mathbf{r}_n$ is the coordinate of the center of the nth unit cell. Here we have made two assumptions: (1) The factors $\exp[i(\mathbf{k}'-\mathbf{k})\cdot\mathbf{r}]$ and $U(\mathbf{r},t)$ vary slowly enough that they can be considered approximately constant over a unit cell. (2) The functions $u_{n\mathbf{k}}^*(\mathbf{r})$ do not change very much with $\mathbf{k}$. That is, $u_{n\mathbf{k}'}(\mathbf{r}) \simeq u_{n\mathbf{k}}(\mathbf{r})$ over the range of values of $|\mathbf{k}'-\mathbf{k}|$, for which

$U_{\mathbf{kk}'}$ is significant. These approximations are inherent in the use of the effective mass equation, since its validity rests on the accuracy of Equation (3.35). Then,

$$\int_{\text{unit cell}} d^3r u^*_{n\mathbf{k}}(\mathbf{r}) u_{n'\mathbf{k}'}(\mathbf{r}) \simeq \int_{\text{unit cell}} d^3r u^*_{n\mathbf{k}}(\mathbf{r}) u_{n'\mathbf{k}}(\mathbf{r}) = \delta_{nn'}(1/N). \tag{3.39}$$

We can now use Equation (3.38) to write Equation (3.27) as

$$\begin{aligned}
\langle \psi_{n\mathbf{k}} | U | \psi_{n'\mathbf{k}'} \rangle &= \int d^3r u^*_{n\mathbf{k}}(\mathbf{r}) u_{n'\mathbf{k}'}(\mathbf{r}) e^{i\mathbf{k}\cdot\mathbf{r}} e^{i(\mathbf{k}'-\mathbf{k})\cdot\mathbf{r}} U(\mathbf{r},t) \\
&= \sum_{n=1}^{N} \int d^3r u^*_{n\mathbf{k}}(\mathbf{r}_n) u_{n'\mathbf{k}'}(\mathbf{r}_n) e^{i\mathbf{k}\cdot\mathbf{r}_n} e^{i(\mathbf{k}'-\mathbf{k})\cdot\mathbf{r}_n} U(\mathbf{r}_n,t) \\
&\simeq \delta_{nn'} \frac{1}{N} \sum_{n=1}^{N} e^{i(\mathbf{k}'-\mathbf{k})\cdot\mathbf{r}_n} U(\mathbf{r}_n,t) \quad \text{[from Eq. (3.38)]} \\
&\simeq \delta_{nn'} \int \frac{d^3r}{V} e^{i(\mathbf{k}'-\mathbf{k})\cdot\mathbf{r}} U(\mathbf{r},t) \\
&= \delta_{nn'} \langle \psi_{\mathbf{k}'} | U | \psi_{\mathbf{k}} \rangle \quad \text{[from Eq. (3.34)].}
\end{aligned} \tag{3.40}$$

This proves Equation (3.35). We can derive a very simple relation between the envelope function $\Psi(\mathbf{r},t)$ and the real wave function $\Psi_o(\mathbf{r},t)$. We can rewrite Equation (3.23) as

$$\begin{aligned}
\Psi_0(\mathbf{r},t) &= \sum_{n,\mathbf{k}} \varphi_{n\mathbf{k}}(t) e^{i\mathbf{k}\cdot\mathbf{r}} u_{n\mathbf{k}}(\mathbf{r}). \\
&\simeq u_{n\mathbf{k}}(\mathbf{r}) \sum_{n,\mathbf{k}} \varphi_{n\mathbf{k}}(t) e^{i\mathbf{k}\cdot\mathbf{r}}.
\end{aligned} \tag{3.41}$$

Using Equation (3.28),

$$\Psi_0(\mathbf{r},t) \simeq \bar{u}_{n\mathbf{k}}(\mathbf{r}) \Psi(\mathbf{r},t). \tag{3.42}$$

The true wave function is thus approximately equal to the product of the envelope function and the periodic part of the Bloch wave function.

3.1.4 The Multiband Effective Mass Equation

The multiband effective mass equation uses a slightly different basis set $\chi_{n\mathbf{k}}$ to the wave function $\Psi_0(\mathbf{r},t)$.

$$\Psi_0(\mathbf{r},t) = \varphi_{n\mathbf{k}}(t)\psi_{n\mathbf{k}}, \tag{3.43}$$

where

$$\psi_{n\mathbf{k}} = \frac{e^{i\mathbf{k}\cdot\mathbf{r}}}{\sqrt{V}}\bar{u}_{n0}(\mathbf{r}) = e^{i\mathbf{k}\cdot\mathbf{r}}u_{n0}(\mathbf{r}). \tag{3.44}$$

Comparing Equation (3.44) with Equation (3.19), we can see that the difference is that we are now using the functions $u_{n\mathbf{k}}(\mathbf{r})$ at $\mathbf{k} = 0$. Consequently, in our derivation, it will not be necessary to assume that the function $u_{n\mathbf{k}}(\mathbf{r})$ does not change with $\mathbf{k}$. However, we encounter the complexity of having to solve multiple coupled differential equations instead of the single differential equation, Equation (3.17). Using this new basis set, the Schrödinger equation is written as before.

$$i\hbar\frac{d}{dt}\varphi_{n,\mathbf{k}}(t) = \sum_{n',\mathbf{k}'}\langle\psi_{n\mathbf{k}}|H_0 + U|\psi_{n'\mathbf{k}'}\rangle\varphi_{n',\mathbf{k}'}(t). \tag{3.45}$$

Consider first the matrix elements of U,

$$\langle\psi_{n\mathbf{k}}|U|\psi_{n'\mathbf{k}'}\rangle = \int d^3r e^{i(\mathbf{k}'-\mathbf{k})\cdot\mathbf{r}}u_{n0}^*(\mathbf{r})u_{n'0}(\mathbf{r})U(\mathbf{r},t). \tag{3.46}$$

Using Equation (3.40),

$$\begin{aligned}\langle\psi_{n\mathbf{k}}|U|\psi_{n'\mathbf{k}'}\rangle &\simeq \delta_{nn'}\frac{1}{N}\sum_{n=1}^{N}e^{i(\mathbf{k}'-\mathbf{k})\cdot\mathbf{r}_n}U(\mathbf{r}_n,t)\\ &\simeq \delta_{nn'}\int\frac{d^3r}{V}e^{i(\mathbf{k}'-\mathbf{k})\cdot\mathbf{r}}U(\mathbf{r},t)\\ &= \delta_{nn'}U_{\mathbf{k}\mathbf{k}'},\end{aligned} \tag{3.47}$$

where

$$U_{\mathbf{k}\mathbf{k}'} = \int\frac{d^3r}{V}e^{i(\mathbf{k}'-\mathbf{k})\cdot\mathbf{r}}U(\mathbf{r},t). \tag{3.48}$$

In the last step, we have used Equation (3.34); $\psi_{\mathbf{k}}$ corresponds to the plane wave states as in Equation (3.28). Note that we do not need the additional assumption that $u_{n\mathbf{k}}(\mathbf{r})$ does not vary much with $\mathbf{k}$. However, the matrix representation of H_0 is no longer diagonal, since the basis functions in Equation (3.44) are not eigenfunctions of H_0, unlike the basis functions in Equation (3.19). It can be shown that

$$\langle \psi_{n\mathbf{k}} | H_0 | \psi_{n'\mathbf{k}'} \rangle = H_0^{nn'}(\mathbf{k}) \delta_{\mathbf{k}'\mathbf{k}} \tag{3.49}$$

where

$$H_0^{nn'}(\mathbf{k}) = \begin{cases} \dfrac{\hbar^2 k^2}{2m_o} + \varepsilon_{n0}, & n' = n \\ \dfrac{\hbar \mathbf{P}_{nn'} \cdot \mathbf{k}}{m_o}, & n' \neq n \end{cases} \tag{3.50}$$

$\mathbf{P}_{nn'}$ is called the momentum matrix element between bands n and n', and is defined by

$$\mathbf{P}_{nn'} = (-i\hbar) \langle u_{n'0}^* | \nabla u_{n0} \rangle \tag{3.51}$$

ε_{n0} is the energy at the edge ($k = 0$, Γ-point) of band n.

The derivation is as follows: We have

$$H_0 \psi_{n0} = \varepsilon_{n0} \psi_{n0} \tag{3.52}$$

since ε_{n0} is the energy eigenvalue of band n at $k = 0$. From Equation (3.44),

$$\psi_{n\mathbf{k}} = e^{i\mathbf{k}\cdot\mathbf{r}} \psi_{n0}. \tag{3.53}$$

Hence,

$$-i\hbar \nabla \psi_{n\mathbf{k}} = e^{i\mathbf{k}\cdot\mathbf{r}} (\hbar \mathbf{k} - i\hbar \nabla) \psi_{n0} \tag{3.54}$$

$$-\hbar^2 \nabla^2 \psi_{n\mathbf{k}} = e^{i\mathbf{k}\cdot\mathbf{r}} (\hbar^2 k^2 - 2i\hbar^2 \mathbf{k} \cdot \nabla - \hbar^2 \nabla^2) \psi_{n0}. \tag{3.55}$$

Using H_0 from Equation (3.15), we have

$$H_0 \psi_{n\mathbf{k}} = e^{i\mathbf{k}\cdot\mathbf{r}} \left(H_0 - \frac{i\hbar^2 \mathbf{k}}{m_o} \cdot \nabla + \frac{\hbar^2 \nabla^2}{2m_o} \right) \psi_{n0} \tag{3.56}$$

$$= \left(\varepsilon_{n0} + \frac{\hbar^2 k^2}{2m_o} \right) \psi_{n\mathbf{k}} + \frac{e^{i\mathbf{k}\cdot\mathbf{r}} \hbar \mathbf{k}}{m_o} \cdot (-\hbar \nabla \psi_{n0}). \tag{3.57}$$

Hence,

$$\langle \psi_{n\mathbf{k}} | H_0 | \psi_{n'\mathbf{k}'} \rangle = \left(\varepsilon_{n0} + \frac{\hbar^2 k^2}{2m_o} \right) \delta_{n'n} \delta_{\mathbf{k}'\mathbf{k}} + \frac{\hbar \mathbf{k}}{m_o} \cdot \langle \psi_{n'0} | e^{i(\mathbf{k}-\mathbf{k}')\cdot\mathbf{r}} (-i\hbar \nabla) | \psi_{n0} \rangle \tag{3.58}$$

$$\simeq\left(\varepsilon_{n0}+\frac{\hbar^2k^2}{2m_o}\right)\delta_{n'n}\delta_{\mathbf{k'k}}+\frac{\hbar\mathbf{k}}{m_o}\cdot\langle\psi_{n'0}|e^{i(\mathbf{k}-\mathbf{k}')\cdot\mathbf{r}}(-i\hbar\nabla)|\psi_{n0}\rangle. \tag{3.59}$$

Here,

$$\begin{aligned}
&\langle\psi_{n'0}|e^{i(\mathbf{k}-\mathbf{k}')\cdot\mathbf{r}}(-i\hbar\nabla)|\psi_{n0}\rangle \\
&\quad=\int d^3r u^*_{n',0}e^{i(\mathbf{k}-\mathbf{k}')\cdot\mathbf{r}}(-i\hbar\nabla)u_{n0} \\
&\quad=\left[\sum_{n=1}^{N}e^{i(\mathbf{k}-\mathbf{k}')\cdot\mathbf{r}_n}\right](-i\hbar)\int_{n\text{th unit cell}}d^3r u^*_{n'0}\nabla u_{n0}\ \text{from Equation (3.22)} \\
&\quad=\frac{1}{N}\left[\sum_{n=1}^{N}e^{i(\mathbf{k}-\mathbf{k}')\cdot\mathbf{r}_n}\right]N(-i\hbar)\int_{n\text{th unit cell}}d^3r u^*_{n'0}\nabla u_{n0} \\
&\quad=\left[\frac{1}{V}\int e^{i(\mathbf{k}-\mathbf{k}')\cdot\mathbf{r}}d^3r\right](-i\hbar)\int d^3r u^*_{n'0}\nabla u_{n0} \\
&\quad=\delta_{\mathbf{kk}'}\mathbf{P}_{nn'}\quad(\text{from Eq. (3.51)})
\end{aligned} \tag{3.60}$$

$$\langle\psi_{n'\mathbf{k}'}|H_0|\psi_{n\mathbf{k}}\rangle=\left(\varepsilon_{n0}+\frac{\hbar^2k^2}{2m_o}\right)\delta_{n'n}\delta_{\mathbf{k'k}}+\frac{\hbar\mathbf{P}_{nn'}\cdot\mathbf{k}}{m_o}\delta_{\mathbf{k'k}}. \tag{3.61}$$

Noting that $\mathbf{P}_{nn'}$ is usually zero.

Using Equations (3.46) and (3.49), we can write Equation (3.45) as

$$i\hbar\frac{d}{dt}\varphi_{n\mathbf{k}}(t)=\sum_{n'}H_0^{nn'}\varphi_{n'\mathbf{k}}+\sum_{\mathbf{k}'}U_{\mathbf{kk}'}\varphi_{n\mathbf{k}'}. \tag{3.62}$$

The corresponding multiband effective mass equation is

$$i\hbar\frac{d}{dt}\Psi_n(\mathbf{r},t)=\sum_{n'}H_0^{nn'}(-i\nabla)\Psi_n(\mathbf{r},t)+U(\mathbf{r},t)\Psi_n(\mathbf{r},t). \tag{3.63}$$

To show that Equation (3.62) is equivalent to Equation (3.63), we just have to write down the matrix representation of Equation (3.63) using a plane wave basis set to expand $\Psi_n(\mathbf{r},t)$.

$$\Psi_n(\mathbf{r},t)=\sum_{\mathbf{k}}\varphi_{n\mathbf{k}}(t)\psi_{\mathbf{k}}. \tag{3.64}$$

The matrix form of Equation (3.63) then follows from our usual prescription.

$$i\hbar\frac{\partial}{\partial t}\varphi_{n\mathbf{k}} = \sum_{\mathbf{k}}\sum_{n'}\langle\psi_{\mathbf{k}}|H_0^{nn'}(-i\nabla)|\psi_{\mathbf{k}'}\rangle\varphi_{n'\mathbf{k}'} + \sum_{\mathbf{k}'}\langle\psi_{\mathbf{k}}|U|\psi_{\mathbf{k}'}\rangle\varphi_{n\mathbf{k}'}. \quad (3.65)$$

Equation (3.62) follows from Equation (3.65) if we note that

$$\langle\psi_{\mathbf{k}}|H_0^{nn'}(-i\nabla)|\psi_{\mathbf{k}'}\rangle = H_0^{nn'}(k)\delta_{\mathbf{kk}'}. \quad (3.66)$$

This relation is proved in exactly the same way we proved Equation (3.49). The relation between the actual wave function and the multiband envelope function follows readily from Equation (3.43).

$$\begin{aligned}\Psi_0(\mathbf{r},t) &= \sum_{n\mathbf{k}}\varphi_{n\mathbf{k}}(t)\frac{e^{i\mathbf{k}\cdot\mathbf{r}}}{\sqrt{V}}\bar{u}_{n,0}(\mathbf{r}). \\ &= \sum_{n}\bar{u}_{n0}(\mathbf{r})\sum_{n\mathbf{k}}\varphi_{n\mathbf{k}}(t)\psi_{\mathbf{k}}.\end{aligned} \quad (3.67)$$

Using Equation (3.64),

$$\Psi_0(\mathbf{r},t) = \sum_{n}\bar{u}_{n0}(\mathbf{r})\Psi_n(\mathbf{r},t). \quad (3.68)$$

Note that this result is exact, and we do not need to assume that $u_{n\mathbf{k}}(\mathbf{r})$ does not change much with $\mathbf{k}$.

3.2 THE LUTTINGER–KOHN HAMILTONIAN

The difficulty with the multiband effective mass equation of Equation (3.63) is that the differential equations for different bands n are all coupled together, unlike the single-band effective mass equation of Equation (3.17). For practical use, we need to limit ourselves to a few bands. However, if we simply truncate the multiband effective mass equation to the desired number of bands, we will run into some problems. One method is to renormalize the elements of the truncated matrix to account for the bands we are neglecting using perturbation theory. On the other hand, this method is difficult because the functions $u_{n,0}(\mathbf{r})$ are often not known accurately for remote bands. Another method is to write each of the matrix elements $H_{n,n'}(\mathbf{k})$ as a second-order polynomial in k, and treat the polynomial coefficients as fitting parameters so that eigenvalues of the matrix ($H(\mathbf{k})$) reproduce the correct energy dispersion relation for the bands. Here, we will review the derivation of the 8×8 Hamiltonian in the valence band following the approach of Luttinger and Kohn [2].

3.2.1 Nondegenerate Bands Without Spin–Orbit Coupling

Let H_0 be the Hamiltonian of the electron in the periodic potential $U_L(\mathbf{r})$. The eigenfunctions of H_0 will be denoted by $\psi_{n\mathbf{k}}$ and the corresponding eivenvalues by $\varepsilon_n(\mathbf{k})$. Here, n labels the band, and $\mathbf{k}$ is the wave vector in the first Brillouin zone. Then, the one-electron Schrödinger equation is given by

$$(H_0 + U)\psi = \varepsilon\psi, \tag{3.69}$$

where

$$\begin{aligned} U(\mathbf{r}) &= U_E(\mathbf{r}) + U_S(\mathbf{r}), \\ H_0 &= -\frac{\hbar^2}{2m_0}\nabla^2 + U_L(\mathbf{r}), \\ H_0\psi_{n\mathbf{k}} &= \varepsilon_n(\mathbf{k})\psi_{n\mathbf{k}}. \end{aligned} \tag{3.70}$$

If we consider only the periodic potential $U_L(\mathbf{r})$, the solutions to the Schrödinger equation can be written in the form of Bloch waves, as discussed in Section 3.1.1. The wave function of an electron in a band n with wave vector $\mathbf{k}$ in the first Brillouin zone of the crystal is written as

$$\psi_{n\mathbf{k}} = e^{i\mathbf{k}\cdot\mathbf{r}} u_{n\mathbf{k}}(\mathbf{r}) \tag{3.71}$$

where $u_{n\mathbf{k}}$ is a periodic function like $U_L(\mathbf{r})$ that is different for each n and each $\mathbf{k}$. Now, it is necessary to choose a complete set of functions in which to expand ψ. The $\psi_{n\mathbf{k}}$ form, of course, is a complete set of functions, in which any wave function can be expanded. However, we consider another set of functions as defined in Equation (3.44)

$$\chi_{n\mathbf{k}} = e^{i\mathbf{k}\cdot\mathbf{r}} u_{n0}. \tag{3.72}$$

These form a complete set if $\psi_{n\mathbf{k}}$ is a complete set, as proved below.

Assume that any function $f(\mathbf{r})$ is expanded in the $\psi_{n\mathbf{k}}$:

$$f(\mathbf{r}) = \sum_n \int d\mathbf{k} g_n(\mathbf{k})\psi_{n\mathbf{k}} = \sum_n \int d\mathbf{k} g_n(\mathbf{k}) e^{i\mathbf{k}\cdot\mathbf{r}} u_{n\mathbf{k}}. \tag{3.73}$$

Also, the periodic function $u_{n\mathbf{k}}$ can be expressed in terms of the Bloch functions at the bottom of the band. That is,

$$u_{n\mathbf{k}} = \sum_{n} b_{nn'}(\mathbf{k}) u_{n'0}. \tag{3.74}$$

Substituting this into Equation (3.73), we obtain the result

$$\begin{aligned} f(\mathbf{r}) &= \sum_{n'} \int d\mathbf{k} g_{n'}(\mathbf{k}) e^{i\mathbf{k}\cdot\mathbf{r}} \sum_{n} b_{n'n}(\mathbf{k}) u_{n0} \\ &= \sum_{n} \int d\mathbf{k} \left(\sum_{n'} g_{n'}(\mathbf{k}) b_{n'n}(\mathbf{k}) \right) \left(e^{i\mathbf{k}\cdot\mathbf{r}} u_{n0} \right) \\ &\equiv \sum_{n} \int d\mathbf{k} \bar{g}_n(\mathbf{k}) \chi_{n\mathbf{k}} \end{aligned} \tag{3.75}$$

with $\bar{g}_n(\mathbf{k}) = \sum_{n'} g_{n'}(\mathbf{k}) b_{n'n}(\mathbf{k})$. The $\chi_{n\mathbf{k}}$ satisfy the orthonomality if the $\psi_{n\mathbf{k}}$ do. That is, the Bloch waves satisfy the following condition

$$\langle \psi_{n\mathbf{k}} | \psi_{n'\mathbf{k}'} \rangle = \int_{entire\ crystal} \psi^*_{n\mathbf{k}} \psi_{n'\mathbf{k}'} d\mathbf{r} = \delta_{nn'} \delta_{\mathbf{k}-\mathbf{k}'}. \tag{3.76}$$

Thus,

$$\int_{entire\ crystal} e^{i(\mathbf{k}'-\mathbf{k})\cdot\mathbf{r}} u_{n\mathbf{k}}(\mathbf{r}) u_{n'\mathbf{k}}(\mathbf{r}) d\mathbf{r} = \delta_{nn'} \frac{1}{(2\pi)^3} \int e^{i(\mathbf{k}'-\mathbf{k})\cdot\mathbf{r}} d\mathbf{r}. \tag{3.77}$$

For $\mathbf{k}' = 0$ and $\mathbf{k} = 0$,

$$\int_{entire\ crystal} u_{n0}(\mathbf{r}) u_{n'0}(\mathbf{r}) d\mathbf{r} = \delta_{nn'} \frac{V}{(2\pi)^3}. \tag{3.78}$$

For the $\chi_{n\mathbf{k}}$,

$$\langle \chi_{n\mathbf{k}} | \chi_{n'\mathbf{k}'} \rangle = \int_{entire\ crystal} e^{i(\mathbf{k}'-\mathbf{k})\cdot\mathbf{r}} u^*_{n0} u_{n'0} d\mathbf{r}. \tag{3.79}$$

Since $u^*_{n0} u_{n'0}$ has the lattice periodicity, we can expand it in a Fourier series as follows:

$$u^*_{n0} u_{n'0} = \sum_{m} B^{nn'}_m e^{-i\mathbf{K}_\mathbf{m}\cdot\mathbf{r}}, \tag{3.80}$$

where the $B^{nn'}_m$ are just numerical coefficients, and $\mathbf{K}_\mathbf{m}$ are the reciprocal lattice vectors. Inserting Equation (3.80) in Equation (3.79), we obtain

$$\langle \chi_{n\mathbf{k}} | \chi_{n'\mathbf{k}'} \rangle = (2\pi)^3 \sum_m B_m^{nn'} \delta(\mathbf{k}' - \mathbf{k} - \mathbf{K_m}), \tag{3.81}$$

from the definition of the delta function

$$\delta(\mathbf{k}' - \mathbf{k} - \mathbf{K_m}) = \frac{1}{(2\pi)^3} \int e^{i(\mathbf{k}' - \mathbf{k} - \mathbf{K_m})\cdot\mathbf{r}} d\mathbf{r}. \tag{3.82}$$

However, since $\mathbf{k}'$ and $\mathbf{k}$ are both in the first Brillouin zone, $\mathbf{k}' - \mathbf{k} = \mathbf{K_m}$ is only possible if $m = 0$.

$$\langle \chi_{n\mathbf{k}} | \chi_{n'\mathbf{k}'} \rangle = (2\pi)^3 B_0^{nn'} \delta(\mathbf{k}' - \mathbf{k}). \tag{3.83}$$

Using Fourier's theorem with Ω as the volume of the unit cell, the $B_m^{nn'}$ are given by

$$B_m^{nn'} = \frac{1}{N\Omega} \int_{entire\ crystal} e^{i(\mathbf{K}_m)\cdot\mathbf{r}} u_{n0}^* u_{n'0} d\mathbf{r} = \frac{1}{\Omega} \int_{cell} e^{i(\mathbf{K}_m)\cdot\mathbf{r}} u_{n0}^* u_{n'0} d\mathbf{r}, \tag{3.84}$$

$$B_0^{nn'} = \frac{1}{\Omega} \int_{cell} u_{n0}^* u_{n'0} d\mathbf{r} = \frac{1}{(2\pi)^3} \delta_{nn'}, \tag{3.85}$$

using Equation (3.78). Thus, from Equation (3.83),

$$\langle \chi_{n\mathbf{k}} | \chi_{n'\mathbf{k}'} \rangle = \delta_{nn'} \delta(\mathbf{k}' - \mathbf{k}) \tag{3.86}$$

which is the required orthonormality. In Equation (3.69), we now make the Ansatz

$$\Psi = \sum_{n'} \int d\mathbf{k}' A_{'n}(\mathbf{k}') \chi_{n'\mathbf{k}'}, \tag{3.87}$$

which gives the equation

$$\sum_{n'} \int d\mathbf{k}' A_{n'}(\mathbf{k}') (H_0 + U) \chi_{n'\mathbf{k}'} = \varepsilon \sum_{n'} \int d\mathbf{k}' A_{n'}(\mathbf{k}') \chi_{n'\mathbf{k}'},$$

$$\sum_{n'} \int d\mathbf{k}' A_{n'}(\mathbf{k}') \int \chi_{n\mathbf{k}} (H_0 + U) \chi_{n'\mathbf{k}'} d\mathbf{r} = \varepsilon \sum_{n'} \int d\mathbf{k}' A_{n'}(\mathbf{k}') \int \chi_{n\mathbf{k}} \chi_{n'\mathbf{k}'} d\mathbf{r},$$

$$\sum_{n'} \int d\mathbf{k}' A_{n'}(\mathbf{k}') \langle \chi_{n\mathbf{k}} | (H_0 + U) | \chi_{n'\mathbf{k}'} \rangle = \varepsilon \sum_{n'} \int d\mathbf{k}' A_{'n}(\mathbf{k}') \langle \chi_{n'\mathbf{k}'} | \chi_{n'\mathbf{k}'} \rangle,$$

$$\sum_{n'} \int d\mathbf{k}' A_{n'}(\mathbf{k}') \langle \chi_{n\mathbf{k}} | (H_0 + U) | \chi_{n'\mathbf{k}'} \rangle = \varepsilon A_n(\mathbf{k}). \tag{3.88}$$

The matrix elements can be evaluated in the following manner. For H_0, we have

$$\begin{aligned} H_0\chi_{n\mathbf{k}} &= \left[-\frac{\hbar^2}{2m_0}\nabla^2 + U_L(\mathbf{r})\right] e^{i\mathbf{k}\cdot\mathbf{r}} u_{n0} \\ &= e^{i\mathbf{k}\cdot\mathbf{r}}\left[\frac{\hbar^2 k^2}{2m_0} + \frac{\hbar}{m_0}\mathbf{k}\cdot\frac{\hbar\nabla}{i} - \frac{\hbar^2}{2m_0}\nabla^2 + U_L(\mathbf{r})\right] u_{n0} \\ &= e^{i\mathbf{k}\cdot\mathbf{r}}\left[H_0 + \frac{\hbar}{m_0}\mathbf{k}\cdot\mathbf{p} + \frac{\hbar^2 k^2}{2m_0}\right] u_{n0}. \end{aligned} \tag{3.89}$$

Thus,

$$\begin{aligned} &\langle \chi_{n\mathbf{k}} | H_0 | \chi_{n'\mathbf{k}'} \rangle \\ &= \int e^{-i\mathbf{k}\cdot\mathbf{r}} u_{n0}^* H_0 e^{i\mathbf{k}'\cdot\mathbf{r}} u_{n'0} d\mathbf{r} \\ &= \int e^{i(\mathbf{k}'-\mathbf{k})\cdot\mathbf{r}} u_{n0}^* \left[H_0 + \frac{\hbar}{m_0}\mathbf{k}\cdot\mathbf{p} + \frac{\hbar^2 k^2}{2m_0}\right] u_{n0} d\mathbf{r} \\ &= \int e^{i(\mathbf{k}'-\mathbf{k})\cdot\mathbf{r}} u_{n0}^* \left[\varepsilon_{n'}(0) + \frac{\hbar}{m_0}\mathbf{k}\cdot\mathbf{p} + \frac{\hbar^2 k^2}{2m_0}\right] u_{n0} d\mathbf{r}, \end{aligned} \tag{3.90}$$

where $\varepsilon_{n'}(0)$ is the energy at the bottom of the nth band, and $\mathbf{P}$ is the momentum operator $(\hbar/i)\nabla$. Since the entire factor multiplying $e^{i(\mathbf{k}'-\mathbf{k})\cdot\mathbf{r}}$ is periodic, the same argument that leads from Equation (3.79) to (3.86) yields

$$\langle \chi_{n\mathbf{k}} | H_0 | \chi_{n'\mathbf{k}'} \rangle = \int e^{i(\mathbf{k}'-\mathbf{k})\cdot\mathbf{r}} u_{n0}^* \left[\varepsilon_{n'}(0) + \frac{\hbar}{m_0}\mathbf{k}\cdot\mathbf{p} + \frac{\hbar^2 k^2}{2m_0}\right] u_{n0} d\mathbf{r}, \tag{3.91}$$

and

$$u_{n0}^* \left[\varepsilon_{n'}(0) + \frac{\hbar}{m_0}\mathbf{k}\cdot\mathbf{p} + \frac{\hbar^2 k^2}{2m_0}\right] u_{n0} = \sum_m C_m^{nn'} e^{-i\mathbf{K}_\mathbf{m}\cdot\mathbf{r}}. \tag{3.92}$$

Thus,

$$\langle \chi_{n\mathbf{k}} | H_0 | \chi_{n'\mathbf{k}'} \rangle = (2\pi)^3 \sum_m C_m^{nn'} \delta(\mathbf{k}' - \mathbf{k} - \mathbf{K}_\mathbf{m}). \tag{3.93}$$

Similarly, since $\mathbf{k}'-\mathbf{k}=\mathbf{K_m}$ is only possible if $m=0$,

$$\langle \chi_{n\mathbf{k}}|H_0|\chi_{n'\mathbf{k}'}\rangle = (2\pi)^3 C_0^{nn'}\delta(\mathbf{k}'-\mathbf{k}). \tag{3.94}$$

Using Fourier's theorem with Ω, the volume of the unit cell, the $C_m^{nn'}$ are given by

$$\begin{aligned} C_m^{nn'} &= \frac{1}{\Omega}\int_{cell} e^{i(\mathbf{K}_m)\cdot\mathbf{r}} u_{n0}^* \left[\varepsilon_{n'}(0)+\frac{\hbar}{m_0}\mathbf{k}\cdot\mathbf{p}+\frac{\hbar^2k^2}{2m_0}\right]u_{n0}d\mathbf{r} \\ C_0^{nn'} &= \frac{1}{\Omega}\int_{cell} u_{n0}^* \left[\varepsilon_{n'}(0)+\frac{\hbar}{m_0}\mathbf{k}\cdot\mathbf{p}+\frac{\hbar^2k^2}{2m_0}\right]u_{n0}d\mathbf{r} \end{aligned} \tag{3.95}$$

Thus,

$$\begin{aligned} \langle \chi_{n\mathbf{k}}|H_0|\chi_{n'\mathbf{k}'}\rangle &= \frac{(2\pi)^3}{\Omega}\delta(\mathbf{k}'-\mathbf{k})\int_{cell} u_{n0}^* \left[\varepsilon_{n'}(0)+\frac{\hbar}{m_0}\mathbf{k}\cdot\mathbf{p}+\frac{\hbar^2k^2}{2m_0}\right]u_{n'0}d\mathbf{r} \\ &= \delta(\mathbf{k}'-\mathbf{k})\left[\left(\varepsilon_n(0)+\frac{\hbar^2k^2}{2m_0}\right)\delta_{nn'}+\frac{\hbar k_\alpha p_{nn'}^\alpha}{m_0}\right] \end{aligned} \tag{3.96}$$

Here, dummy indices mean a summation over $\alpha = x, y, z$, and the momentum matrix element $p_{nn'}^\alpha$ at the bottom of the band is defined by

$$p_{nn'}^\alpha = \frac{(2\pi)^3}{\Omega}\int_{cell} u_{n0}^*\left(\frac{\hbar}{i}\nabla_\alpha\right)u_{n'0}d\mathbf{r}. \tag{3.97}$$

Also, the expectation value $p_{nn}^\alpha(\mathbf{k})$ of the momentum is given by $p_{nn}^\alpha(\mathbf{k}) = m\partial\varepsilon_n(\mathbf{k})/\partial k_\alpha$. Thus,

$$p_{nn}^\alpha = \frac{(2\pi)^3}{\Omega}\int_{cell} u_{n0}^*\left(\frac{\hbar}{i}\nabla_\alpha\right)u_{n0}d\mathbf{r} = \frac{(2\pi)^3}{\Omega}\int_{cell} \psi_{n0}^*\left(\frac{\hbar}{i}\nabla_\alpha\right)\psi_{n0}d\mathbf{r} \tag{3.98}$$

is the expectation value at the bottom of the band and becomes zero at the minimum. In addition, they have the following properties:

$$p_{nn'}^\alpha = p_{n'n}^\alpha = (p_{n'n}^\alpha)^* \tag{3.99}$$

which follows if a center of symmetry exists in the crystal. For the matrix elements of U, we proceed similarly:

$$\begin{aligned}
&\langle \chi_{n\mathbf{k}}|U|\chi_{n'\mathbf{k}'}\rangle \\
&\quad= \int e^{-i\mathbf{k}\cdot\mathbf{r}} u_{n0}^{*} U e^{i\mathbf{k}'\cdot\mathbf{r}} u_{n'0} d\mathbf{r} \\
&\quad= \int e^{i(\mathbf{k}'-\mathbf{k})\cdot\mathbf{r}} u_{n0}^{*} u_{n0} U d\mathbf{r} \\
&\quad= \sum_m \int e^{i(\mathbf{k}'-\mathbf{k})\cdot\mathbf{r}} B_m^{nn'} e^{-i\mathbf{K_m}\cdot\mathbf{r}} U d\mathbf{r} \\
&\quad= (2\pi)^3 \sum_m \int B_m^{nn'} \bar{U}(\mathbf{k}-\mathbf{k}'+\mathbf{K}_m),
\end{aligned} \tag{3.100}$$

where

$$\bar{U}(\mathbf{k}-\mathbf{k}'+\mathbf{K}_m) = \frac{1}{(2\pi)^3} \int d\mathbf{r} e^{-i(\mathbf{k}-\mathbf{k}'+\mathbf{K}_m)\cdot\mathbf{r}} U(\mathbf{r}). \tag{3.101}$$

The right-hand side requires us to integrate over the entire sample volume V. But first, let us consider the effect of integrating over a single unit cell, say the nth one.

$$\begin{aligned}
\frac{1}{(2\pi)^3} \int_{nth\ unit\ cell} d\mathbf{r} e^{-i(\mathbf{k}-\mathbf{k}'+\mathbf{K}_m)\cdot\mathbf{r}} U(\mathbf{r}) &\simeq \frac{1}{(2\pi)^3} U(\mathbf{r}_n) \int_{nth\ unit\ cell} d\mathbf{r} e^{-i(\mathbf{k}-\mathbf{k}'+\mathbf{K}_m)\cdot\mathbf{r}} \\
&= \frac{1}{N} U(\mathbf{r}_n) \delta(\mathbf{k}-\mathbf{k}'+\mathbf{K}_m)
\end{aligned} \tag{3.102}$$

where $\mathbf{r}_n$ is the coordinate of the center of the nth unit cell. Here we assumed that $U(\mathbf{r},t)$ varies slowly enough that it can be considered approximately constant over a unit cell. Similarly, $\mathbf{k}'-\mathbf{k}=\mathbf{K_m}$ is only possible if $m=0$.

$$\langle \chi_{n\mathbf{k}}|U|\chi_{n'\mathbf{k}'}\rangle = (2\pi)^3 B_0^{nn'} \bar{U}(\mathbf{k}-\mathbf{k}') = \delta_{nn'} \bar{U}(\mathbf{k}-\mathbf{k}') \tag{3.103}$$

from Equation (3.85). From Equations (3.91) and (3.103), Equation (3.90) becomes

$$\begin{aligned}
&\left(\varepsilon_n(0) + \frac{\hbar^2 k^2}{2m_0}\right) A_n(\mathbf{k}) + \sum_{n'\neq n} \frac{\hbar k_\alpha p_{nn'}^{\alpha}}{m_0} A_{n'}(\mathbf{k}) + \int d\mathbf{k}' \bar{U}(\mathbf{k}-\mathbf{k}') A_n(\mathbf{k}) \\
&= \varepsilon A_n(\mathbf{k}).
\end{aligned} \tag{3.104}$$

Equation (3.104) still contains terms involving $p_{nn'}^{\alpha}$, which are proportional to $\mathbf{k}$, and represent coupling between bands. Since the interband matrix elements $p^{\alpha}_{nn'}$ are causing the trouble, we try to find a transformation to yield equations with no interband elements to the first order in $\mathbf{k}$. To do so, we remove them to the first order by a canonical transformation T. That is, we put

$$A_n(\mathbf{k}) = \sum_{n'} \int d\mathbf{k} \langle \chi_{n\mathbf{k}} | T | \chi_{n'\mathbf{k}'} \rangle B_{n'}(\mathbf{k}). \tag{3.105}$$

It is convenient to write this as

$$HA = \varepsilon A, \tag{3.106}$$

and

$$A = TA \equiv e^S B. \tag{3.107}$$

Substituting Equation (3.107) into Equation (3.106), we obtain

$$\begin{aligned} HA = He^S B = \varepsilon A = \varepsilon e^S B \\ e^{-S} He^S B = \varepsilon B \equiv \bar{H} B. \end{aligned} \tag{3.108}$$

Clearly,

$$\bar{H} = e^{-S} He^S = H + [H,S] + \frac{1}{2}[[H,S],S] + \ldots \tag{3.109}$$

Let us write $H = H^{(0)} + H^{(1)} + U$, where

$$\begin{aligned} \langle n\mathbf{k} | H^{(0)} | n'\mathbf{k}' \rangle &= \left(\varepsilon_n(0) + \frac{\hbar^2 k^2}{2m_0} \right) \delta_{nn'} \delta(\mathbf{k}' - \mathbf{k}) \\ \langle n\mathbf{k} | H^{(1)} | n'\mathbf{k}' \rangle &= \frac{\hbar k_\alpha p^{\alpha}_{nn'}}{m_0} \delta(\mathbf{k}' - \mathbf{k}) \\ \langle n\mathbf{k} | U | n'\mathbf{k}' \rangle &= \bar{U}(\mathbf{k}' - \mathbf{k}) \delta_{nn'}. \end{aligned} \tag{3.110}$$

Then,

$$\begin{aligned} \bar{H} = H^{(0)} + H^{(1)} + U + [H^{(0)},S] + [H^{(1)},S] + [U,S] \\ + \frac{1}{2}[[H^{(0)},S],S] + \frac{1}{2}[[U,S],S] + \ldots, \end{aligned} \tag{3.111}$$

the omitted terms being of order S^3 or more. By choosing S such that

$$H^{(1)} + [H^{(0)}, S] = 0, \tag{3.112}$$

we succeeded in eliminating the interband transitions to the first order in $\mathbf{k}$. Going back to the χ_{nk} representation, we see at once that

$$\begin{aligned} &\langle \chi_{nk} | H^{(1)} | \chi_{n'k'} \rangle + \langle \chi_{nk} | H^{(0)} S - S H^{(0)} | \chi_{n'k'} \rangle = 0 \\ &\frac{\hbar k_\alpha p^\alpha_{nn'}}{m_0} \delta(\mathbf{k}' - \mathbf{k}) + (\varepsilon_n - \varepsilon_{n'}) \langle \chi_{nk} | S | \chi_{n'k'} \rangle = 0 \end{aligned} \tag{3.113}$$

$$\langle \chi_{nk} | S | \chi_{n'k'} \rangle = \begin{cases} -\dfrac{k_\alpha p^\alpha_{nn'}}{m_0 \omega_{nn'}} \delta(\mathbf{k}' - \mathbf{k}), & n \neq n' \\ 0, & n = n' \end{cases}, \tag{3.114}$$

where $\hbar\omega_{n,n'} = \varepsilon_n - \varepsilon_{n'}$. Consequently, the second-order terms from $H^{(0)}$ and $H^{(1)}$ are

$$[H^{(1)}, S] + \frac{1}{2}[[H^{(0)}, S], S] = \frac{1}{2}[H^{(1)}, S], \tag{3.115}$$

which becomes

$$\begin{aligned} \frac{1}{2}\langle \chi_{nk} | [H^{(1)}, S] | \chi_{n'k'} \rangle &= \frac{1}{2} \sum_{n''} \langle \chi_{nk} | H^{(1)} | \chi_{n''k''} \rangle \langle \chi_{n''k''} | S | \chi_{n'k'} \rangle \\ &\quad - \frac{1}{2} \sum_{n''} \langle \chi_{nk} | S | \chi_{n''k''} \rangle \langle \chi_{n''k''} | H^{(1)} | \chi_{n'k'} \rangle \\ &= \frac{1}{2} \sum_{n''} \frac{\hbar k_\alpha p^\alpha_{nn''}}{m_0} \delta(\mathbf{k} - \mathbf{k}'') \left[-\frac{k''_\beta p^\beta_{n''n'}}{m_0 \omega_{n''n'}} \delta(\mathbf{k}'' - \mathbf{k}') \right] \\ &\quad - \left[-\frac{k_\beta p^\beta_{nn''}}{m_0 \omega_{nn''}} \delta(\mathbf{k} - \mathbf{k}'') \right] \frac{1}{2} \sum_{n''} \frac{\hbar k''_\alpha p^\alpha_{n''n'}}{m_0} \delta(\mathbf{k}'' - \mathbf{k}') \end{aligned} \tag{3.116}$$

Using δ function,

$$\begin{aligned} &\frac{1}{2}\langle \chi_{nk} | [H^{(1)}, S] | \chi_{n'k'} \rangle \\ &= \frac{\hbar k_\alpha k_\beta}{2m_0^2} \left[\sum_{n'' \neq n, n'} p^\alpha_{nn''} p^\beta_{n''n'} \left(\frac{1}{\omega_{nn''}} + \frac{1}{\omega_{n'n''}} \right) \right] \delta(\mathbf{k} - \mathbf{k}'), \end{aligned} \tag{3.117}$$

in the $\chi_{n\mathbf{k}}$ representation. The correction terms to U are also easily found. Consider first $\langle \chi_{nk} \| [U,S] \| \chi_{n'k'} \rangle$. Using Equations (3.110) and (3.114), we obtain

$$\begin{aligned} &\langle \chi_{nk} \| [U,S] \| \chi_{n'k'} \rangle \\ &= \sum_{n''} \langle \chi_{nk} | U | \chi_{n''k''} \rangle \langle \chi_{n''k''} | S | \chi_{n'k'} \rangle - \sum_{n''} \langle \chi_{nk} | S | \chi_{n''k''} \rangle \langle \chi_{n''k''} | U | \chi_{n'k'} \rangle \\ &= \sum_{n''} \delta_{nn''} U(\mathbf{k}-\mathbf{k}'') \left[-\frac{k''_\alpha p^\alpha_{n''n'}}{m_0 \omega_{n''n'}} \delta(\mathbf{k}''-\mathbf{k}') \right] \\ &\quad - \sum_{n''} \left[-\frac{k_\alpha p^\alpha_{nn''}}{m_0 \omega_{nn''}} \delta(\mathbf{k}-\mathbf{k}'') \right] \delta_{n'n''} U(\mathbf{k}'-\mathbf{k}''). \end{aligned} \tag{3.118}$$

Thus,

$$\langle \chi_{nk} \| [U,S] \| \chi_{n'k'} \rangle = \begin{cases} (k_\alpha - k'_\alpha) U(\mathbf{k}-\mathbf{k}'') p^\alpha_{nn'} / (m_0 \omega_{nn'}), & n \neq n' \\ 0, & n = n' \end{cases}. \tag{3.119}$$

Here, $k \sim 1/a_i$, where a_i is the extent of the impurity state when the U is the potential due to the impurities. $p_{nn'} \sim 1/a$ and $\omega_{nn'} \sim 1/(m_o a^2)$, where a is the lattice spacing. Equation (3.119) is of the order of the square of the ratio of a/a_i, and can be neglected. Similarly, the next term, $\frac{1}{2}[[U,S],S]$, is also omitted because it is reduced over $[U,S]$ by a factor S, which is of the squared order of a/a_i. In the limit of very extended impurity states, Equation (3.104) becomes

$$\begin{aligned} &\left(\varepsilon_n(0) + \frac{\hbar^2 k^2}{2m_0} + \frac{\hbar^2 k_\alpha k_\beta}{m_0^2} \left[\sum_{n'',\neq n} p^\alpha_{nn''} p^\beta_{n''n} \left(\frac{1}{\omega_{nn''}} \right) \right] \right) B_n(\mathbf{k}) \\ &\quad + \int d\mathbf{k}' U(\mathbf{k}-\mathbf{k}') B_n(\mathbf{k}') = \varepsilon B_n(\mathbf{k}) \end{aligned} \tag{3.120}$$

where we have neglected terms of the order k^3 and interband elements of order k^2. Since the coefficient of $B_n(\mathbf{k})$ is exactly the expansion of $\varepsilon_n(\mathbf{k})$ to second order, we can write

$$\varepsilon_n(\mathbf{k}) B_n(\mathbf{k}) + \int d\mathbf{k}' U(\mathbf{k}-\mathbf{k}') B_n(\mathbf{k}') = \varepsilon B_n(\mathbf{k}). \tag{3.121}$$

Equation (3.121) is the well-known *effective mass equation* written in momentum space. To get the more usual formulation, we introduce a function

$$F_n(\mathbf{r}) = \int e^{i\mathbf{k}\cdot\mathbf{r}} B_n(\mathbf{k}) d\mathbf{k}, \tag{3.122}$$

where the integration is over the first Brillouin zone. Inverting Equation (3.122), we obtain

$$B_n(\mathbf{k}) = \frac{\Omega}{(2\pi)^3} \sum_m e^{-i\mathbf{k}\cdot\mathbf{R}_m} F_n(\mathbf{R}_m), \tag{3.123}$$

since $B_n(\mathbf{k})$ is periodic in $\mathbf{k}$, $\mathbf{R}_m$ being the vector to the lattice point m. Then, Equation (3.121) becomes simply

$$\varepsilon_n(-i\nabla)F_n(\mathbf{r}) + \int d\mathbf{r}'\Delta(\mathbf{r}-\mathbf{r}')U(\mathbf{r}')F_n(\mathbf{r}') = \varepsilon F_n(\mathbf{r}), \tag{3.124}$$

where

$$\Delta(\mathbf{r}-\mathbf{r}') = \frac{1}{(2\pi)^3} \int d\mathbf{k} e^{i\mathbf{k}\cdot(\mathbf{r}-\mathbf{r}')}. \tag{3.125}$$

Since $\Delta\mathbf{r}$ is a δ-like function of extension $\sim a$, we have

$$[\varepsilon_n(-i\nabla) + U(\mathbf{r})]F_n(\mathbf{r}) = \varepsilon F_n(\mathbf{r}). \tag{3.126}$$

From Equation (3.87), the leading term in the wave function becomes

$$\Psi = \sum_n \int d\mathbf{k} B_n(\mathbf{k}) e^{i\mathbf{k}\cdot\mathbf{r}} u_{n0}(\mathbf{r}) = \sum_n F_n(\mathbf{r}) u_{n0}(\mathbf{r}) = \sum_n F_n(\mathbf{r}) \Psi_{n0}(\mathbf{r}). \tag{3.127}$$

3.2.2 Degenerate Bands Without Spin–Orbit Coupling

We now consider the case of a degenerate band neglecting spin–orbit coupling. For simplicity, we shall limit ourselves to the case where the degeneracy occurs at the point $\mathbf{k} = 0$. Let us call the degenerate functions at this point $\phi_1, \phi_2, \ldots, \phi_r$, a typical one being denoted by ϕ_j. The degeneracy implies that they all have the same energy:

$$H_0\phi_j = \varepsilon_0\phi_j. \tag{3.128}$$

Let us further make the convention that the wave functions at $\mathbf{k} = 0$ for the other bands be denoted by ϕ_i $(i \neq j)$.

We now introduce a complete set of functions $\phi_{n\mathbf{k}}$ defined by

$$\phi_{n\mathbf{k}} = e^{i\mathbf{k}\cdot\mathbf{r}}\phi_n, \tag{3.129}$$

the index n running over both j and i. We then make the Ansatz

$$\Psi = \sum_n \int d\mathbf{k} A_n(\mathbf{k})\phi_{n\mathbf{k}}. \tag{3.130}$$

For the impurity scattering problem,

$$\sum_{n'} \int d\mathbf{k}' \langle n\mathbf{k}|H_0 + U|n'\mathbf{k}'\rangle A_{n'}(\mathbf{k}') = \varepsilon A_n(\mathbf{k}'). \tag{3.131}$$

The analysis leading from Equation (3.90) to Equation (3.104) is again valid due to the orthonormality of the $\phi_{n\mathbf{k}}$. On specializing n to j in Equation (3.131), we obtain

$$\left(\varepsilon_0 + \frac{\hbar^2 k^2}{2m_0}\right) A_j(\mathbf{k}) + \sum_i \frac{\hbar k_\alpha p_{ji}^\alpha}{m_0} A_i(\mathbf{k}) + \int d\mathbf{k}' \bar{U}(\mathbf{k}-\mathbf{k}') A_j(\mathbf{k}') = \varepsilon A_j(\mathbf{k}) \tag{3.132}$$

This is the equation that replaces Equation (3.104) for the degenerate case. We again transform the interband elements p_{ji}^α by a transformation analogous to Equations (3.107) and (3.114). Here, the matrix elements of $\mathbf{p}$ vanish between states of the same parity, so that we can write

$$p_{jj'}^\alpha = 0. \tag{3.133}$$

Then,

$$\langle \chi_{nk}|S|\chi_{n'k'}\rangle = \begin{cases} -\dfrac{k_\alpha p_{nn'}^\alpha}{m_0 \omega_{nn'}} \delta(\mathbf{k}'-\mathbf{k}), & n = j, n' \neq j \\ 0, & n = j, n' = j \end{cases}. \tag{3.134}$$

Since we are interested in degenerate bands, putting $n = j$ and $n' = j'$ in Equation (3.105) leads to the result, to the second-order terms in k:

$$\sum_{j'} \left(\varepsilon_0 \delta_{jj'} + \frac{\hbar^2 k^2}{2m_0} \delta_{jj'} + \frac{\hbar^2 k_\alpha k_\beta}{m_0^2} \sum_i \frac{p_{ji}^\alpha p_{ij'}^\beta}{\varepsilon_0 - \varepsilon_i} \right) B_{j'}(\mathbf{k}) + \int d\mathbf{k}' U(\mathbf{k}-\mathbf{k}') B_j(\mathbf{k}') = \varepsilon B_j(\mathbf{k}). \tag{3.135}$$

If we set the energy zero at ε_0, then we can write

$$\sum_{\alpha\beta} \sum_{j'=1}^{r} D_{jj'}^{\alpha\beta} k_\alpha k_\beta B_{j'}(\mathbf{k}) + \int d\mathbf{k}' U(\mathbf{k}-\mathbf{k}') B_j(\mathbf{k}') = \varepsilon B_j(\mathbf{k}), \tag{3.136}$$

where

$$D_{jj'}^{\alpha\beta} \equiv \frac{\hbar^2}{2m_0}\delta_{jj'}\delta_{\alpha\beta} + \frac{\hbar^2}{m_0^2}\sum_i \frac{p_{ji}^{\alpha} p_{ij'}^{\beta}}{\varepsilon_0 - \varepsilon_i}. \tag{3.137}$$

This set of numbers $D_{jj'}^{\alpha\beta}$ plays the same role in the theory of degenerate bands that the effective masses do for a simple band. We again introduce a function $F_j(\mathbf{r})$ as Equation (3.122),

$$F_j(\mathbf{r}) = \int e^{i\mathbf{k}\cdot\mathbf{r}} B_j(\mathbf{r}) d\mathbf{k}. \tag{3.138}$$

Then, Equation (3.136) becomes

$$\sum_{\alpha\beta}\sum_{j'=1}^{r}\left[D_{jj'}^{\alpha\beta}\left(\frac{1}{i}\nabla_\alpha\right)\left(\frac{1}{i}\nabla_\beta\right) + U(\mathbf{r})\delta_{jj'}\right]F_{j'}(\mathbf{r}) = \varepsilon F_j(\mathbf{r}), \tag{3.139}$$

and the leading term in the wave function is

$$\Psi = \sum_{j=1}^{r} F_j(\mathbf{r})\phi_j(\mathbf{r}). \tag{3.140}$$

The Hamitonian can be written as the sum of a band-edge contribution and a **k**-dependent contribution

$$H(\mathbf{k})_{jj'} = E_j(\mathbf{k}=0) + \sum_{\alpha\beta} D_{jj'}^{\alpha\beta} k_\alpha k_\beta, \tag{3.141}$$

where $E_j(\mathbf{k}=0) = \varepsilon_0\delta_{jj'}$.

3.2.3 Spin–Orbit Coupling

In this section, we explain the spin–orbit coupling effect and introduce the concept of the spinor for the wave function. So far, we have treated the electronic wave function Ψ as a complex scalar. However, Ψ is neither a scalar nor a vector; it is a spinor. We know that we need one component to specify a scalar, while we need three components along the three coordinate axes to specify a vector. On the other hand, to specify a spinor, we need two components along the "up" and "down" directions. The two spin components are often uncoupled, and we can treat Ψ as a scalar and simply multiply the results by 2 at the end to account for the two spin polarizations. However, this is not always true. In a magnetic field, for example, the energies of two polarizations are different. Moreover, an electron moving in a strong electric field sees an effective magnetic field due to relativistic effects even if there are no external magnetic fields. This is known as the "spin–orbit interaction." We include effects of spin–orbit coupling as follows:

$$\tilde{H}_0 = H_0 + H_{so}, \tag{3.142}$$

where

$$H_{so} = \frac{\hbar}{4m_0^2c^2}(\sigma \times \nabla V)\cdot \mathbf{p}, \tag{3.143}$$

and $\sigma_x, \sigma_y, \sigma_z$ are the Pauli spin matrices defined as

$$\sigma_x = \begin{pmatrix} 0 & 1 \\ 1 & 0 \end{pmatrix}, \sigma_y = \begin{pmatrix} 0 & -i \\ i & 0 \end{pmatrix}, \sigma_y = \begin{pmatrix} 1 & 0 \\ 0 & -1 \end{pmatrix}. \tag{3.144}$$

Since V is periodic and $\mathbf{p}$ invariant under translations, the total unperturbed Hamiltonian $\bar{H}_0$ will still have Bloch waves as solutions. Call these $\tilde{\psi}_{n\mathbf{k}} = e^{i\mathbf{k}\cdot\mathbf{r}}\tilde{u}_{n\mathbf{k}}$, with the energy $\tilde{\varepsilon}_n(\mathbf{k})$. We may again introduce a set of functions analogous to the $\chi_{n\mathbf{k}}$, say $\tilde{\chi}_{n\mathbf{k}}$, defined by

$$\tilde{\chi}_{n\mathbf{k}} = e^{i\mathbf{k}\cdot\mathbf{r}}\tilde{u}_{n0}. \tag{3.145}$$

Then, the Ansatz

$$\psi = \sum_n \int d\mathbf{k} A_n(\mathbf{k})\tilde{\chi}_{n\mathbf{k}} \tag{3.146}$$

leads to the same Schrödinger equation for $A_n(\mathbf{k})$ as the one we had without spin–orbit coupling, except that the matrix elements $p^{\alpha}_{nn'}$ are replaced by $\pi^{\alpha}_{nn'}$, defined by

$$\pi^{\alpha}_{nn'} = \frac{(2\pi)^3}{\Omega}\int_{cell} \tilde{u}^*_{n0}\left(\frac{\hbar}{i}\nabla_{\alpha} + \frac{\hbar}{4m_0^2c^2}(\sigma \times \nabla V)_{\alpha}\right)\tilde{u}_{n'0}d\mathbf{r} \tag{3.147}$$

This extra term arises from the fact that the spin–orbit coupling contains the differential operation $\mathbf{p}$.

Degenerate bands are split, in general, into several bands by the spin–orbit coupling. If the spin–orbit interaction were absent, the six valence band wave functions would simply be

$$\begin{pmatrix} |X\rangle \\ 0 \end{pmatrix}, \begin{pmatrix} 0 \\ |X\rangle \end{pmatrix}, \begin{pmatrix} |Y\rangle \\ 0 \end{pmatrix}, \begin{pmatrix} 0 \\ |Y\rangle \end{pmatrix}, \begin{pmatrix} |Z\rangle \\ 0 \end{pmatrix}, \begin{pmatrix} 0 \\ |Z\rangle \end{pmatrix}, \tag{3.148}$$

and we could continue to treat Ψ as a scalar and multiply at the end by 2. In many common semiconductors like gallium arsenide (GaAs), due to crystal symmetry, the functions $u_{n0}(\mathbf{r})$ for the conduction and valence bands can be expressed in terms of these functions $|S\rangle, |X\rangle, |Y\rangle$, and $|Z\rangle$. Electrons in the conduction band have completely symmetric wave functions $|S\rangle$. Here, $|S\rangle$ is completely symmetric in x, y, and z [like $(x^2+y^2+z^2)^{-1/2}$]; $|X\rangle$ is antisymmetric in x but symmetric in y and z [like $x(x^2+y^2+z^2)^{-1/2}$]:$|Y\rangle$ is antisymmetric in y but symmetric in z and x; and $|Z\rangle$ is antisymmetric in z but symmetric in x and y. In the absence of spin–orbit coupling, we shall denote three degenerate space functions at $\mathbf{k}=0$ by $|X\rangle, |Y\rangle$, and $|Z\rangle$. Each has an additional double degeneracy due to spin. We will construct a matrix D defined by

$$D_{jj'} = \sum_{\alpha,\beta} k_\alpha k_\beta D_{jj'}^{\alpha\beta}. \tag{3.149}$$

using these functions.

Then, as can be seen from Equation (3.137), the symmetry of the diamond lattice requires that D have the form

$$D = \begin{pmatrix} Ak_x^2 + B(k_y^2+k_z^2) & Ck_xk_y & Ck_xk_z \\ Ck_xk_y & Ak_y^2 + B(k_x^2+k_z^2) & Ck_yk_z \\ Ck_xk_z & Ck_yk_z & Ak_z^2 + B(k_y^2+k_x^2) \end{pmatrix} \tag{3.150}$$

for each spin, where A, B, C are three real constants defined by

$$\begin{aligned} A &= \frac{\hbar^2}{2m_0} + \frac{\hbar^2}{m_0^2}\sum_i \frac{p_{Xi}^x p_{iX}^x}{\varepsilon_0 - \varepsilon_i} \\ B &= \frac{\hbar^2}{2m_0} + \frac{\hbar^2}{m_0^2}\sum_i \frac{p_{Xi}^y p_{iX}^y}{\varepsilon_0 - \varepsilon_i} \\ C &= \frac{\hbar^2}{m_0^2}\sum_i \frac{p_{Xi}^x p_{iY}^y + p_{Xi}^y p_{iY}^x}{\varepsilon_0 - \varepsilon_i}. \end{aligned} \tag{3.151}$$

As an example of the calculation, consider D_{XX},

$$D_{XX} = \sum_{\alpha\beta}\left(\frac{\hbar^2}{2m_0}\delta_{XX}\delta_{\alpha\beta} + \frac{\hbar^2}{m_0^2}\sum_i \frac{p_{Xi}^{\alpha}p_{iX}^{\beta}}{\varepsilon_0 - \varepsilon_i}\right)k_\alpha k_\beta, \tag{3.152}$$

where

$$p_{Xi}^{\alpha}p_{iX}^{\beta} = \left\langle X\left|\frac{\hbar}{i}\nabla_\alpha\right|i\right\rangle\left\langle i\left|\frac{\hbar}{i}\nabla_\beta\right|X\right\rangle. \tag{3.153}$$

Equation (3.153) is zero except for $\alpha = \beta = X, Y, Z$, because of its antisymmetry. That is, for $\alpha = X$ and $\beta = Y$,

$$p_{Xi}^{X}p_{iX}^{Y} = \left\langle X\left|\frac{\hbar}{i}\nabla_x\right|i\right\rangle\left\langle i\left|\frac{\hbar}{i}\nabla_y\right|X\right\rangle, \tag{3.154}$$

which is antisymmetric in x or y, and becomes zero. Thus,

$$D_{XX} = \left(\frac{\hbar^2}{2m_0} + \frac{\hbar^2}{m_0^2}\sum_i \frac{p_{Xi}^{X}p_{iX}^{X}}{\varepsilon_0 - \varepsilon_i}\right)k_x k_x + \left(\frac{\hbar^2}{2m_0} + \frac{\hbar^2}{m_0^2}\sum_i \frac{p_{Xi}^{Y}p_{iX}^{Y}}{\varepsilon_0 - \varepsilon_i}\right)k_y k_y$$
$$+ \left(\frac{\hbar^2}{2m_0} + \frac{\hbar^2}{m_0^2}\sum_i \frac{p_{Xi}^{Z}p_{iX}^{Z}}{\varepsilon_0 - \varepsilon_i}\right)k_z k_z. \tag{3.155}$$

Using $p_{Xi}^{y}p_{iX}^{y} = p_{Xi}^{z}p_{iX}^{z}$,

$$\begin{aligned} D_{XX} &= \left(\frac{\hbar^2}{2m_0} + \frac{\hbar^2}{m_0^2}\sum_i \frac{p_{Xi}^{X}p_{iX}^{X}}{\varepsilon_0 - \varepsilon_i}\right)k_x^2 + \left(\frac{\hbar^2}{2m_0} + \frac{\hbar^2}{m_0^2}\sum_i \frac{p_{Xi}^{Y}p_{iX}^{Y}}{\varepsilon_0 - \varepsilon_i}\right)(k_y^2 + k_z^2) \\ &\equiv Ak_x^2 + B(k_y^2 + k_z^2) \end{aligned} \tag{3.156}$$

For D_{XY}, only terms with $\alpha = X, \beta = Y$ or $\alpha = Z, \beta = X$ are nonzero. Similarly, we can obtain the following relations for nonzero terms:

$$\begin{aligned} &p_{Xi}^{\alpha}p_{iX}^{\alpha}, p_{Yi}^{\alpha}p_{iY}^{\alpha}, p_{Zi}^{\alpha}p_{iZ}^{\alpha} \quad \mathit{for}\ \alpha = X, Y, Z \\ &p_{Xi}^{Y}p_{iY}^{X}, p_{Yi}^{X}p_{iX}^{Y}, p_{Xi}^{Z}p_{iZ}^{X}, p_{Zi}^{X}p_{iX}^{Z}, p_{Yi}^{Z}p_{iZ}^{Y}, p_{Zi}^{Y}p_{iY}^{Z}, \end{aligned} \tag{3.157}$$

where

$$\begin{aligned} &p_{Xi}^{X}p_{iX}^{X} = p_{Yi}^{Y}p_{iY}^{Y} = p_{Zi}^{Z}p_{iZ}^{Z} \\ &p_{Xi}^{Y}p_{iX}^{Y} = p_{Yi}^{Z}p_{iY}^{Z} = p_{Xi}^{Z}p_{iX}^{Z} = \ldots \text{etc.} \end{aligned} \tag{3.158}$$

3.2.4 Wave Functions Modified by Spin–Orbit Coupling

The introduction of spin–orbit coupling splits some of the degeneracy of the band at $\mathbf{k} = 0$. Treating H_{so} as a perturbation, we can obtain the correct wave functions. The wave function $\Psi(\mathbf{r},t)$ for electrons is actually not a scalar but is a two-component spinor

$$\begin{pmatrix} \alpha(\mathbf{r},t) \\ \beta(\mathbf{r},t) \end{pmatrix}. \tag{3.159}$$

Thus, the Schrödinger equation is not a single differential equation, but two coupled differential equations as follows.

$$i\hbar\frac{\partial}{\partial t}\alpha = H_{11}\alpha + H_{12}\beta \tag{3.160}$$

$$i\hbar\frac{\partial}{\partial t}\beta = H_{21}\alpha + H_{22}\beta. \tag{3.161}$$

Or, in matrix form,

$$i\hbar\frac{\partial}{\partial t}\begin{pmatrix} \alpha \\ \beta \end{pmatrix} = \begin{bmatrix} H_{11} & H_{12} \\ H_{21} & H_{22} \end{bmatrix}\begin{pmatrix} \alpha \\ \beta \end{pmatrix}. \tag{3.162}$$

In general, we assume that

$$H_{12} = H_{21} = 0 \tag{3.163}$$

$$H_{11} = H_{12} = -\frac{\hbar^2}{2m_o}\nabla^2 + U_L \equiv H_o. \tag{3.164}$$

Equation (3.162) then decouples into two identical Schrödinger equations—one for α and one for β. Consequently, we can just solve one Schrödinger equation for a scalar and multiply the results at the end by 2 to account for the other spin component. However, this simplicity is lost if H_{12} and H_{21} are nonzero.

Let us now consider how the wave functions and energy levels in a semiconductor are modified by the spin–orbit interaction. The Schrödinger equation reads

$$i\hbar\frac{\partial\Psi}{\partial t} = (H + H_{so})\Psi, \tag{3.165}$$

where H is the 2×2 matrix operator defined in Equations (3.163) and (3.164), Ψ is a 2×1 spinor, and H_{so} is the 2×2 spin–orbit operator defined below. The spin–orbit coupling in Equation (3.143) can be written as

$$H_{so} = -i\frac{\hbar^2}{4m_0^2c^2}\sigma\cdot[\nabla U \times \nabla]. \tag{3.166}$$

H_{so} is a 2×2 matrix operator whose individual elements can be written out with differential operators using the definition of the Pauli matrices $\sigma_{x,y,z}$ from Equation (3.144).

$$(H_{so})_{11} = -iC\left(\frac{\partial U}{\partial x}\frac{\partial}{\partial y} - \frac{\partial U}{\partial y}\frac{\partial}{\partial x}\right) \tag{3.167}$$

$$(H_{so})_{12} = -iC\left\{\left(\frac{\partial U}{\partial y}\frac{\partial}{\partial z} - \frac{\partial U}{\partial z}\frac{\partial}{\partial y}\right) - i\left(\frac{\partial U}{\partial z}\frac{\partial}{\partial x} - \frac{\partial U}{\partial x}\frac{\partial}{\partial z}\right)\right\} \tag{3.168}$$

$$(H_{so})_{21} = -iC\left\{\left(\frac{\partial U}{\partial y}\frac{\partial}{\partial z} - \frac{\partial U}{\partial z}\frac{\partial}{\partial y}\right) + i\left(\frac{\partial U}{\partial z}\frac{\partial}{\partial x} - \frac{\partial U}{\partial x}\frac{\partial}{\partial z}\right)\right\} \tag{3.169}$$

$$(H_{so})_{22} = iC\left(\frac{\partial U}{\partial x}\frac{\partial}{\partial y} - \frac{\partial U}{\partial y}\frac{\partial}{\partial x}\right), \tag{3.170}$$

where $C = \dfrac{\hbar^2}{4m_0^2c^2}$.

We are interested in the energy eigenfunctions and eigenvalues at $k = 0$. Writing $\Psi(\mathbf{r},t) = \xi(\mathbf{r})\exp(-iEt/\hbar)$ [since $k = 0$, $\exp(i\mathbf{k}\cdot\mathbf{r}) = 1$], we have the eigenvalue equation.

$$E\xi = (H + H_{so})\xi \tag{3.171}$$

This is actually two coupled differential equations that can be written out using Equations (3.163) and (3.164) as follows, assuming $\xi(\mathbf{r}) = \begin{pmatrix}\alpha(\mathbf{r})\\ \beta(\mathbf{r})\end{pmatrix}$.

$$E\alpha = [H_0 + (H_{so})_{11}]\alpha + (H_{so})_{12}\beta \tag{3.172}$$

$$E\beta = [(H_{so})_{21}]\alpha + (H_0 + H_{so})_{22}\beta. \tag{3.173}$$

The functions $|S\rangle$, $|X\rangle$, $|Y\rangle$, and $|Z\rangle$ are eigenfunctions of H_0. That is,

$$H_0|S\rangle = E_{c0}|S\rangle, H_0|X\rangle = E'_{v0}|X\rangle, H_0|Y\rangle = E'_{v0}|Y\rangle, H_0|Z\rangle = E'_{v0}|Z\rangle. \tag{3.174}$$

If the spin–orbit interaction H_{so} were absent, the energy levels and the wave functions at the zone center ($k=0$) will be written as shown below.

$$\begin{aligned} v1 &: \alpha = |X\rangle, \beta = 0 \\ v2 &: \alpha = |Y\rangle, \beta = 0 \\ v3 &: \alpha = |Z\rangle, \beta = 0 \\ v4 &: \alpha = |0\rangle, \beta = |X\rangle \\ v5 &: \alpha = |0\rangle, \beta = |Y\rangle \\ v6 &: \alpha = |0\rangle, \beta = |Z\rangle. \end{aligned} \tag{3.175}$$

Now, let us use these functions to expand $\alpha(\mathbf{r})$ and $\beta(\mathbf{r})$.

$$\alpha(\mathbf{r}) = \alpha_s|S\rangle + \alpha_x|X\rangle + \alpha_y|Y\rangle + \alpha_z|Z\rangle \equiv \sum_{i=S,X,Y,Z} \alpha_i|i\rangle \tag{3.176}$$

$$\beta(\mathbf{r}) = \sum_{i=S,X,Y,Z} \beta_i|i\rangle. \tag{3.177}$$

We can write the differential equations in Equations (3.172) and (3.173) as a matrix equation.

$$E\alpha_i = \sum_j \langle i|H_o + (H_{so})_{11}|j\rangle\alpha_j + \sum_j \langle i|(H_{so})_{12}|j\rangle\beta_j. \tag{3.178}$$

$$E\beta_i = \sum_j \langle i|(H_{so})_{21}|j\rangle\alpha_j + \sum_j \langle i|H_o + (H_{so})_{22}|j\rangle\beta_j. \tag{3.179}$$

Since the functions $|i\rangle$, $i = S, X, Y, Z$ are eigenfunctions of H_o, it is easy to write down the matrix elements of H_o.

$$\langle i|H_0|j\rangle = \begin{cases} E_{c0}\delta_{ij}, & \text{if} \quad i = S \\ E'_{v0}\delta_{ij}, & \text{if} \quad i = X, Y, Z \end{cases}. \tag{3.180}$$

Using the fact that the atomic potential U has cubic symmetry, we can show that $\langle i|(H_{SO})_{11}|j\rangle$ or $\langle i|(H_{SO})_{22}|j\rangle$ is nonzero only if $i = X$, $j = Y$ or $i = Y$, $j = X$. Then,

$$\langle X|(H_{SO})_{11}|Y\rangle = -\langle X|(H_{SO})_{22}|Y\rangle \equiv -i\Delta/3 \tag{3.181}$$

where

$$\Delta = 3C\langle X|\left(\frac{\partial U}{\partial y}\frac{\partial}{\partial x} - \frac{\partial U}{\partial x}\frac{\partial}{\partial y}\right)|Y\rangle. \tag{3.182}$$

Similarly, it can be shown that

$$\begin{aligned}\langle Y|(H_{SO})_{12}|Z\rangle &= \langle Y|(H_{SO})_{21}|Z\rangle = -i\Delta/3 \\ \langle X|(H_{SO})_{12}|Z\rangle &= -\langle X|(H_{SO})_{21}|Z\rangle = \Delta/3.\end{aligned} \tag{3.183}$$

The other matrix elements are all zero. Also, in general,

$$\langle i|(H_{SO})_{nm}|j\rangle = -\langle j|(H_{SO})_{mn}|i\rangle. \tag{3.184}$$

Using the matrix elements in Equations (3.181), (3.183), and (3.184), we can write out Equations (3.178) and (3.179) as follows:

$$E\begin{pmatrix}\alpha_s\\ \alpha_x\\ \alpha_y\\ \alpha_z\\ \beta_s\\ \beta_x\\ \beta_y\\ \beta_z\end{pmatrix} = \begin{pmatrix} E_{c0} & 0 & 0 & 0 & 0 & 0 & 0 & 0\\ 0 & E'_{v0} & -i\Delta/3 & 0 & 0 & 0 & 0 & \Delta/3\\ 0 & i\Delta/3 & E'_{v0} & 0 & 0 & 0 & 0 & -i\Delta/3\\ 0 & 0 & 0 & E'_{v0} & 0 & -\Delta/3 & i\Delta/3 & 0\\ 0 & 0 & 0 & 0 & E_{c0} & 0 & 0 & 0\\ 0 & 0 & 0 & -\Delta/3 & 0 & E'_{v0} & i\Delta/3 & 0\\ 0 & 0 & 0 & -i\Delta/3 & 0 & -i\Delta/3 & E'_{v0} & 0\\ 0 & \Delta/3 & i\Delta/3 & 0 & 0 & 0 & 0 & E'_{v0}\end{pmatrix}\begin{pmatrix}\alpha_s\\ \alpha_x\\ \alpha_y\\ \alpha_z\\ \beta_s\\ \beta_x\\ \beta_y\\ \beta_z\end{pmatrix}. \tag{3.185}$$

Here, we neglected the coupling between the conduction and the valence bands, and will consider it in next section. By rearranging the rows and columns in Equation (3.185), we can get the matrix in block-diagonalized form.

$$E\begin{pmatrix}\alpha_s\\ \beta_s\\ \alpha_x\\ \alpha_y\\ \beta_z\\ \beta_x\\ \beta_y\\ \alpha_z\end{pmatrix} = \begin{pmatrix} E_{c0} & 0 & 0 & 0 & 0 & 0 & 0 & 0\\ 0 & E_{c0} & 0 & 0 & 0 & 0 & 0 & 0\\ 0 & 0 & E'_{v0} & -i\Delta/3 & \Delta/3 & 0 & 0 & 0\\ 0 & 0 & i\Delta/3 & E'_{v0} & -i\Delta/3 & 0 & 0 & 0\\ 0 & 0 & \Delta/3 & i\Delta/3 & E'_{v0} & 0 & 0 & 0\\ 0 & 0 & 0 & 0 & 0 & E'_{v0} & i\Delta/3 & -\Delta/3\\ 0 & 0 & 0 & 0 & 0 & -i\Delta/3 & E'_{v0} & -i\Delta/3\\ 0 & 0 & 0 & 0 & 0 & -\Delta/3 & i\Delta/3 & E'_{v0}\end{pmatrix}\begin{pmatrix}\alpha_s\\ \beta_s\\ \alpha_x\\ \alpha_y\\ \beta_z\\ \beta_x\\ \beta_y\\ \alpha_z\end{pmatrix}. \tag{3.186}$$

The eigenvalues and eigenfunctions for each block can be calculated individually. Note that the conduction band levels are unaffected by the spin–orbit interaction. The eigenvalues and eigenvetors of the upper 3×3 block are given by

$$\begin{aligned} E_h &= E'_{v0} + \frac{\Delta}{3} = E_{v0} \quad \text{(heavy hole}, h1) \\ \alpha_x &= -1/\sqrt{2}, \alpha_y = -i/\sqrt{2}, \beta_z = 0 \end{aligned} \tag{3.187}$$

$$\begin{aligned} E_l &= E'_{v0} + \frac{\Delta}{3} = E_{v0} \quad \text{(light hole}, l2) \\ \alpha_x &= 1/\sqrt{6}, \alpha_y = -i/\sqrt{6}, \beta_z = \sqrt{2/3} \end{aligned} \tag{3.188}$$

$$\begin{aligned} E_s &= E'_{v0} - \frac{2\Delta}{3} = E_{v0} - \Delta \quad \text{(split-off}, s2) \\ \alpha_x &= 1/\sqrt{3}, \alpha_y = -i/\sqrt{3}, \beta_z = -1/\sqrt{3}. \end{aligned} \tag{3.189}$$

Note that, at $k = 0$, the first two eigenvalues are equal, so that any linear combination of the heavy- and light-hole eigenvectors is also an eigenvector. In fact, there are numerous other combinations possible for Equations (3.187) and (3.188). The reason for using this choice is that if we consider electrons in motion along the z-direction ($k_x = k_y = 0, k_z \neq 0$), the heavy- and light-hole eigenvectors are decoupled. The lower 3×3 block in Equation (3.186) gives the same eigenvalues, but slightly different eigenvectors.

$$\begin{aligned} E_h &= E'_{v0} + \frac{\Delta}{3} = E_{v0} \quad \text{(heavy hole}, h2) \\ \beta_x &= 1/\sqrt{2}, \beta_y = -i/\sqrt{2}, \alpha_z = 0 \end{aligned} \tag{3.190}$$

$$\begin{aligned} E_l &= E'_{v0} + \frac{\Delta}{3} = E_{v0} \quad \text{(light hole}, l1) \\ \beta_x &= -1/\sqrt{6}, \beta_y = -i/\sqrt{6}, \alpha_z = \sqrt{2/3} \end{aligned} \tag{3.191}$$

$$\begin{aligned} E_s &= E'_{v0} - \frac{2\Delta}{3} = E_{v0} - \Delta \quad \text{(split-off}, s1) \\ \beta_x &= 1/\sqrt{3}, \beta_y = i/\sqrt{3}, \alpha_z = 1/\sqrt{3}. \end{aligned} \tag{3.192}$$

The introduction of spin–orbit coupling splits some of the degeneracy of the band at $k = 0$. The wave functions fall into a group of four and a group of two:

$$
\begin{aligned}
h1:&\left|\frac{3}{2},\frac{3}{2}\right\rangle = -\frac{1}{\sqrt{2}}[|X\rangle + i|Y\rangle]\uparrow \\
l1:&\left|\frac{3}{2},\frac{1}{2}\right\rangle = -\frac{1}{\sqrt{6}}[(|X\rangle + i|Y\rangle)\downarrow -2|Z\rangle\uparrow] \\
l2:&\left|\frac{3}{2},-\frac{1}{2}\right\rangle = \frac{1}{\sqrt{6}}[(|X\rangle - i|Y\rangle)\uparrow +2|Z\rangle\downarrow] \\
h2:&\left|\frac{3}{2},-\frac{3}{2}\right\rangle = \frac{1}{\sqrt{2}}[|X\rangle - i|Y\rangle]\downarrow \\
s1:&\left|\frac{1}{2},\frac{1}{2}\right\rangle = \frac{1}{\sqrt{3}}[(|X\rangle + i|Y\rangle)\downarrow +|Z\rangle\uparrow] \\
s2:&\left|\frac{1}{2},-\frac{1}{2}\right\rangle = \frac{1}{\sqrt{3}}[(|X\rangle - i|Y\rangle)\uparrow -|Z\rangle\downarrow].
\end{aligned}
\tag{3.193}
$$

where $\uparrow$ and $\downarrow$ are the spin functions corresponding to spin "up" and "down," respectively.

3.3 THE ZINC BLENDE HAMILTONIAN

3.3.1 Matrix Elements for the 6×6 Hamiltonian

We shall consider the situation in which the spin–orbit coupling is sufficiently small so that the zeroth-order functions of Equation (3.193) may be used. Considering both the $J = 3/2$ and 1/2 bands together, and using Equation (3.137) having p instead π, and using Equation (3.141), we find

$$
H = -\begin{pmatrix}
h1:\left|\frac{3}{2},\frac{3}{2}\right\rangle \\
l1:\left|\frac{3}{2},\frac{1}{2}\right\rangle \\
l2:\left|\frac{3}{2},-\frac{1}{2}\right\rangle \\
h2:\left|\frac{3}{2},-\frac{3}{2}\right\rangle \\
s1:\left|\frac{1}{2},\frac{1}{2}\right\rangle \\
s2:\left|\frac{1}{2},-\frac{1}{2}\right\rangle
\end{pmatrix}
\begin{pmatrix}
P+Q & -S & R & 0 & -S/\sqrt{2} & \sqrt{2}R \\
-S^{\dagger} & P-Q & 0 & R & -\sqrt{2}Q & \sqrt{3/2}S \\
R^{\dagger} & 0 & P-Q & S & \sqrt{3/2}S^{\dagger} & \sqrt{2}Q \\
0 & R^{\dagger} & S^{\dagger} & P+Q & -\sqrt{2}R^{\dagger} & -S^{\dagger}/\sqrt{2} \\
-S^{\dagger}/\sqrt{2} & -\sqrt{2}Q^{\dagger} & \sqrt{3/2}S & -\sqrt{2}R & P+\Delta & 0 \\
\sqrt{2}R^{\dagger} & \sqrt{3/2}S^{\dagger} & \sqrt{2}Q^{\dagger} & -S/\sqrt{2} & 0 & P+\Delta
\end{pmatrix},
\tag{3.194}
$$

where

$$\begin{aligned}
P &= \frac{\hbar^2}{2m_0}\gamma_1(k_x^2+k_y^2+k_z^2),\\
Q &= \frac{\hbar^2}{2m_0}\gamma_1(k_x^2+k_y^2-2k_z^2),\\
R &= \frac{\hbar^2}{2m_0}[-\sqrt{3}\gamma_2(k_x^2-k_y^2)+i2\sqrt{3}\gamma_3 k_x k_y],\\
S &= \frac{\hbar^2}{2m_0}2\sqrt{3}\gamma_3(k_x-ik_y)k_z,\\
\Delta &= \text{spin-orbit splitting at } \mathbf{k}=0,
\end{aligned} \tag{3.195}$$

where we put $E_{v0}=0$ as follows:

$$E_j(\mathbf{k}=0)=\begin{cases} E'_{v0}+\dfrac{\Delta}{3}=E_{v0}=0 & \text{for } l=h1,l1,h2,l2 \\ E'_{v0}-\dfrac{2\Delta}{3}=E_{v0}-\Delta=-\Delta & \text{for } l=s1,s2 \end{cases}. \tag{3.196}$$

As an example of the calculation, we consider $H_{h1,h1}$.

$$H_{h1,h1}=\sum_{\alpha\beta}\left(\frac{\hbar^2}{2m_0}\delta_{h1h1}\delta_{\alpha\beta}+\frac{\hbar}{m_0^2}\sum_i\frac{p_{h1i}^{\alpha}p_{ih1}^{\beta}}{\varepsilon_0-\varepsilon_i}\right)k_\alpha k_\beta, \tag{3.197}$$

where

$$\begin{aligned}
p_{h1i}^{\alpha}p_{ih1}^{\beta} &= \left\langle h1\left|\frac{\hbar}{i}\nabla_\alpha\right|i\right\rangle\left\langle i\left|\frac{\hbar}{i}\nabla_\beta\right|h1\right\rangle\\
&= \frac{1}{2}\left\langle X-iY\left|\frac{\hbar}{i}\nabla_\alpha\right|i\right\rangle\left\langle i\left|\frac{\hbar}{i}\nabla_\beta\right|X+iY\right\rangle\\
&= \frac{1}{2}(p_{Xi}^{\alpha}p_{iX}^{\beta}-ip_{Yi}^{\alpha}p_{iX}^{\beta}+p_{Xi}^{\alpha}p_{iY}^{\beta}+p_{Yi}^{\alpha}p_{iY}^{\beta})\\
&= \frac{1}{2}(p_{Xi}^{\alpha}p_{iX}^{\beta}+p_{Yi}^{\alpha}p_{iY}^{\beta}) \text{ from Equations (3.157) and (3.158).}
\end{aligned} \tag{3.198}$$

This becomes zero except for $\alpha=\beta$.

$$
\begin{aligned}
H_{h1,h1} &= k_x^2\left(\frac{\hbar^2}{2m_0}+\frac{\hbar}{2m_0^2}\sum_i\frac{p_{Xi}^X p_{iX1}^X + p_{Yi}^X p_{iY}^X}{\varepsilon_0-\varepsilon_i}\right)\\
&+k_y^2\left(\frac{\hbar^2}{2m_0}+\frac{\hbar}{2m_0^2}\sum_i\frac{p_{Xi}^Y p_{iX}^Y + p_{Yi}^Y p_{iY}^Y}{\varepsilon_0-\varepsilon_i}\right)\\
&+k_z^2\left(\frac{\hbar^2}{2m_0}+\frac{\hbar}{2m_0^2}\sum_i\frac{p_{Xi}^Z p_{iX}^Z + p_{Yi}^Z p_{iY}^Z}{\varepsilon_0-\varepsilon_i}\right)\\
&=\frac{1}{2}k_x^2(A+B)+\frac{1}{2}k_y^2(A+B)+k_z^2 B \text{ from Equation (3.151).}
\end{aligned}
\tag{3.199}
$$

Now, it is convenient to introduce the dimensionless Luttinger effective mass parameters γ_1, γ_2, and γ_3 by

$$
\begin{aligned}
A &= \frac{\hbar^2}{2m_0}(\gamma_1+4\gamma_2)\\
B &= \frac{\hbar^2}{2m_0}(\gamma_1-2\gamma_2)\\
C &= -\frac{\hbar^2}{2m_0}(6\gamma_3).
\end{aligned}
\tag{3.200}
$$

Then, Equation (3.199) becomes

$$
\begin{aligned}
H_{h1,h1} &= \frac{\hbar^2}{2m_0}(\gamma_1+\gamma_2)k_x^2+\frac{\hbar^2}{2m_0}(\gamma_1+\gamma_2)k_y^2+\frac{\hbar^2}{2m_0}(\gamma_1-2\gamma_2)k_z^2\\
&=\frac{\hbar^2}{2m_0}\gamma_1(k_x^2+k_y^2+k_z^2)+\frac{\hbar^2}{2m_0}\gamma_2(k_x^2+k_y^2-2k_z^2)\equiv P+Q
\end{aligned}
\tag{3.201}
$$

We always have to ensure that whatever $H(\mathbf{k})$ we use, its eigenvalues should give us the correct dispersion laws for the bands. Thus, note that the Luttinger parameters γ_1, γ_2, and γ_3 will be slightly different depending on whether we use the 4×4 matrix, the 6×6 matrix, or the 8×8 matrix, including the spin–orbit band.

3.3.2 Matrix Elements of the Hamiltonian between Conduction and Valence Bands

Matrix elements between conduction and valence bands can be derived by Equation (3.50) when we consider the coupling between conduction and valence bands. That is, we first calculate momentum matrix elements $p_{nn'}^{\alpha}$ between conduction and valence bands, then substitute them into Equation (3.50). Since the functions $u_{n0}(\mathbf{r})$ are spinors of the form

$$\begin{pmatrix} \alpha_{n,0} \\ \beta_{n,0} \end{pmatrix}, \tag{3.202}$$

Equation (3.50) becomes

$$\mathbf{p}_{nn'}^{\alpha} = -i\hbar\langle\alpha_{n,0}|\nabla\alpha_{n',0}\rangle - i\hbar\langle\beta_{n,0}|\nabla\beta_{n',0}\rangle. \tag{3.203}$$

In many common semiconductors like GaAs, due to crystal symmetry, the functions $\bar{u}_{n,0}(\mathbf{r})$ for the conduction and valence bands can be expressed in terms of these functions $|S\rangle$, $|X\rangle$, $|Y\rangle$, and $|Z\rangle$. Electrons in the conduction band have completely symmetric wave functions $|S\rangle$. The up-spin and down-spin wave functions corresponding to $c1$ and $c2$ are written as

$$u_{c1,0}(\mathbf{r}) = \begin{pmatrix} |S\rangle \\ 0 \end{pmatrix} \tag{3.204}$$

$$u_{c2,0}(\mathbf{r}) = \begin{pmatrix} 0 \\ |S\rangle \end{pmatrix}. \tag{3.205}$$

To calculate momentum matrix elements $p_{nn'}^{\alpha}$ between conduction and valence bands, we substitute the functions α and β from Equations (3.193) and (3.205). Due to the symmetry of the functions $|S\rangle, |X\rangle, |Y\rangle$, and $|Z\rangle$, only the following integrals are nonzero.

$$\left\langle S\left|-i\hbar\frac{\partial}{\partial x}\right|X\right\rangle = \left\langle S\left|-i\hbar\frac{\partial}{\partial y}\right|Y\right\rangle = \left\langle S\left|-i\hbar\frac{\partial}{\partial z}\right|Z\right\rangle \equiv P_0. \tag{3.206}$$

All other integrals, such as

$$\left\langle S\left|-i\hbar\frac{\partial}{\partial x}\right|Y\right\rangle, \tag{3.207}$$

are zero. Using Equation (3.206), it is straightforward to evaluate $p_{nn'}$. For example, suppose $n = c1$ and $n' = h1$. Using Equation (3.193), we have from Equation (3.203),

$$\mathbf{p}_{c1h1} = -\frac{i\hbar}{\sqrt{2}}(\langle S|\nabla|X\rangle + i\langle S|\nabla|Y\rangle). \tag{3.208}$$

Using Equation (3.206),

$$\mathbf{p}_{c1h1} = \frac{P_0}{\sqrt{2}}(\hat{e}_x + i\hat{e}_y), \tag{3.209}$$

where $\hat{e}_x$ and $\hat{e}_y$ are unit vectors along the x, y, and z directions. Then, using Equation (3.50), matrix elements of Hamiltonian between conduction and valence bands are given by

$$\begin{pmatrix} & h1 & h2 & l1 & l2 & s1 & s2 \\ c1 & -\frac{\pi}{\sqrt{2}}(k_x + ik_y) & 0 & \sqrt{\frac{2}{3}}\pi k_z & \frac{\pi}{\sqrt{6}}(k_x - ik_y) & \frac{\pi}{\sqrt{3}}k_z & \frac{\pi}{\sqrt{3}}(k_x - ik_y) \\ c2 & 0 & \frac{\pi}{\sqrt{2}}(k_x - ik_y) & \frac{\pi}{\sqrt{6}}(k_x + ik_y) & \sqrt{\frac{2}{3}}\pi k_z & \frac{\pi}{\sqrt{3}}(k_x + ik_y) & -\frac{\pi}{\sqrt{3}}k_z \end{pmatrix}, \tag{3.210}$$

where $\pi = \hbar P_0/m_0 = \sqrt{(\hbar^2/2m_0)E_p}$, and E_p is a parameter that controls mixing of the valence and conduction bands.

3.3.3 The 8 × 8 Hamiltonian

Combining Equations (3.194) and (3.210), we can obtain the 8×8 Hamiltonian given by

$$H=\left(\begin{matrix}
A_c & 0 & -\sqrt{3}V^\dagger & \sqrt{2}U & V & 0 & U & \sqrt{2}V \\
0 & A_c & 0 & -V^\dagger & \sqrt{2}U & \sqrt{3}V & \sqrt{2}V^\dagger & -U \\
-\sqrt{3}V & 0 & -(P+Q) & S & -R & 0 & \frac{1}{\sqrt{2}}S & -\sqrt{2}R \\
\sqrt{2}U^\dagger & -V & S^\dagger & -(P-Q) & 0 & -R & \sqrt{2}Q & -\sqrt{3/2}S \\
V^\dagger & \sqrt{2}U^\dagger & -R^\dagger & 0 & -(P-Q) & -S & -\sqrt{3/2}S^\dagger & -\sqrt{2}Q \\
0 & \sqrt{3}V^\dagger & 0 & -R^\dagger & -S^\dagger & -(P+Q) & \sqrt{2}R^\dagger & \frac{1}{\sqrt{2}}S^\dagger \\
U^\dagger & \sqrt{2}V & \frac{1}{\sqrt{2}}S^\dagger & \sqrt{2}Q^\dagger & -\sqrt{3/2}S & \sqrt{2}R & -(P+\Delta) & 0 \\
\sqrt{2}V^\dagger & -U^\dagger & -\sqrt{2}R^\dagger & -\sqrt{3/2}S^\dagger & -\sqrt{2}Q^\dagger & \frac{1}{\sqrt{2}}S & 0 & -(P+\Delta)
\end{matrix}\right)\begin{matrix} c1 \\ c2 \\ h1 \\ l1 \\ l2, \\ h2 \\ s1 \\ s2 \end{matrix} \tag{3.211}$$

with

$$\begin{aligned}
A_c &= E_g + \frac{\hbar^2}{2m_0}(k_x^2+k_y^2+k_z^2), \\
V &= \frac{1}{\sqrt{6}}\pi(k_x - ik_y), \\
U &= \frac{1}{\sqrt{3}}\pi k_z.
\end{aligned} \tag{3.212}$$

Here, the basis set is summarized as follows:

$$\begin{aligned}
c1 &: |S\rangle\uparrow \\
c2 &: |S\rangle\downarrow \\
h1 &: \left|\frac{3}{2},\frac{3}{2}\right\rangle = -\frac{1}{\sqrt{2}}[|X\rangle + i|Y\rangle]\uparrow \\
l1 &: \left|\frac{3}{2},\frac{1}{2}\right\rangle = -\frac{1}{\sqrt{6}}[(|X\rangle + i|Y\rangle)\downarrow - 2|Z\rangle\uparrow] \\
l2 &: \left|\frac{3}{2},-\frac{1}{2}\right\rangle = \frac{1}{\sqrt{6}}[(|X\rangle - i|Y\rangle)\uparrow + 2|Z\rangle\downarrow] \\
h2 &: \left|\frac{3}{2},-\frac{3}{2}\right\rangle = \frac{1}{\sqrt{2}}[|X\rangle - i|Y\rangle]\downarrow \\
s1 &: \left|\frac{1}{2},\frac{1}{2}\right\rangle = \frac{1}{\sqrt{3}}[(|X\rangle + i|Y\rangle)\downarrow + |Z\rangle\uparrow] \\
s2 &: \left|\frac{1}{2},-\frac{1}{2}\right\rangle = \frac{1}{\sqrt{3}}[(|X\rangle - i|Y\rangle)\uparrow - |Z\rangle\downarrow].
\end{aligned} \tag{3.213}$$

3.3.4 The Strained Hamiltonian

In this section, we will introduce the strained 8×8 Hamiltonian. We defined the unit vectors $\hat{x}$, $\hat{y}$, and $\hat{z}$ in the unstrained system. Under the influence of a uniform deformation, these axes are distorted to $\mathbf{x}$, $\mathbf{y}$, and $\mathbf{z}$. The new axes are related to the old one by

$$\begin{aligned}\mathbf{x} &= (1+\varepsilon_{xx})\hat{x} + \varepsilon_{xy}\hat{y} + \varepsilon_{xz}\hat{z},\\ \mathbf{y} &= \varepsilon_{yx}\hat{x} + (1+\varepsilon_{yy})\hat{y} + \varepsilon_{yz}\hat{z},\\ \mathbf{z} &= \varepsilon_{zx}\hat{x} + \varepsilon_{zy}\hat{y} + (1+\varepsilon_{zz})\hat{z}.\end{aligned} \tag{3.214}$$

The coefficients are strain tensors and define the deformation in the system. The new axes are not orthogonal in general. To label a position A (or atom A) in the unstrained system, we have

$$\mathbf{r} = x\hat{x} + y\hat{y} + z\hat{z}. \tag{3.215}$$

After the distortion, the same atom can be labeled as

$$\mathbf{r}' = x\mathbf{x}' + y\mathbf{y}' + z\mathbf{z}'. \tag{3.216}$$

Note that the coefficients x, y, z of the vectors are unchanged. The displacement of the deformation is given by

$$\begin{aligned}\mathbf{R} &= \mathbf{r}' - \mathbf{r}'\\ &= x(\mathbf{x}' - \hat{x}) + y(\mathbf{y}' - \hat{y}) + z(\mathbf{z}' - \hat{z})\\ &= (x\varepsilon_{xx} + y\varepsilon_{yx} + z\varepsilon_{zx})\hat{x} + (x\varepsilon_{xy} + y\varepsilon_{yy} + z\varepsilon_{zy})\hat{y} + (x\varepsilon_{xz} + y\varepsilon_{yz} + z\varepsilon_{zz})\hat{z}\\ &= u(\mathbf{r})\hat{x} + v(\mathbf{r})\hat{y} + w(\mathbf{r})\hat{z}.\end{aligned} \tag{3.217}$$

For a nonuniform distortion, one must define a position-dependent strain. Taking the origin of $\mathbf{r}$ close to the point and assuming a small strain, we have

$$\varepsilon_{xx} = \frac{\partial u}{\partial x}; \varepsilon_{yx} = \frac{\partial u}{\partial y}; \varepsilon_{zx} = \frac{\partial u}{\partial z}, \text{etc.} \tag{3.218}$$

We assume a homogeneous strain and $\varepsilon_{ij} = \varepsilon_{ji}$. We define six strain components as

$$
\begin{aligned}
e_1 &= \varepsilon_{xx}, e_2 = \varepsilon_{yy}, e_3 = \varepsilon_{zz} \\
e_4 &= \varepsilon_{xy} + \varepsilon_{yx} \\
e_5 &= \varepsilon_{yz} + \varepsilon_{zy} \\
e_6 &= \varepsilon_{zx} + \varepsilon_{xz},
\end{aligned} \tag{3.219}
$$

keeping only the linear terms in strain. It is useful to define the net fractional change in the volume produced by the distortion. This quantity is called dilation. The initially cubic volume of unity after distortion has a volume

$$
V' = \mathbf{x}' \cdot (\mathbf{y}' \times \mathbf{z}') = \begin{vmatrix} 1+\varepsilon_{xx} & \varepsilon_{xy} & \varepsilon_{xz} \\ \varepsilon_{yx} & 1+\varepsilon_{yy} & \varepsilon_{yz} \\ \varepsilon_{zx} & \varepsilon_{zy} & 1+\varepsilon_{xz} \end{vmatrix} \simeq 1 + \varepsilon_{xx} + \varepsilon_{yy} + \varepsilon_{zz}. \tag{3.220}
$$

The dilation is then

$$
\delta = \frac{V' - V}{V} \simeq \varepsilon_{xx} + \varepsilon_{yy} + \varepsilon_{zz}. \tag{3.221}
$$

The force acting on a unit area in the solid is called stress, and Hooke's law states that the strain is linearly proportional to the stress, or, conversely, that the stress is linearly proportional to the strain. In general,

$$
T_{ij} = c_{ijkl} S_{kl}, i, j, k, l = x, y, z, \tag{3.222}
$$

with summation over the repeated subscript k and l. The first subscript letter indicates the direction of the force, and the second subscript letter is the direction of the normal to the plane on which the stress is acting. The number of stress components decreases to six if we impose the condition on a cubic system, since there is no torque in the system. Then, $T_{xy} = T_{yx}$; $T_{yz} = T_{zx}; T_{xz} = T_{zx}$. We then have the six independent components: $T_{xx}, T_{yy}, T_{zz}, T_{xy}, T_{yz}, T_{zx}$. The c_{ijkl} are called elastic stiffness constants. If the coordinate axes x, y, z are chosen to coincide with the cubic crystal axes, the nonzero elastic stiffness constants are

$$
\begin{aligned}
& c_{xxxx} = c_{yyyy} = c_{zzzz} \\
& c_{xxyy} = c_{yyxx} = c_{xxzz} = c_{yyzz} = c_{zzyy} \\
& c_{yxyx} = c_{yxxy} = c_{zyyz} = c_{zyzy} \\
& \quad = c_{xzxz} = c_{xzzx} = c_{zxxz} = c_{zxzx} \\
& \quad = c_{xyxy} = c_{xyyx} = c_{yxxy} = c_{yxyx}.
\end{aligned} \tag{3.223}
$$

We use abbreviated subscript notation: $1 = xx$; $2 = yy$; $3 = zz$; $4 = yz$; $5 = zx$; $6 = xy$. Then, the stiffness Equation (3.222) reduce to

$$T_I = c_{IJ} S_J, \tag{3.224}$$

where $I, J = 1,2,3,4,5,6$. In abbreviated subscript notation, the nonzero elastic stiffness constants are

$$\begin{aligned} c_{11} &= c_{22} = c_{33} \\ c_{12} &= c_{21} = c_{13} = c_{31} = c_{23} = c_{32} \\ c_{44} &= c_{55} = c_{66}, \end{aligned} \tag{3.225}$$

and the stiffness matrix is

$$[c] = \begin{pmatrix} c_{11} & c_{12} & c_{12} & 0 & 0 & 0 \\ c_{12} & c_{11} & c_{12} & 0 & 0 & 0 \\ c_{12} & c_{12} & c_{11} & 0 & 0 & 0 \\ 0 & 0 & 0 & c_{44} & 0 & 0 \\ 0 & 0 & 0 & 0 & c_{44} & 0 \\ 0 & 0 & 0 & 0 & 0 & c_{44} \end{pmatrix}. \tag{3.226}$$

The strain energy of the system is a quadratic function of the strain, and can be written as

$$U = \frac{1}{2} \sum_{i=1}^{6} \sum_{j=1}^{6} c_{ij} e_i e_j. \tag{3.227}$$

For cubic systems, U is given by

$$U = \frac{1}{2} c_{11}(e_1^2 + e_2^2 e_3^2) + C_{44}(e_4^2 + e_5^2 + e_6^2) + C_{12}(e_1 e_2 + e_2 e_3 + e_3 e_2). \tag{3.228}$$

Using the relations between the deformed coordinates and the undeformed coordinates, the Hamiltonian for a strained semiconductor can be derived. The strained Hamiltonian introduces extra terms that can be obtained by using the following correspondences for the 8×8 Luttinger–Kohn Hamiltonian given by Equation (3.211)

$$\begin{aligned} &k_\alpha k_\beta \leftrightarrow \varepsilon_{\alpha\beta} \\ &\frac{\hbar^2}{2m_e^*} \leftrightarrow a_c \\ &\frac{\hbar^2 \gamma_1}{2m_0} \leftrightarrow -a_v \end{aligned}$$

$$\frac{\hbar^2\gamma_2}{2m_0} \leftrightarrow -\frac{b}{2}$$
$$\frac{\hbar^2\gamma_3}{2m_0} \leftrightarrow -\frac{d}{2\sqrt{3}}$$
$$k_\alpha \leftrightarrow -\sum_\beta \varepsilon_{\alpha\beta}k_\beta$$

$$H_\varepsilon = \begin{pmatrix} A_{c\varepsilon} & 0 & -\sqrt{3}V_\varepsilon^\dagger & \sqrt{2}U_\varepsilon & V_\varepsilon & 0 & U_\varepsilon & \sqrt{2}V_\varepsilon \\ 0 & A_{c\varepsilon} & 0 & -V_\varepsilon^\dagger & \sqrt{2}U_\varepsilon & \sqrt{3}V_\varepsilon & \sqrt{2}V_\varepsilon^\dagger & -U_\varepsilon \\ -\sqrt{3}V_\varepsilon & 0 & -(P_\varepsilon+Q_\varepsilon) & S_\varepsilon & -R_\varepsilon & 0 & \frac{1}{\sqrt{2}}S_\varepsilon & -\sqrt{2}R_\varepsilon \\ \sqrt{2}U_\varepsilon^\dagger & -V_\varepsilon & S_\varepsilon^\dagger & -(P_\varepsilon-Q_\varepsilon) & 0 & -R_\varepsilon & \sqrt{2}Q_\varepsilon & -\sqrt{3/2}S_\varepsilon \\ V_\varepsilon^\dagger & \sqrt{2}U_\varepsilon^\dagger & -R_\varepsilon^\dagger & 0 & -(P_\varepsilon-Q_\varepsilon) & -S_\varepsilon & -\sqrt{3/2}S_\varepsilon^\dagger & -\sqrt{2}Q_\varepsilon \\ 0 & \sqrt{3}V_\varepsilon^\dagger & 0 & -R_\varepsilon^\dagger & -S_\varepsilon^\dagger & -(P_\varepsilon+Q_\varepsilon) & \sqrt{2}R_\varepsilon^\dagger & \frac{1}{\sqrt{2}}S_\varepsilon^\dagger \\ U_\varepsilon^\dagger & \sqrt{2}V_\varepsilon & \frac{1}{\sqrt{2}}S_\varepsilon^\dagger & \sqrt{2}Q_\varepsilon^\dagger & -\sqrt{3/2}S_\varepsilon & \sqrt{2}R_\varepsilon & -(P_\varepsilon) & 0 \\ \sqrt{2}V_\varepsilon^\dagger & -U_\varepsilon^\dagger & -\sqrt{2}R_\varepsilon^\dagger & -\sqrt{3/2}S_\varepsilon^\dagger & -\sqrt{2}Q_\varepsilon^\dagger & \frac{1}{\sqrt{2}}S\varepsilon & 0 & -P\varepsilon \end{pmatrix} \begin{matrix} c1 \\ c2 \\ h1 \\ l1 \\ l2 \\ h2 \\ s1 \\ s2 \end{matrix}, \tag{3.229}$$

with

$$\begin{aligned} A_{c\varepsilon} &= a_c(\varepsilon_{xx}+\varepsilon_{yy}+\varepsilon_{zz}), \\ V_\varepsilon &= -\frac{1}{\sqrt{6}}\pi\sum_j(\varepsilon_{xj}-i\varepsilon_{yj})k_j, \\ U_\varepsilon &= -\frac{1}{\sqrt{3}}\pi\sum_j\varepsilon_{zj}k_j, \\ P_\varepsilon &= -a_v(\varepsilon_{xx}+\varepsilon_{yy}+\varepsilon_{zz}), \\ Q_\varepsilon &= -\frac{b}{2}(\varepsilon_{xx}+\varepsilon_{yy}-2\varepsilon_{zz}), \\ S_\varepsilon &= -d(\varepsilon_{xz}-\varepsilon_{yz}), \\ R_\varepsilon &= \frac{\sqrt{3}}{2}b(\varepsilon_{xx}-\varepsilon_{yy})-id\varepsilon_{xy}. \end{aligned} \tag{3.230}$$

3.4 THE WURTZITE HAMILTONIAN

In the section, we consider the $k \cdot p$ method for strained wurtzite semiconductors. Similar to Section 3.2.4, we can write the matrix element of H_o:

$$\begin{aligned}
\langle S|H_0|S\rangle &= E_{c0},\\
\langle X|H_0|X\rangle &= \langle Y|H_0|Y\rangle = E_{v0}+\Delta_1,\\
\langle Z|H_0|Z\rangle &= E_{v0},\\
\langle X|(H_{SO})_{11}|Y\rangle &= -\langle X|(H_{SO})_{22}|Y\rangle \equiv -i\Delta_2,\\
\langle Y|(H_{SO})_{12}|Z\rangle &= \langle Y|(H_{SO})_{21}|Z\rangle = -i\Delta_3,\\
\langle X|(H_{SO})_{12}|Z\rangle &= -\langle X|(H_{SO})_{21}|Z\rangle = \Delta_3,
\end{aligned} \tag{3.231}$$

where

$$\Delta_2 = C\left\langle X\left|\left(\frac{\partial U}{\partial y}\frac{\partial}{\partial x}-\frac{\partial U}{\partial x}\frac{\partial}{\partial y}\right)\right|Y\right\rangle \tag{3.232}$$

and

$$\Delta_3 = C\left\langle Y\left|\left(\frac{\partial U}{\partial y}\frac{\partial}{\partial z}-\frac{\partial U}{\partial z}\frac{\partial}{\partial y}\right)\right|Z\right\rangle = C\left\langle X\left|\left(\frac{\partial U}{\partial x}\frac{\partial}{\partial z}-\frac{\partial U}{\partial z}\frac{\partial}{\partial x}\right)\right|Z\right\rangle. \tag{3.233}$$

In the above, we used the sixfold symmetry on the $x-y$ plane. For example, (x, y) mapping to $\pi/3$ rotation around the z-axis leaves the Hamiltonian invariant, and the wave functions $|X\rangle$ and $|Y\rangle$ transform to

$$\begin{aligned}
|X'\rangle &= \frac{1}{2}|X\rangle + \frac{\sqrt{3}}{2}|Y\rangle,\\
|Y'\rangle &= -\frac{\sqrt{3}}{2}|X\rangle + \frac{1}{2}|Y\rangle.
\end{aligned} \tag{3.234}$$

Here,

$$\langle Y|H_0|Y\rangle = \langle Y'|H_0|Y'\rangle = \frac{3}{4}\langle X|H_0|X\rangle + \frac{1}{4}\langle Y|H_0|Y\rangle. \tag{3.235}$$

Hence, $\langle Y|H_0|Y\rangle = \langle X|H_0|X\rangle$. Similarly, we obtain $\langle X|H_0|Y\rangle = 0$.

Using the matrix elements in Equation (3.231) for wurtzite semiconductors, we can write Equations (3.178) and (3.179) as follows:

$$E\begin{pmatrix}\alpha_s\\ \alpha_x\\ \alpha_y\\ \alpha_z\\ \beta_s\\ \beta_x\\ \beta_y\\ \beta_z\end{pmatrix}=\begin{pmatrix}E_{c0} & 0 & 0 & 0 & 0 & 0 & 0 & 0\\ 0 & E_{v0}+\Delta_1 & -i\Delta_2 & 0 & 0 & 0 & 0 & \Delta_3\\ 0 & i\Delta_2 & E_{v0}+\Delta_1 & 0 & 0 & 0 & 0 & -i\Delta_3\\ 0 & 0 & 0 & E_{v0} & 0 & -\Delta_3 & i\Delta_3 & 0\\ 0 & 0 & 0 & 0 & E_{c0} & 0 & 0 & 0\\ 0 & 0 & 0 & -\Delta_3 & 0 & E_{v0}+\Delta_1 & i\Delta_2 & 0\\ 0 & 0 & 0 & -i\Delta_3 & 0 & -i\Delta_2 & E_{v0}+\Delta_1 & 0\\ 0 & \Delta_3 & i\Delta_3 & 0 & 0 & 0 & 0 & E_{v0}\end{pmatrix}\begin{pmatrix}\alpha_s\\ \alpha_x\\ \alpha_y\\ \alpha_z\\ \beta_s\\ \beta_x\\ \beta_y\\ \beta_z\end{pmatrix}. \tag{3.236}$$

By using a method similar to that used for zinc blende structure, we can get the matrix in block-diagonalized form

$$E\begin{pmatrix}\alpha_s\\ \beta_s\\ \alpha_x\\ \alpha_y\\ \beta_z\\ \beta_x\\ \beta_y\\ \alpha_z\end{pmatrix}=\begin{pmatrix}E_{c0} & 0 & 0 & 0 & 0 & 0 & 0 & 0\\ 0 & E_{c0} & 0 & 0 & 0 & 0 & 0 & 0\\ 0 & 0 & E_{v0}+\Delta_1 & -i\Delta_2 & \Delta_3 & 0 & 0 & 0\\ 0 & 0 & i\Delta_2 & E_{v0}+\Delta_1 & -i\Delta_3 & 0 & 0 & 0\\ 0 & 0 & \Delta_3 & i\Delta_3 & E_{v0} & 0 & 0 & 0\\ 0 & 0 & 0 & 0 & 0 & E_{v0}+\Delta_1 & i\Delta_2 & -\Delta_3\\ 0 & 0 & 0 & 0 & 0 & -i\Delta_2 & E_{v0}+\Delta_1 & -i\Delta_3\\ 0 & 0 & 0 & 0 & 0 & -\Delta_3 & i\Delta_3 & E_{v0}\end{pmatrix}\begin{pmatrix}\alpha_s\\ \beta_s\\ \alpha_x\\ \alpha_y\\ \beta_z\\ \beta_x\\ \beta_y\\ \alpha_z\end{pmatrix}. \tag{3.237}$$

The eigenvalues and eigenvectors of the upper 3×3 block are given by

$$\begin{aligned}&E_h = E_{v0}+\Delta_1+\Delta_2 \quad \text{(heavy hole, } h1\text{)}\\ &\alpha_x = -1/\sqrt{2}, \alpha_y = -i/\sqrt{2}, \beta_z = 0\end{aligned} \tag{3.238}$$

$$\begin{aligned}&E_l = E_{v0}+\frac{\Delta_1-\Delta_2}{2}+\sqrt{\left(\frac{\Delta_1-\Delta_2}{2}\right)^2+2\Delta_3^2} \quad \text{(light hole, } l2\text{)}\\ &\alpha_x = \Delta_3/\sqrt{E_c^2+2\Delta_3^2}, \alpha_y = -i\Delta_3/\sqrt{E_c^2+2\Delta_3^2}, \beta_z = -E_c/\sqrt{E_c^2+2\Delta_3^2}\end{aligned} \tag{3.239}$$

$$\begin{aligned}&E_c = E_{v0}+\frac{\Delta_1-\Delta_2}{2}-\sqrt{\left(\frac{\Delta_1-\Delta_2}{2}\right)^2+2\Delta_3^2} \quad \text{(crystal-field split-off hole, } c2\text{)}\\ &\alpha_x = \Delta_3/\sqrt{E_l^2+2\Delta_3^2}, \alpha_y = -i\Delta_3/\sqrt{E_l^2+2\Delta_3^2}, \beta_z = -E_l/\sqrt{E_l^2+2\Delta_3^2}\end{aligned} \tag{3.240}$$

The lower 3×3 block in Equation (3.237) gives the same eigenvalues and eigenvectors.

If the spin–orbit interaction H_{so} were absent, $\Delta_2 = \Delta_3 = 0$. Then, the energy levels and the wave functions at the zone center ($k = 0$) are written as shown below.

$$\begin{aligned}
&E_h = E_{v0} + \Delta_1 \quad \text{(heavy hole, } h1\text{)} \\
&\alpha_x = -1/\sqrt{2}, \alpha_y = -i/\sqrt{2}, \beta_z = 0 \\
&E_l = E_{v0} + \Delta_1 \quad \text{(light hole, } l2\text{)} \\
&\alpha_x = 1/\sqrt{2}, \alpha_y = -i/\sqrt{2}, \beta_z = 0 \\
&E_c = E_{v0} \quad \text{(crystal-field split-off hole, } c2\text{)} \\
&\alpha_x = 0, \alpha_y = 0, \beta_z = 1
\end{aligned} \tag{3.241}$$

Those for the lower 3×3 block can be obtained by using a similar process. Here, the first two eigenvalues are equal so that any linear combination of the heavy- and light-hole eigenvectors is also an eigenvector. In the wurtzite Hamitonian, we will use the following basis functions:

$$\begin{aligned}
c1 &: |iS\rangle \uparrow \\
c2 &: |iS\rangle \downarrow \\
h1 : |u_1\rangle &= -\frac{1}{\sqrt{2}}[|X\rangle + i|Y\rangle] \uparrow \\
l1 : |u_2\rangle &= \frac{1}{\sqrt{2}}[(|X\rangle - i|Y\rangle] \uparrow \\
c1 : |u_3\rangle &= |Z\rangle \uparrow \\
l2 : |u_4\rangle &= \frac{1}{\sqrt{2}}[(|X\rangle - i|Y\rangle] \downarrow \\
h2 : |u_5\rangle &= -\frac{1}{\sqrt{2}}[|X\rangle + i|Y\rangle] \downarrow \\
c2 : |u_6\rangle &= |Z\rangle \downarrow
\end{aligned} \tag{3.242}$$

Similarly to a zinc blende case, we will construct a matrix D defined by

$$D_{jj'} = \sum_{\alpha\beta} k_\alpha k_\beta D_{jj'}^{\alpha\beta}, \tag{3.243}$$

where $j, j' = X, Y, Z$.

Then, the 3×3 matrix has the form

$$D=\begin{pmatrix} L_1k_x^2+M_1k_y^2+M_2k_z^2 & N_1k_xk_y & N_2k_xk_z \\ N_1k_xk_y & M_1k_x^2+L_1k_y^2+M_2k_z^2 & N_2k_yk_z \\ N_2k_xk_z & N_2k_yk_z & M_3k_x^2+M_3k_y^2+L_2k_z^2 \end{pmatrix}, \tag{3.244}$$

for each spin, where the band-structure parameters L, M, N are three real constants, which are similar to the Luttinger–Kohn parameters γ_1, γ_2, and γ_3 for zinc blende structures,

$$\begin{aligned}
L_1 &= \frac{\hbar^2}{2m_0}+\frac{\hbar^2}{m_0^2}\sum_i \frac{p_{Xi}^x p_{iX}^x}{\varepsilon_0-\varepsilon_i} = \frac{\hbar^2}{2m_0}+\frac{\hbar^2}{m_0^2}\sum_i \frac{p_{Yi}^y p_{iY}^y}{\varepsilon_0-\varepsilon_i} \\
L_2 &= \frac{\hbar^2}{2m_0}+\frac{\hbar^2}{m_0^2}\sum_i \frac{p_{Zi}^z p_{iZ}^z}{\varepsilon_0-\varepsilon_i} \\
M_1 &= \frac{\hbar^2}{2m_0}+\frac{\hbar^2}{m_0^2}\sum_i \frac{p_{Xi}^y p_{iX}^y}{\varepsilon_0-\varepsilon_i} = \frac{\hbar^2}{2m_0}+\frac{\hbar^2}{m_0^2}\sum_i \frac{p_{Yi}^x p_{iY}^x}{\varepsilon_0-\varepsilon_i} \\
M_2 &= \frac{\hbar^2}{2m_0}+\frac{\hbar^2}{m_0^2}\sum_i \frac{p_{Xi}^z p_{iX}^z}{\varepsilon_0-\varepsilon_i} = \frac{\hbar^2}{2m_0}+\frac{\hbar^2}{m_0^2}\sum_i \frac{p_{Yi}^z p_{iY}^z}{\varepsilon_0-\varepsilon_i} \\
M_3 &= \frac{\hbar^2}{2m_0}+\frac{\hbar^2}{m_0^2}\sum_i \frac{p_{Zi}^x p_{iZ}^x}{\varepsilon_0-\varepsilon_i} = \frac{\hbar^2}{2m_0}+\frac{\hbar^2}{m_0^2}\sum_i \frac{p_{Zi}^y p_{iZ}^y}{\varepsilon_0-\varepsilon_i} \\
N_1 &= \frac{\hbar^2}{m_0^2}\sum_i \frac{p_{Xi}^x p_{iY}^y + p_{Xi}^y p_{iY}^x}{\varepsilon_0-\varepsilon_i}
\end{aligned} \tag{3.245}$$

as can be seen from Equation (3.137). Using the results in Equation (3.244), the 6×6 Hamiltonian in the bases $\{u_1, u_2, \ldots, u_6\}$ can be easily derived using

$$D_{6\times 6}=\begin{pmatrix} D_{11} & D_{21}^* & -D_{23}^* & & & \\ D_{21} & D_{11} & D_{23} & & 0 & \\ -D_{23} & D_{23}^* & D_{33} & & 0 & \\ & & & D_{11} & D_{21} & D_{23} \\ & 0 & & D_{21}^* & D_{11} & -D_{23}^* \\ & & & D_{23}^* & -D_{23} & D_{33} \end{pmatrix}, \tag{3.246}$$

where

$$
\begin{aligned}
D_{11} &= \left(\frac{L_1 + M_1}{2}\right)(k_x^2 + k_y^2) + M_2 k_z^2 \\
D_{21} &= -\frac{1}{2}[(L_1 - M_1)(k_x^2 - k_y^2) + 2iN_1 k_x k_y] = -\frac{1}{2} N_1 (k_x + ik_y)^2, \\
D_{23} &= \frac{1}{\sqrt{2}} N_2 (k_x + ik_y) k_z, \\
D_{33} &= M_2 (k_x^2 + k_y^2) + L_2 k_z^2.
\end{aligned}
\tag{3.247}
$$

Also, we obtain the following relation from symmetry consideration:

$$
L_1 - M_1 = N_1. \tag{3.248}
$$

The 6×6 Hamiltonian at the zone center in the bases $\{u_1, u_2, \ldots, u_6\}$ can be obtained from Equation (3.236):

$$
H_{6\times6}(\mathbf{k}=0) = \begin{pmatrix}
E_{v0} + \Delta_1 + \Delta_2 & 0 & 0 & 0 & 0 & 0 \\
0 & E_{v0} + \Delta_1 \Delta_2 & 0 & 0 & 0 & \sqrt{2}\Delta_3 \\
0 & 0 & E_{v0} & 0 & \sqrt{2}\Delta_3 & 0 \\
0 & 0 & 0 & E_{v0} + \Delta_1 + \Delta_2 & 0 & 0 \\
0 & 0 & \sqrt{2}\Delta_3 & 0 & E_{v0} + \Delta_1 - \Delta_2 & E_{v0} \\
0 & \sqrt{2}\Delta_3 & 0 & 0 & 0 & 0
\end{pmatrix}
\begin{pmatrix} u_1 \\ u_2 \\ u_3 \\ u_4 \\ u_5 \\ u_6 \end{pmatrix}. \tag{3.249}
$$

Hence, the full Hamiltonian, $H = H_{6\times6}(\mathbf{k}=0) + D_{6\times6}$, can be written as

$$
H_{6\times6}(\mathbf{k}) = \begin{pmatrix}
F & -K^* & -H^* & 0 & 0 & 0 \\
-K & G & H & 0 & 0 & \Delta \\
-H & H^* & \lambda & 0 & \Delta & 0 \\
0 & 0 & 0 & F & -K & H \\
0 & 0 & \Delta & -K^* & G & -H^* \\
0 & \Delta & 0 & H^* & -H & \lambda
\end{pmatrix}
\begin{pmatrix} u_1 \\ u_2 \\ u_3 \\ u_4 \\ u_5 \\ u_6 \end{pmatrix}, \tag{3.250}
$$

where

$$
\begin{aligned}
F &= \Delta_1 + \Delta_2 + \lambda + \theta, \\
G &= \Delta_1 - \Delta_2 + \lambda + \theta, \\
\lambda &= \frac{\hbar^2}{2m_o}[A_1 k_z^2 + A_2(k_x^2 + k_y^2)], \\
\theta &= \frac{\hbar^2}{2m_o}[A_3 k_z^2 + A_4(k_x^2 + k_y^2)], \\
K &= \frac{\hbar^2}{2m_o} A_5 (k_x + ik_y)^2, \\
H &= \frac{\hbar^2}{2m_o} A_6 (k_x + ik_y) k_z, \\
\Delta &= \sqrt{2}\Delta_3.
\end{aligned}
\tag{3.251}
$$

The relations between the band-structure parameters L, M, N derived from the $k \cdot p$ method and A_i parameters used in the Bir–Pikusmodel are given by

$$
\begin{aligned}
&\frac{\hbar^2}{2m_o} A_1 = L_2, \frac{\hbar^2}{2m_o} A_2 = M_3, \frac{\hbar^2}{2m_o} A_3 = M_2 - L_2, \\
&\frac{\hbar^2}{2m_o} A_4 = \frac{L_1 + M_1}{2} - M_3, \frac{\hbar^2}{2m_o} A_5 = \frac{N_1}{2}, \frac{\hbar^2}{2m_o} A_6 = \frac{N_2}{\sqrt{2}}.
\end{aligned}
\tag{3.252}
$$

The Hamiltonian for a strained semiconductor can be easily derived by replacing A_i with the deformation potentials D_i and a straightforward addition of corresponding terms:

$$
k_\alpha k_\beta \to \varepsilon_{\alpha\beta}. \tag{3.253}
$$

$$
\begin{aligned}
F &= \Delta_1 + \Delta_2 + \lambda + \theta, \\
G &= \Delta_1 - \Delta_2 + \lambda + \theta, \\
\lambda &= \frac{\hbar^2}{2m_o}[A_1 k_z^2 + A_2(k_x^2 + k_y^2)] + \lambda_\varepsilon, \\
\theta &= \frac{\hbar^2}{2m_o}[A_3 k_z^2 + A_4(k_x^2 + k_y^2)] + \theta_\varepsilon, \\
K &= \frac{\hbar^2}{2m_o} A_5 (k_x + ik_y)^2 + D_5 \varepsilon_+, \\
H &= \frac{\hbar^2}{2m_o} A_6 (k_x + ik_y) k_z + D_6 \varepsilon_{z+}, \\
\lambda_\varepsilon &= D_1 \varepsilon_{zz} + D_2(\varepsilon_{xx} + \varepsilon_{yy}), \\
\theta_\varepsilon &= D_3 \varepsilon_{zz} + D_4(\varepsilon_{xx} + \varepsilon_{yy}), \\
\Delta &= \sqrt{2}\Delta_3,
\end{aligned}
\tag{3.254}
$$

where

$$\begin{aligned} \varepsilon_{\pm} &= \varepsilon_{xx} \pm 2i\varepsilon_{xy} - \varepsilon_{yy}, \\ \varepsilon_{z\pm} &= \varepsilon_{zx} \pm i\varepsilon_{yz}. \end{aligned} \tag{3.255}$$

Matrix elements between conduction and valence bands can be derived by using a process similar to the one we have used for a zinc blende case in Section 3.3.2: Here, we use two Kane's parameters, P_1 and P_2, which are defined as

$$\begin{aligned} \left\langle iS \left| -i\hbar\frac{\partial}{\partial z} \right| Z \right\rangle &= \frac{m_0}{\hbar} P_1, \\ \left\langle iS \left| -i\hbar\frac{\partial}{\partial x} \right| X \right\rangle = \left\langle iS \left| -i\hbar\frac{\partial}{\partial y} \right| Y \right\rangle &= \frac{m_0}{\hbar} P_2. \end{aligned} \tag{3.256}$$

Then, from Equation (3.203), matrix elements of Hamiltonian between conduction and valence bands under bases functions of Equation (3.242) are given by

$$\begin{pmatrix} & h1 & h2 & l1 & l2 & s1 & s2 \\ c1 & -\frac{P_2}{\sqrt{2}}(k_x+ik_y) & 0 & \frac{P_2}{\sqrt{2}}(k_x-ik_y) & 0 & P_1k_z & 0 \\ c2 & 0 & -\frac{P_2}{\sqrt{2}}(k_x+ik_y) & 0 & \frac{P_2}{\sqrt{2}}(k_x-ik_y) & 0 & P_1k_z \end{pmatrix}, \tag{3.257}$$

For example, suppose $n = c1$ and $n' = h1$. Then, we have

$$\mathbf{p}_{c1h1} = \frac{i\hbar}{\sqrt{2}}(\langle iS|\nabla|X\rangle + i\langle iS|\nabla|Y\rangle). \tag{3.258}$$

Using Equation (3.256),

$$H_{c1h1} = -\frac{P_2}{\sqrt{2}}(k_x + ik_y). \tag{3.259}$$

Combining Equations (3.250) and (3.257), we can obtain the 8×8 Hamiltonian given by

$$H=\begin{pmatrix} A_c & 0 & V & 0 & -V^* & 0 & U & 0 \\ 0 & A_c & 0 & V & 0 & -V^* & 0 & U \\ V^* & 0 & F & -K^* & -H^* & 0 & 0 & 0 \\ 0 & V^* & -K & G & H & 0 & 0 & \Delta \\ -V & 0 & -H & H^* & \lambda & 0 & \Delta & 0 \\ 0 & -V & 0 & 0 & 0 & F & -K & H \\ U^* & 0 & 0 & 0 & \Delta & -K^* & G & -H^* \\ 0 & U^* & 0 & \Delta & 0 & H^* & -H & \lambda \end{pmatrix}\begin{matrix} c1 \\ c2 \\ h1 \\ l1 \\ c1 \\ l2 \\ h2 \\ c2 \end{matrix}, \quad (3.260)$$

with

$$\begin{aligned} A_c &= E_g + \frac{\hbar^2}{2m_0}(k_x^2 + k_y^2 + k_z^2) + a_c(\varepsilon_{xx} + \varepsilon_{yy} + \varepsilon_{zz}), \\ V &= -\frac{P_2}{\sqrt{2}}(k_x + ik_y) + \frac{P_2}{\sqrt{2}}\sum_j(\varepsilon_{xj} + i\varepsilon_{yj})k_j, \\ U &= P_1 k_z - P_1 \sum_j \varepsilon_{zj} k_j, \end{aligned} \quad (3.261)$$

where P_1 and P_2 are the conduction band transverse (longitudinal) deformation potentials. Here, the basis set for a wurtzite crystal is summarized as follows:

$$\begin{aligned} &c1 : |iS\rangle\uparrow \\ &c2 : |iS\rangle\downarrow \\ &h1 : |u_1\rangle = -\frac{1}{\sqrt{2}}[|X\rangle + i|Y\rangle]\uparrow \\ &l1 : |u_2\rangle = \frac{1}{\sqrt{2}}[(|X\rangle - i|Y\rangle]\uparrow \\ &c1 : |u_3\rangle = |Z\rangle\uparrow \\ &l2 : |u_4\rangle = \frac{1}{\sqrt{2}}[(|X\rangle - i|Y\rangle]\downarrow \\ &h2 : |u_5\rangle = -\frac{1}{\sqrt{2}}[|X\rangle + i|Y\rangle]\downarrow \\ &c2 : |u_6\rangle = |Z\rangle\downarrow. \end{aligned} \quad (3.262)$$

3.5 BAND STRUCTURE OF ZINC BLENDE AND WURTZITE SEMICONDUCTORS

Usually, the coupling between the conduction and the valence bands can be neglected for a semiconductor with a large bandgap, and, in that case, we can use the 6×6 Hamiltonian for the valence band. In this section, we show the block-diagonalized 3×3 Hamiltonian for zinc blende and wurtzite bulk crystals [3–8]. Also, the Hamiltonian for zinc blende bulk crystal is written using the basis functions of wurtzite crystals for a consistent block diagonalization into two 3×3 Hamiltonians using almost the same transformation.

3.5.1 The Hamiltonian of a Zinc Blende Bulk Semiconductor

As shown in Section 3.2, the valence-band structure of a strained (001)-oriented zinc blende semiconductor can be described by the following 6×6 Hamiltonian in the envelope-function space:

$$H^{(001)}_{\text{zinc blende}} = -\begin{pmatrix} P+Q & R & \sqrt{2}R & -S/\sqrt{2} & -S & 0 \\ R^* & P-Q & \sqrt{2}Q & \sqrt{3/2}S^* & 0 & S \\ \sqrt{2}R^* & \sqrt{2}Q & P+\Delta & 0 & \sqrt{3/2}S^* & -S/\sqrt{2} \\ -S^*/\sqrt{2} & \sqrt{3/2}S & 0 & P+\Delta & -\sqrt{2}Q & -\sqrt{2}R \\ -S^* & 0 & \sqrt{3/2}S & -\sqrt{2}Q & P-Q & R \\ 0 & S^* & -S^*/\sqrt{2} & -\sqrt{2}R^* & R^* & P+Q \end{pmatrix} \begin{matrix} \left|\frac{3}{2},\frac{3}{2}\right\rangle \\ \left|\frac{3}{2},-\frac{1}{2}\right\rangle \\ \left|\frac{1}{2},-\frac{1}{2}\right\rangle \\ \left|\frac{1}{2},\frac{1}{2}\right\rangle \\ \left|\frac{3}{2},\frac{1}{2}\right\rangle \\ \left|\frac{3}{2},-\frac{3}{2}\right\rangle \end{matrix}, \tag{3.263}$$

where

$$P = P_k + P_\varepsilon$$
$$Q = Q_k + Q_\varepsilon$$
$$R = R_k + R_\varepsilon$$

$$
\begin{aligned}
S &= S_k + S_\varepsilon \\
P_k &= \frac{\hbar^2}{2m_o}\gamma_1(k_x^2 + k_y^2 + k_z^2) \\
Q_k &= \frac{\hbar^2}{2m_o}\gamma_2(k_x^2 + k_y^2 - 2k_z^2) \\
R_k &= \frac{\hbar^2}{2m_o}\sqrt{3}[-\gamma_2(k_x^2 - k_y^2) + 2i\gamma_3 k_x k_y] \\
S_k &= \frac{\hbar^2}{2m_o}2\sqrt{3}\gamma_3(k_x - ik_y)k_z \\
P_\varepsilon &= -a_v(\varepsilon_{xx} + \varepsilon_{yy} + \varepsilon_{zz}) \\
Q_\varepsilon &= -b/2(\varepsilon_{xx} + \varepsilon_{yy} - 2\varepsilon_{zz}) \\
R_\varepsilon &= \frac{\sqrt{3}}{2}b(\varepsilon_{xx} - \varepsilon_{yy}) - id\varepsilon_{xy} \\
S_\varepsilon &= -d(\varepsilon_{zx} - i\varepsilon_{yz}),
\end{aligned}
\tag{3.264}
$$

where $k_i(i = x, y, z)$ is the wave vector component in the unprimed (x, y, z) coordinate system. The zinc blende basis set is given in Equation (3.213). The biaxial strain components for the (001)-oriented zinc blende crystal are given by

$$
\begin{aligned}
\varepsilon_{xx} &= \varepsilon_{yy} = \varepsilon_{\parallel}^c = \frac{a_s^c - a_e^c}{a_e^c}, \\
\varepsilon_{zz} &= -2\frac{C_{12}^c}{C_{11}^c}\varepsilon_{\parallel}^c, \\
\varepsilon_{xy} &= \varepsilon_{yz} = \varepsilon_{zx} = 0,
\end{aligned}
\tag{3.265}
$$

where a_s^c and a_e^c are the lattice constants of the substrate and epilayer materials, and C_{11}^c and C_{12}^c are the stiffness constants in the strained epilayers. The superscript c means those for zinc blende (cubic) crystals.

The 6×6 Hamiltonian matrix Equation (3.263) can be block diagonalized using the unitary transformation:

$$
\bar{H} = UHU^\dagger = \begin{pmatrix} H^U & 0 \\ 0 & H^L, \end{pmatrix}, \tag{3.266}
$$

where the upper and lower blocks, H^U and H^L, are given by

$$H^U = -\begin{pmatrix} P_k + Q_k & |R_k| - i|S_k| & \sqrt{2}|R_k| + i\frac{1}{\sqrt{2}}|S_k| \\ |R_k| + i|S_k| & P_k - Q_k & \sqrt{2}Q_k + i\frac{\sqrt{3}}{\sqrt{2}}|S_k| \\ \sqrt{2}|R_k| - i\frac{1}{\sqrt{2}}|S_k| & \sqrt{2}Q_k - i\frac{\sqrt{3}}{\sqrt{2}}|S_k| & P_k + \Delta \end{pmatrix}, \tag{3.267}$$

and

$$H^L = -\begin{pmatrix} P_k + \Delta & \sqrt{2}Q_k + i\frac{\sqrt{3}}{\sqrt{2}}|S_k| & \sqrt{2}|R_k| + i\frac{1}{\sqrt{2}}|S_k| \\ \sqrt{2}Q_k - i\frac{\sqrt{3}}{\sqrt{2}}|S_k & P_k - Q_k & |R_k| - i|S_k| \\ \sqrt{2}|R_k| - i\frac{1}{\sqrt{2}}|S_k| & |R_k| - i|S_k| & P_k + Q_k \end{pmatrix}, \tag{3.268}$$

where

$$\begin{aligned} |R_k| &= \frac{\hbar^2}{2m_o}\sqrt{3}\left(\frac{\gamma_2 + \gamma_3}{2}\right)k_t^2 \\ |S_k| &= \frac{\hbar^2}{2m_o}2\sqrt{3}\gamma_3 k_t k_z \\ k_t^2 &= k_x^2 + k_y^2. \end{aligned} \tag{3.269}$$

The unitary transformation U is defined by

$$T = \begin{pmatrix} \alpha^* & 0 & 0 & 0 & 0 & -\alpha \\ 0 & \beta^* & 0 & 0 & -\beta & 0 \\ 0 & 0 & \gamma^* & \gamma & 0 & 0 \\ 0 & 0 & \gamma^* & -\gamma & 0 & 0 \\ 0 & \beta^* & 0 & 0 & \beta & 0 \\ \alpha^* & 0 & 0 & 0 & 0 & \alpha \end{pmatrix}, \tag{3.270}$$

where

$$\alpha = \frac{1}{\sqrt{2}}e^{i\left(\frac{3\pi}{4} - \frac{3\phi}{2}\right)}, \quad \beta = \gamma = \frac{1}{\sqrt{2}}e^{i\left(-\frac{\pi}{4} + \frac{\phi}{2}\right)}, \tag{3.271}$$

and ϕ is the angle defined by $\phi = \tan^{-1}(k_y/k_x)$. For the strain potential, H_ε^σ is given by

$$H_\varepsilon^U = -\begin{pmatrix} P_\varepsilon + Q_\varepsilon & 0 & 0 \\ 0 & P_\varepsilon + Q_\varepsilon & \sqrt{2}Q_\varepsilon \\ 0 & \sqrt{2}Q_\varepsilon & P_\varepsilon \end{pmatrix}, \tag{3.272}$$

and

$$H_\varepsilon^L = -\begin{pmatrix} P_\varepsilon & \sqrt{2}Q_\varepsilon & 0 \\ \sqrt{2}Q_\varepsilon & P_\varepsilon - Q_\varepsilon & 0 \\ 0 & 0 & P_\varepsilon + Q_\varepsilon \end{pmatrix}. \tag{3.273}$$

Under the unitary transformation, the upper- and lower-block Hamiltonians, H^U and H^L, are decoupled. Let us denote the upper- and lower-block envelope functions by

$$\Psi_{m\mathbf{k}_\parallel}^\sigma(\mathbf{r}) = \sum_\nu g_m^\nu(\mathbf{k}_\parallel, z)e^{i\mathbf{k}\cdot\mathbf{r}}|\nu\rangle, \tag{3.274}$$

where $g_m^\nu(\mathbf{k}_\parallel, z)$ are the envelope functions, $\{|\nu\rangle\}$ denotes the transformed Bloch bases at the zone center, m is the quantum well subband index, $\mathbf{k}_\parallel = k_x\hat{x} + k_y\hat{y}, \mathbf{r}_\parallel = x\hat{x} + y\hat{y}$, and $\sigma = U$ (or L) refers to the upper (or lower) blocks, respectively. We note that the summation is for $\nu = 1,2,3$ for $\sigma = U$, and for $\nu = 4,5,6$ for $\sigma = L$. The transformed Bloch bases are given by

$$\begin{aligned}
|1\rangle &= \alpha^*\left|\frac{3}{2},\frac{3}{2}\right\rangle - \alpha\left|\frac{3}{2},-\frac{3}{2}\right\rangle \\
|2\rangle &= \beta^*\left|\frac{3}{2},-\frac{1}{2}\right\rangle - \beta\left|\frac{3}{2},\frac{1}{2}\right\rangle \\
|3\rangle &= \gamma^*\left|\frac{1}{2},-\frac{1}{2}\right\rangle + \gamma\left|\frac{1}{2},\frac{1}{2}\right\rangle \\
|4\rangle &= \gamma^*\left|\frac{1}{2},-\frac{1}{2}\right\rangle - \gamma\left|\frac{1}{2},\frac{1}{2}\right\rangle \\
|5\rangle &= \beta^*\left|\frac{3}{2},-\frac{1}{2}\right\rangle + \beta\left|\frac{3}{2},\frac{1}{2}\right\rangle \\
|6\rangle &= \alpha^*\left|\frac{3}{2},\frac{3}{2}\right\rangle + \alpha\left|\frac{3}{2},-\frac{3}{2}\right\rangle.
\end{aligned} \tag{3.275}$$

3.5.2 Zinc Blende Hamiltonians in the Wurtzite Basis Functions

Sometimes, it is convenient to work on a wurtzite basis for zinc blende Hamiltonians. We define a matrix M_{zb}, which relates the zinc blende bases given in Equation (3.213) to $(X\uparrow, Y\uparrow, Z\uparrow, X\downarrow, Y\downarrow, Z\downarrow)^t$.

$$\text{(zinc blende bases)} = M_{zb}(X\uparrow, Y\uparrow, Z\uparrow, X\downarrow, Y\downarrow, Z\downarrow)^t, \tag{3.276}$$

where

$$M_{zb} = \begin{pmatrix} -1/\sqrt{2} & -i/\sqrt{2} & 0 & 0 & 0 & 0 \\ 0 & 0 & \sqrt{2}/\sqrt{3} & -1/\sqrt{6} & -i/\sqrt{6} & 0 \\ 1/\sqrt{6} & -i/\sqrt{6} & 0 & 0 & 0 & \sqrt{2}/\sqrt{3} \\ 0 & 0 & 0 & 1/\sqrt{2} & -i/\sqrt{2} & 0 \\ 0 & 0 & 1/\sqrt{3} & 1/\sqrt{3} & i/\sqrt{3} & 0 \\ 1/\sqrt{3} & -i/\sqrt{3} & 0 & 0 & 0 & -1/\sqrt{3} \end{pmatrix}. \tag{3.277}$$

Similarly, we define another matrix M_{wz}, which relates the wurtzite bases to $(X\uparrow, Y\uparrow, Z\uparrow, X\downarrow, Y\downarrow, Z\downarrow)^t$.

$$(U_1, U_2, U_3, U_4, U_5, U_6)^t = M_{wz}(X\uparrow, Y\uparrow, Z\uparrow, X\downarrow, Y\downarrow, Z\downarrow)^t, \tag{3.278}$$

where the wurtzite bases are given by

$$\begin{aligned} |U_1\rangle &= -\frac{1}{\sqrt{2}}|(X+iY)\uparrow\rangle \\ |U_2\rangle &= \frac{1}{\sqrt{2}}|(X-iY)\uparrow\rangle \\ |U_3\rangle &= |Z\uparrow\rangle, \\ |U_4\rangle &= \frac{1}{\sqrt{2}}|(X-iY)\downarrow\rangle \\ |U_5\rangle &= -\frac{1}{\sqrt{2}}|(X+iY)\downarrow\rangle \\ |U_6\rangle &= |Z\downarrow\rangle, \end{aligned} \tag{3.279}$$

and the matrix is

$$M_{wz}=\begin{pmatrix} -1/\sqrt{2} & -i/\sqrt{2} & 0 & 0 & 0 & 0 \\ 1/\sqrt{2} & -i/\sqrt{2} & 0 & 0 & 0 & 0 \\ 0 & 0 & 1 & 0 & 0 & 0 \\ 0 & 0 & 0 & 1/\sqrt{2} & -i/\sqrt{2} & 0 \\ 0 & 0 & 0 & -1/\sqrt{2} & -i/\sqrt{2} & 0 \\ 0 & 0 & 0 & 0 & 0 & 1 \end{pmatrix}. \tag{3.280}$$

The zinc blende Hamiltonian in the wurtzite bases can be obtained by the relation

$$H^{(001)} = M^{*}_{\text{bases}} H^{(001)}_{\text{zinc blende}} \tilde{M}_{\text{bases}}, \tag{3.281}$$

where

$$M_{bases} = M_{wz} M_{zb}^{-1}. \tag{3.282}$$

We can obtain (001)-oriented zinc blende Hamiltonian with wurtzite bases:

$$H^{(001)}=\begin{pmatrix} H_{11}^{(1)} & -H_{21}^{(1)*} & -H_{23}^{(1)*} & 0 & 0 & 0 \\ -H_{21}^{(1)} & H_{22}^{(1)} & H_{23}^{(1)} & 0 & 0 & \Delta^{(1)} \\ -H_{23}^{(1)} & H_{23}^{(1)*} & H_{33}^{(1)} & 0 & \Delta^{(1)} & 0 \\ 0 & 0 & 0 & H_{11}^{(1)} & -H_{21}^{(1)} & H_{23}^{(1)} \\ 0 & 0 & \Delta^{(1)} & -H_{21}^{(1)*} & H_{22}^{(1)} & -H_{23}^{(1)*} \\ 0 & \Delta^{(1)} & 0 & H_{23}^{(1)*} & -H_{23}^{(1)} & H_{33}^{(1)} \end{pmatrix} \begin{matrix} |U_1\rangle \\ |U_2\rangle \\ |U_3\rangle \\ |U_4\rangle \\ |U_5\rangle \\ |U_6\rangle \end{matrix}, \tag{3.283}$$

where the superscript (1) means (001)-orientated zinc blende crystals in the wurtzite bases, and the matrix elements are:

$$\begin{aligned} H_{11}^{(1)} &= -\frac{\hbar^2}{2m_o}\left[(\gamma_1+\gamma_2)(k_x^2+k_y^2)+(\gamma_1-2\gamma_2)k_z^2\right]-P_\varepsilon-Q_\varepsilon, \\ H_{22}^{(1)} &= H_{11}^{(1)}-\frac{2}{3}\Delta, \\ H_{33}^{(1)} &= -\frac{\Delta}{3}-\frac{\hbar^2}{2m_o}\left[(\gamma_1-2\gamma_2)(k_x^2+k_y^2)+(\gamma_1+4\gamma_2)k_z^2\right]-P_\varepsilon+2Q_\varepsilon, \\ H_{21}^{(1)} &= -3\frac{\hbar^2}{2m_o}\left[\gamma_2(k_x^2-k_y^2)+2i\gamma_3k_xk_y\right]-\frac{3}{2}b(\varepsilon_{xx}^{(1)}-\varepsilon_{yy}^{(1)}), \\ H_{23}^{(1)} &= -\frac{\hbar^2}{2m_o}3\sqrt{2}\gamma_3(k_x+ik_y)k_z, \\ \Delta^{(1)} &= \frac{\sqrt{2}}{3}\Delta. \end{aligned} \tag{3.284}$$

3.5.3 The Hamiltonian of Wurtzite Bulk Semiconductors

From Section 3.4, we have the full Hamiltonian for (0001)-orientated wurtzite crystal:

$$H^{(0001)} = \begin{pmatrix} H_{11}^{(2)} & -H_{21}^{(2)*} & -H_{23}^{(2)*} & 0 & 0 & 0 \\ -H_{21}^{(2)} & H_{22}^{(2)} & H_{23}^{(2)} & 0 & 0 & \Delta^{(2)} \\ -H_{23}^{(2)} & H_{23}^{(2)*} & H_{33}^{(2)} & 0 & \Delta^{(2)} & 0 \\ 0 & 0 & 0 & H_{11}^{(2)} & -H_{21}^{(2)} & H_{23}^{(2)} \\ 0 & 0 & \Delta^{(2)} & -H_{21}^{(2)*} & H_{22}^{(2)} & -H_{23}^{(2)*} \\ 0 & \Delta^{(2)} & 0 & H_{23}^{(2)*} & -H_{23}^{(2)} & H_{33}^{(2)} \end{pmatrix} \begin{matrix} |U_1\rangle \\ |U_2\rangle \\ |U_3\rangle \\ |U_4\rangle \\ |U_5\rangle \\ |U_6\rangle \end{matrix}, \tag{3.285}$$

where the superscript (2) stands for (0001)-oriented wurtzite crystals. The matrix elements are:

$$\begin{aligned} H_{11}^{(2)} &= \Delta_1 + \Delta_2 + \lambda + \theta, \\ H_{22}^{(2)} &= \Delta_1 - \Delta_2 + \lambda + \theta, \\ H_{33}^{(2)} &= \lambda + \lambda_\varepsilon, \\ \lambda &= \frac{\hbar^2}{2m_o}[A_1 k_z^2 + A_2(k_x^2 + k_y^2)] + \lambda_\varepsilon, \\ \theta &= \frac{\hbar^2}{2m_o}[A_3 k_z^2 + A_4(k_x^2 + k_y^2)] + \theta_\varepsilon, \\ H_{21}^{(2)} &= \frac{\hbar^2}{2m_o} A_5 (k_x + ik_y)^2 + D_5 \varepsilon_+^{(2)}, \\ H_{23}^{(2)} &= \frac{\hbar^2}{2m_o} A_6 (k_x + ik_y) k_z + D_6 \varepsilon_{z+}^{(2)}, \\ \lambda_\varepsilon &= D_1 \varepsilon_{zz}^{(2)} + D_2(\varepsilon_{xx}^{(2)} + \varepsilon_{yy}^{(2)}), \\ \theta_\varepsilon &= D_3 \varepsilon_{zz}^{(2)} + D_4(\varepsilon_{xx}^{(2)} + \varepsilon_{yy}^{(2)}), \\ \varepsilon_+^{(2)} &= \varepsilon_{xx}^{(2)} - \varepsilon_{yy}^{(2)} + 2i\varepsilon_{xy}^{(2)}, \\ \varepsilon_{z+}^{(2)} &= \varepsilon_{xz}^{(2)} + i\varepsilon_{yz}^{(2)}, \\ \Delta^{(2)} &= \sqrt{2}\Delta_3. \end{aligned} \tag{3.286}$$

Here, the A_is are the valence-band effective mass parameters, which are similar to the Luttinger parameters in a zinc blende crystal, the D_is are the deformation potentials for wurtzite crystals, Δ_1 is the crystal field split energy, and Δ_2 and Δ_3 account for spin–orbit interactions in wurtzite crystals. For the biaxial strain, the strain tensor has only the following nonvanishing diagonal elements:

$$\begin{aligned} \varepsilon_{xx}^{(2)} &= \varepsilon_{yy}^{(2)} = \frac{a_s^w - a_e^w}{a_e^w}, \\ \varepsilon_{zz}^{(2)} &= -\frac{2C_{13}^w}{C_{33}^w}\varepsilon_{xx}^{(2)}, \end{aligned} \tag{3.287}$$

where C_{13}^w and C_{33}^w are the stiffness constants for wurtzite structure.

3.5.4 Block Diagonalization of Zinc Blende and Wurtzite Hamiltonians

The full 6×6 Hamiltonian matrix can be block diagonalized using the basis transformation. The off-diagonal terms such as $H_{21}^{(n)}$ and $H_{23}^{(n)}$ contain the $\phi^{(n)}$ dependence, and we can write

$$\begin{aligned} H_{21}^{(n)} &= K_{21}^{(n)} e^{i2\phi^{(n)}}, \\ H_{23}^{(n)} &= K_{23}^{(n)} e^{i\phi^{(n)}}, \end{aligned} \tag{3.288}$$

where $k_x + ik_y = k_t \exp(i\phi^{(n)})$ for $n = 1, 2$. Here, the $K_{21}^{(n)}$ and $K_{23}^{(n)}$ matrix elements for two Hamiltonians are given by

$$\begin{aligned} K_{21}^{(1)} &= \frac{\hbar^2}{2m_o}\left[-\frac{3}{2}(\gamma_2+\gamma_3)\right]k_t^2, \\ K_{23}^{(1)} &= \frac{\hbar^2}{2m_o}\left[-3\sqrt{2}\gamma_3\right]k_t k_z, \end{aligned} \tag{3.289}$$

for the first Hamiltonian $H^{(001)}$, and

$$\begin{aligned} K_{21}^{(2)} &= \frac{\hbar^2}{2m_o}A_5 k_t^2, \\ K_{23}^{(2)} &= \frac{\hbar^2}{2m_o}A_6 k_t k_z. \end{aligned} \tag{3.290}$$

for the second Hamiltonian $H^{(0001)}$. We define the transformation matrix $T^{(n)}$ as

$$T^{(n)} = \begin{pmatrix} \alpha^{(n)*} & 0 & 0 & \alpha^{(n)} & 0 & 0 \\ 0 & \beta^{(n)} & 0 & 0 & \beta^{(n)*} & 0 \\ 0 & 0 & \beta^{(n)*} & 0 & 0 & \beta^{(n)} \\ \alpha^{(n)*} & 0 & 0 & -\alpha^{(n)} & 0 & 0 \\ 0 & \beta^{(n)} & 0 & 0 & -\beta^{(n)*} & 0 \\ 0 & 0 & -\beta^{(n)*} & 0 & 0 & \beta^{(n)} \end{pmatrix}, \tag{3.291}$$

where

$$\alpha^{(n)} = \frac{1}{\sqrt{2}} e^{i\left(\frac{3\pi}{4}+\frac{3\phi^{(n)}}{2}\right)}, \qquad \beta^{(n)} = \frac{1}{\sqrt{2}} e^{i\left(\frac{\pi}{4}+\frac{\phi^{(n)}}{2}\right)}. \tag{3.292}$$

The block-diagonalized Hamiltonian, $H_B^{(n)} = T^{(n)*} H T^{(n)t}$, is given by

$$H_B^{(n)} = \begin{pmatrix} H^{(n)U} & 0 \\ 0 & H^{(n)L} \end{pmatrix}, \tag{3.293}$$

where the upper 3×3 Hamiltonian is

$$H^{(n)U} = \begin{pmatrix} H_{11}^{(n)} & K_{21}^{(n)} & -iK_{23}^{(n)} \\ K_{21}^{(n)} & H_{22}^{(n)} & \Delta^{(n)} - iK_{23}^{(n)} \\ iK_{23}^{(n)} & \Delta^{(n)} + iK_{23}^{(n)} & H_{33}^{(n)} \end{pmatrix}, \tag{3.294}$$

and the lower 3×3 Hamiltonian is

$$H^{(n)L} = \begin{pmatrix} H_{11}^{(n)} & K_{21}^{(n)} & iK_{23}^{(n)} \\ K_{21}^{(n)} & H_{22}^{(n)} & \Delta^{(n)} + iK_{23}^{(n)} \\ -iK_{23}^{(n)} & \Delta^{(n)} - iK_{23}^{(n)} & H_{33}^{(n)} \end{pmatrix}, \tag{3.295}$$

where $n = 1$ and 2. The lower Hamiltonian is the complex conjugate of the upper Hamiltonian. Therefore, they have exactly the same eigenvalues, since the eigenenergies are real. For the wurtzite quantum well structure, there will be a lifting of Kramers degeneracy in the in-plane dispersion relations because of the internal field due to piezoelectric and spontaneous polarizations. The wave functions of the lower Hamiltonian are the complex conjugates of the corresponding wave functions of the upper Hamiltonian.

3.5.5 Analytical Solutions for Valence Band Energies and Wave Functions in Bulk Semiconductors

The eigenequation for the valence band energy $E^{(n)}$ at the band edge is given by

$$\begin{pmatrix} H_{11}^{(n)} - E^{(n)} & K_{21}^{(n)} & -iK_{23}^{(n)} \\ K_{21}^{(n)} & H_{22}^{(n)} - E^{(n)} & \Delta^{(n)} - iK_{23}^{(n)} \\ iK_{23}^{(n)} & \Delta^{(n)} + iK_{23} & H_{33}^{(n)} - E^{(n)} \end{pmatrix} \begin{pmatrix} g_1^{(n)} \\ g_2^{(n)} \\ g_3^{(n)} \end{pmatrix} = 0. \tag{3.296}$$

The eigenequation is a third-order polynomial equation that can always be solved analytically. The complete Bloch wave functions for the valence bands are given by

$$\Psi_v^{(n)}(\mathbf{r},\mathbf{k}) = \frac{1}{\sqrt{V}} e^{i\mathbf{k}\cdot\mathbf{r}} \sum_{\nu=1}^{3} g_\nu(\mathbf{k}) |\nu\rangle^{(n)}, \tag{3.297}$$

where

$$\begin{aligned} |1\rangle^{(n)} &= \alpha^{(n)*}|U_1\rangle + \alpha^{(n)}|U_4\rangle, \\ |2\rangle^{(n)} &= \beta^{(n)}|U_2\rangle + \beta^{(n)*}|U_5\rangle, \\ |3\rangle^{(n)} &= \beta^{(n)*}|U_3\rangle + \beta^{(n)}|U_6\rangle. \end{aligned} \tag{3.298}$$

Similar expressions hold for the lower Hamiltonian (Eq. (3.295)) with bases $|4\rangle^{(n)}, |5\rangle^{(n)}$, and $|6\rangle^{(n)}$.

The conduction bands can be characterized by a parabolic band model, and the effective mass Hamiltonians for the three crystal structures are

$$\begin{aligned} H_{\text{cond}}^{(1)}(\mathbf{k}) &= \frac{\hbar^2}{2m_e^c}(k_x^2 + k_y^2 + k_z^2) + E_g^c + a_c(\varepsilon_{xx}^{(1)} + \varepsilon_{yy}^{(1)} + \varepsilon_{zz}^{(1)}), \\ H_{\text{cond}}^{(2)}(\mathbf{k}') &= \frac{\hbar^2}{2}\left(\frac{k_x'^2 + k_y'^2}{m_{et}^w} + \frac{k_z'^2}{m_{ez}^w}\right) + \Delta_1 + \Delta_2 + E_g^w + a_{cz}^w \varepsilon_{zz}^{(2)} + a_{ct}^w(\varepsilon_{xx}^{(2)} + \varepsilon_{yy}^{(2)}), \end{aligned} \tag{3.299}$$

where m_e^c is the effective mass in the conduction band for zinc blende structure, m_{ez}^w and m_{et}^w are the longitudinal and transverse effective masses for wurtzite structure, a_c is the conduction band deformation potential for zinc blende structure, and a_{cz}^w and a_{ct}^w are the conduction band deformation potentials along the c-axis and perpendicular to the c-axis, respectively. The conduction band wave function is of the form

$$\Psi_{c\eta}^{(n)}(\mathbf{r},\mathbf{k}) = \frac{1}{\sqrt{V}} e^{i\mathbf{k}\cdot\mathbf{r}} |S\eta\rangle, \tag{3.300}$$

where η is the electron spin $\uparrow$ and $\downarrow$, and S is spherically symmetric wave function.

3.5.6 Interband Optical Momentum Matrix Elements in Bulk Semiconductors

The optical momentum matrix elements are given by

$$\left|\hat{\mathbf{e}}\cdot\mathbf{M}^{\eta(n)}\right|^2 = \left|\langle\Psi_{c\eta}^{(n)}|\hat{\mathbf{e}}\cdot\mathbf{p}|\Psi_{v}^{(n)}\rangle\right|^2. \tag{3.301}$$

Using the expressions given in Equations (3.297) and (3.300), and taking the $\phi^{(n)}$ integration of the momentum matrix elements, we obtain the following momentum matrix elements:

$$\text{TE-polarization}\,(\hat{\mathbf{e}} = \hat{\mathbf{x}}\;or\;\hat{\mathbf{y}}):\quad \left|M_x^{\eta(n)}\right|^2 = \frac{M_{px}^{(n)}}{4}\left(g_1^{(n)2} + g_2^{(n)2}\right) \tag{3.302}$$

$$\text{TM-polarization}\,(\hat{\mathbf{e}} = \hat{\mathbf{z}}):\quad \left|M_z^{\eta(n)}\right|^2 = \frac{M_{pz}^{(n)}}{2} g_3^{(n)2} \tag{3.303}$$

where

$$M_{px}^{(1)} = M_{pz}^{(1)} = \left|\langle S|p_x|X\rangle^{(1)}\right|^2 = \left|\langle S|p_z|Z\rangle^{(1)}\right|^2 = \frac{m_o}{2}\left(\frac{m_o}{m_e^c} - 1\right)\frac{(E_g^c + \Delta)E_g^c}{(E_g^c + 2\Delta/3)}, \tag{3.304}$$

and

$$M_{px}^{(2)} = \left|\langle S|p_x|X\rangle^{(2)}\right|^2 = \frac{m_o}{2}\left(\frac{m_o}{m_{et}^w} - 1\right)\frac{E_g[(E_g + \Delta_1 + \Delta_2)(E_g + 2\Delta_2) - 2\Delta_3^2]}{(E_g + \Delta_1 + \Delta_2)(E_g + \Delta_2) - \Delta_3^2},$$

$$M_{pz}^{(2)} = \left|\langle S|p_z|Z\rangle^{(3)}\right|^2 = \frac{m_o}{2}\left(\frac{m_o}{m_{ez}^w} - 1\right)\frac{(E_g + \Delta_1 + \Delta_2)(E_g + 2\Delta_2) - 2\Delta_3^2}{E_g + 2\Delta_2}. \tag{3.305}$$

Substituting $E^{(n)} = E_1^{(n)}$, $E_2^{(n)}$, and $E_3^{(n)}$ into the wave function components, $g_i, i = 1,2,3$, we can the obtain moment matrix element for the interband optical transitions between the conduction (C) band and the HH, LH, and SO or CH bands, respectively.

3.5.7 Interband Optical Momentum Matrix Elements in a Quantum Well Structure

For a quantum well structure, the band-edge discontinuities have to be added properly in the effective mass equation. The valence subband structure for the upper Hamiltonian (Eq. (3.294)) can be determined by solving

$$\sum_{j=1}^{3}\left(H_{ij}^{(n)U}\left(k_z=-i\frac{\partial}{\partial z}\right)+\delta_{ij}V_v^{(n)0}(z)\right)g_m^{(n)(j)}(z;k_t)=E_m^{(n)U}g_m^{(n)(i)}(z;k_t),\tag{3.306}$$

where $g_m^{(n)(i)}$, $i=1,2,\ldots,6$ are the envelope functions of the mth valence subbands, and $V_v^{(n)0}(z)$ is the QW potential profile of the unstrained valence band reference energy. If we choose the reference energy for the well to be zero, the reference energy for the barrier is given by the band-edge discontinuity $\Delta E_v^0=Q_v\Delta E_g$, where ΔE_g is the bandgap difference between the unstrained barrier and the well, and Q_v is the partition ratio. The hole wave function in a quantum well can be written as

$$\Psi_m^{(n)U}(z;k_t)=\frac{e^{i\mathbf{k}_t\cdot\mathbf{r}_t}}{\sqrt{A}}\left(g_m^{(n)(1)}(z;k_t)|1\rangle^{(n)}+g_m^{(n)(2)}(z;k_t)|2\rangle^{(n)}+g_m^{(n)(3)}(z;k_t)|3\rangle^{(n)}\right)\tag{3.307}$$

$$\Psi_m^{(n)L}(z;k_t)=\frac{e^{i\mathbf{k}_t\cdot\mathbf{r}_t}}{\sqrt{A}}\left(g_m^{(n)(4)}(z;k_t)|4\rangle^{(n)}+g_m^{(n)(5)}(z;k_t)|5\rangle^{(n)}+g_m^{(n)(6)}(z;k_t)|6\rangle^{(n)}\right)\tag{3.308}$$

where $g_m^{(n)(i)}$ follow the normalization rules

$$\begin{aligned}\sum_{i=1}^{3}\int dz\left|g_m^{(i)}(z;k_t)\right|^2&=1,\\ \sum_{i=4}^{6}\int dz\left|g_m^{(i)}(z;k_t)\right|^2&=1.\end{aligned}\tag{3.309}$$

Similarly, the valence band structure for the lower Hamiltonian, $E_m^{(n)L}$, can be obtained using $H_{ij}^{(n)L}+\delta_{ij}V_v^{(n)0}(z)$.

3.6 CRYSTAL ORIENTATION EFFECTS ON A ZINC BLENDE HAMILTONIAN

3.6.1 Zinc Blende Hamiltonian with General Crystal Orientation under Transformation to Wurtzite Bases

The Hamiltonian for an arbitrary crystal orientation [9–11] can be obtained using a rotation matrix

$$U = \begin{pmatrix} \cos\theta\cos\phi & \cos\theta\sin\phi & -\sin\theta \\ -\sin\phi & \cos\phi & 0 \\ \sin\theta\cos\phi & \sin\theta\sin\phi & \cos\theta \end{pmatrix}. \tag{3.310}$$

The rotation with the Euler angles θ and ϕ transforms physical quantities from the (x, y, z) coordinates to the (x', y', z') coordinates. Figure 3.1 shows configuration of coordinate systems in (hkl)-oriented QW structures. The relation between the coordinate systems for vectors and tensors is expressed as

$$\begin{aligned} k'_i &= U_{i\alpha}k_\alpha \quad \text{or} \quad k_\alpha = U_{i\alpha}k'_i, \\ \varepsilon'_{ij} &= U_{i\alpha}U_{j\beta}\varepsilon_{\alpha\beta} \quad \text{or} \quad \varepsilon_{\alpha\beta} = U_{i\alpha}U_{j\beta}\varepsilon'_{ij}, \\ C'_{ijkl} &= U_{i\alpha}U_{j\beta}U_{i\gamma}U_{j\delta}C_{\alpha\beta\gamma\delta} \quad \text{or} \quad C_{\alpha\beta\gamma\delta} = U_{i\alpha}U_{j\beta}U_{i\gamma}U_{j\delta}C'_{ijkl}, \end{aligned} \tag{3.311}$$

where $i, j, \alpha, \beta, \gamma, \delta = x, y, z$. The rotation matrix for the two spins is given by

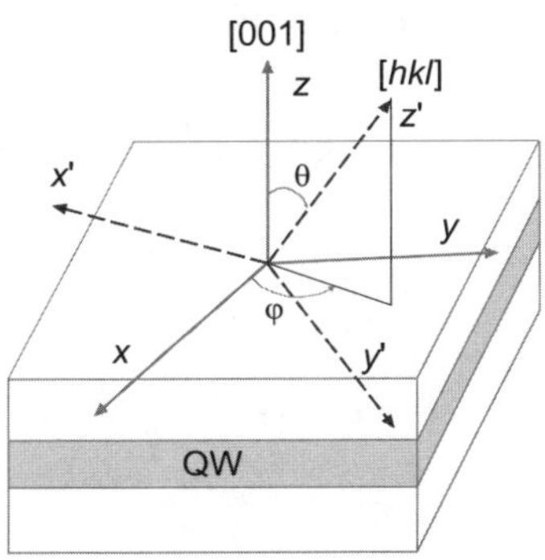

Figure 3.1. Configuration of coordinate systems in (hkl)-oriented QW structures.

$$M_s = \begin{pmatrix} e^{-i\frac{\phi}{2}}\cos\frac{\theta}{2} & e^{i\frac{\phi}{2}}\sin\frac{\theta}{2} \\ -e^{-i\frac{\phi}{2}}\sin\frac{\theta}{2} & e^{i\frac{\phi}{2}}\cos\frac{\theta}{2} \end{pmatrix}. \tag{3.312}$$

Combining the matrices U and M_s, we can obtain the matrix M_c that relates $(X'\uparrow', Y'\uparrow', Z'\uparrow', X'\downarrow', Y'\downarrow', Z'\downarrow')^t$ and $(X\uparrow, Y\uparrow, Z\uparrow, X\downarrow, Y\downarrow, Z\downarrow)^t$:

$$M_c = \begin{pmatrix}
\cos\phi\cos\theta\cos\frac{\theta}{2}e^{-i\frac{\phi}{2}} & \sin\phi\cos\theta\cos\frac{\theta}{2}e^{-i\frac{\phi}{2}} & -\sin\theta\cos\frac{\theta}{2}e^{-i\frac{\phi}{2}} & \cos\phi\cos\theta\sin\frac{\theta}{2}e^{i\frac{\phi}{2}} & \sin\phi\cos\theta\sin\frac{\theta}{2}e^{i\frac{\phi}{2}} & -\sin\theta\sin\frac{\theta}{2}e^{i\frac{\phi}{2}} \\
-\sin\phi\cos\frac{\theta}{2}e^{-i\frac{\phi}{2}} & \cos\phi\cos\frac{\theta}{2}e^{-i\frac{\phi}{2}} & 0 & -\sin\phi\sin\frac{\theta}{2}e^{i\frac{\phi}{2}} & \cos\phi\sin\frac{\theta}{2}e^{i\frac{\phi}{2}} & 0 \\
\cos\phi\sin\theta\cos\frac{\theta}{2}e^{-i\frac{\phi}{2}} & \sin\phi\sin\theta\cos\frac{\theta}{2}e^{-i\frac{\phi}{2}} & \cos\theta\cos\frac{\theta}{2}e^{-i\frac{\phi}{2}} & \cos\phi\sin\theta\sin\frac{\theta}{2}e^{i\frac{\phi}{2}} & \sin\phi\sin\theta\sin\frac{\theta}{2}e^{i\frac{\phi}{2}} & \cos\theta\sin\frac{\theta}{2}e^{i\frac{\phi}{2}} \\
-\cos\phi\cos\theta\sin\frac{\theta}{2}e^{-i\frac{\phi}{2}} & -\sin\phi\cos\theta\sin\frac{\theta}{2}e^{-i\frac{\phi}{2}} & \sin\theta\sin\frac{\theta}{2}e^{-i\frac{\phi}{2}} & \cos\phi\cos\theta\cos\frac{\theta}{2}e^{i\frac{\phi}{2}} & \sin\phi\cos\theta\cos\frac{\theta}{2}e^{i\frac{\phi}{2}} & -\sin\theta\cos\frac{\theta}{2}e^{i\frac{\phi}{2}} \\
\sin\phi\sin\frac{\theta}{2}e^{-i\frac{\phi}{2}} & -\cos\phi\sin\frac{\theta}{2}e^{-i\frac{\phi}{2}} & 0 & -\sin\phi\cos\frac{\theta}{2}e^{i\frac{\phi}{2}} & \cos\phi\cos\frac{\theta}{2}e^{i\frac{\phi}{2}} & 0 \\
-\cos\phi\sin\theta\sin\frac{\theta}{2}e^{-i\frac{\phi}{2}} & -\sin\phi\sin\theta\sin\frac{\theta}{2}e^{-i\frac{\phi}{2}} & -\cos\theta\sin\frac{\theta}{2}e^{-i\frac{\phi}{2}} & \cos\phi\sin\theta\cos\frac{\theta}{2}e^{i\frac{\phi}{2}} & -\sin\phi\sin\theta\cos\frac{\theta}{2}e^{i\frac{\phi}{2}} & \cos\theta\cos\frac{\theta}{2}e^{i\frac{\phi}{2}},
\end{pmatrix} \tag{3.313}$$

where the superscript t means taking the transpose of the vector or matrix.

The zinc blende Hamiltonian in the wurtzite bases for an arbitrary orientation can be obtained by the relation

$$H^{(lmn)} = M^{*}_{\text{total}} H^{(001)}_{\text{zinc blende}} \tilde{M}_{\text{total}}, \tag{3.314}$$

where

$$M_{\text{total}} = M_{wz} M_c M_{zb}^{-1}, \tag{3.315}$$

using matrices M_{wz} and M_{zb} defined in Equations (3.277) and (3.280). If we want to get the Hamiltonian with the zinc blende bases given by Equation (3.263), $M_{total} = M_{wz} M_c M_{zb}^{-1}$ should be replaced with $M_{\text{total}} = M_{zb} M_c M_{zb}^{-1}$. The above method to obtain the Hamiltonian for general crystal orientation is the most general and straightforward. However, the expression of Hamiltonian elements is not usually simplified. Thus, an invariant method to give the same result as the above is often used for general crystal orientation. Below, we first treat 4×4 Hamiltonian with a strain effect, and then extend the result to the 6×6 Hamiltonian.

3.6.2 Invariant Method to Obtain Zinc Blende Hamiltonians for General Crystal Orientation

The 4×4 Hamiltonian first proposed by Luttinger to describe the interaction between the heavy-hole and light-hole bands along with the Bir–Pikus Hamiltonian for strain can be formally written as

$$H_t^v = H_0^v(\mathbf{k}) + H_s^v, \tag{3.316}$$

with $H_0^v(\mathbf{k})$ representing the valence band Hamiltonian, and H_s^v the valence band strain or Bir–Pikus Hamiltonian.

The Luttinger formulation of the most general Hamiltonian for the (001) crystal orientation is given by

$$\begin{aligned}
H_o^v &= \frac{\hbar^2}{2m_o}(\gamma_1 k^2 I_4 - 2\gamma_2[(J_x^2 - 1/3J^2)k_x^2 + (J_y^2 - 1/3J^2)k_y^2 + (J_z^2 - 1/3J^2)k_z^2] \\
&\quad - 4\gamma_3[\{J_xJ_y\}\{k_xk_y\} + \{J_yJ_z\}\{k_yk_z\} + \{J_zJ_x\}\{k_zk_x\}]), \\
H_s^v &= -a_v(\varepsilon_{xx} + \varepsilon_{yy} + \varepsilon_{zz})I_4 + b[(J_x^2 - 1/3J^2)\varepsilon_{xx} + (J_y^2 - 1/3J^2)\varepsilon_{yy} \\
&\quad + (J_z^2 - 1/3J^2)\varepsilon_{zz}] + 2d/\sqrt{3}[\{J_xJ_y\}\{\varepsilon_{xy}\} + \{J_yJ_z\}\{\varepsilon_{yz}\} + \{J_zJ_x\}\{\varepsilon_{zx}\}]),
\end{aligned} \tag{3.317}$$

where the J_is are the angular momentum matrices for a state with spin 3/2, I_4 is a 4×4 identity matrix, and $\{k_xk_y\} = (k_xk_y + k_yk_x)/2$. Here,

$k_i(i=x,y,z)$ is the wave vector component in the unprimed (x,y,z) coordinate system, γ_i (i = 1, 2, 3) are Luttinger parameters, and a_v, b, and d are deformation potentials in the valence band. Note that the valence-band strain Hamiltonian H_s^v was obtained by replacing k_ik_j in H_o^v by ε_{ij}, with γ_1, γ_2, and γ_3 being replaced by $-a_v$, $-b/2$, and $-d/(2\sqrt{3})$, respectively.

The angular momentum matrices are any three matrices, (J_xJ_y,J_z), which satisfy the rules of commutation for angular momentum. These matrices given by Luttinger are as follows:

$$J_x=\begin{pmatrix}0 & 0 & \sqrt{3}/2 & 0\\ 0 & 0 & 1 & \sqrt{3}/2\\ \sqrt{3}/2 & 1 & 0 & 0\\ 0 & \sqrt{3}/2 & 0 & 0\end{pmatrix}, J_y=\begin{pmatrix}0 & 0 & -i\sqrt{3}/2 & 0\\ 0 & 0 & i & -i\sqrt{3}/2\\ i\sqrt{3}/2 & -i & 0 & 0\\ 0 & i\sqrt{3}/2 & 0 & 0\end{pmatrix},$$

$$J_z=\begin{pmatrix}3/2 & 0 & 0 & 0\\ 0 & -1/2 & 0 & 0\\ 0 & 0 & 1/2 & 0\\ 0 & 0 & 0 & -3/2\end{pmatrix}, I_4=\begin{pmatrix}1 & 0 & 0 & 0\\ 0 & 1 & 0 & 0\\ 0 & 0 & 1 & 0\\ 0 & 0 & 0 & 1\end{pmatrix}. \tag{3.318}$$

We can obtain the (001)-oriented Hamiltonian substituting Equation (3.318) into Equation (3.317):

$$H_0^v=\begin{pmatrix}P+Q & R & -S & 0\\ R^* & P-Q & 0 & S^*\\ -S^* & 0 & P-Q & R\\ 0 & S^* & R^* & P+Q\end{pmatrix}. \tag{3.319}$$

This is exactly the same form as the 4×4 Hamiltonian part given in Equation (3.263). Note that the ordering of the basis functions for the choice of J_is is $|\frac{3}{2},\frac{3}{2}\rangle$, $|\frac{3}{2},-\frac{1}{2}\rangle$, $|\frac{3}{2},\frac{1}{2}\rangle$, and $|\frac{3}{2},-\frac{3}{2}\rangle$. The z'-axis, or, equivalently, the (lmn)-axis, is the direction normal to the growth surface. Using the relation $k_\alpha=U_{i\alpha}k_i'$ in Equation (3.311), the most general valence band Hamiltonian obtained from these considerations can be written as

$$P = -\frac{\hbar^2}{2m_o}\gamma_1\left[k_{x'}^2 + k_{y'}^2 + k_{z'}^2\right] \tag{3.320}$$

$$\begin{aligned}
Q = -\frac{\hbar^2}{2m_o}\Bigg\{&\Bigg[\gamma_2\left(\frac{\eta^6 - 4\eta^6 n^2 - 6k^4 n^2 + \eta^2(6k^2 n^2 + n^4)}{\eta^2(\eta^2+n^2)^2}\right)\\
&+ 6\gamma_3\left(\frac{(\eta^4 - \eta^2 k^2 + k^4)n^2}{\eta^2(\eta^2+n^2)^2}\right)\Bigg]k_{x'}^2\\
&+\left[\gamma_1 + \gamma_2\left(\frac{\eta^4 + 6k^4 + \eta^2(-6k^2 + n^2)}{\eta^2(\eta^2+n^2)^2}\right) + 6\gamma_3\left(\frac{(\eta^2 - k^2)k^2}{\eta^2(\eta^2+n^2)^2}\right)\right]k_{y'}^2\\
&+\left[\gamma_1 - 2\gamma_2\left(\frac{\eta^4 + 3k^4 + n^4 - \eta^2(3k^2 + n^2)}{(\eta^2+n^2)^2}\right) + 6\gamma_3\left(\frac{k^4 - \eta^2(k^2 + n^2)}{(\eta^2+n^2)^2}\right)\right]k_{z'}^2\\
&+\left[\left(\frac{6kn(\eta^2 - 2k^2)\sqrt{\eta^2 - k^2}}{\eta^2(\eta^2+n^2)^{3/2}}\right)(\gamma_2 - \gamma_3)\right]k_{x'}k_{y'}\\
&+\left[\left(\frac{6\eta k(\eta^2 - 2k^2)\sqrt{\eta^2 - k^2}}{\eta^2(\eta^2+n^2)^{3/2}}\right)(\gamma_2 - \gamma_3)\right]k_{y'}k_{z'}\\
&+\left[\left(\frac{6\eta n(\eta^4 + 2k^4 - \eta^2(2k^2 + n^2))}{\eta^2(\eta^2+n^2)^2}\right)(\gamma_3 - \gamma_2)\right]k_{x'}k_{z'}\Bigg\}
\end{aligned} \tag{3.321}$$

$$\begin{aligned}
R = \Bigg[&-\sqrt{3}\gamma_2\Bigg(\frac{\eta^6 + 4k^4 n^4 + 2k^2 n^2(k^2 - 2m^2) + \eta^2(-2k^2 n^2 + n^4)}{\eta^2(\eta^2+n^2)^2}\\
&+ i\frac{2k(\eta^2 - 2k^2)n^3\sqrt{\eta^2 - k^2}\sqrt{\eta^2 + n^2}}{\eta^4(\eta^2+n^2)^2}\Bigg)\\
&- 2\sqrt{3}n^2\gamma_3\Bigg(\frac{\eta^6 + \eta^4 k^2 - \eta^2(k^4 - 2k^2 n^2) - 2k^4 n^2}{\eta^4(\eta^2+n^2)^2}\\
&- i\frac{km\sqrt{\eta^2 - k^2}\sqrt{\eta^2 + n^2}(\eta^2 - 2k^2)}{\eta^4(\eta^2+n^2)^2}\Bigg)\Bigg]k_{x'}^2\\
&+\Bigg[\gamma_2\Bigg(\frac{\eta^6 + 2\eta^2(\eta^4 - 3\eta^2 k^2 + 3k^4) + 3n(\eta^2 - 2k^2)(\eta^2 n - 2k^2 n)}{\sqrt{3}\eta^4(\eta^2+n^2)}\\
&+ i\frac{6kn(\eta^2 - 2k^2)\sqrt{\eta^2 - k^2}\sqrt{\eta^2 + n^2}}{\sqrt{3}\eta^4(\eta^2+n^2)}\Bigg)
\end{aligned}$$

$$
\begin{aligned}
&+2\sqrt{3}k\sqrt{\eta^2-k^2}\gamma_3\left(\frac{\eta^2k\sqrt{\eta^2-k^2}+2k\sqrt{\eta^2-k^2}n^2}{\eta^4(\eta^2+n^2)}\right. \\
&\left.\left.-i\frac{\eta^2n\sqrt{\eta^2+n^2}-2k^2n\sqrt{\eta^2+n^2}}{\eta^4(\eta^2+n^2)}\right)\right]k_{y'}^2 \\
&+\left[-2\sqrt{3}(\gamma_2-\gamma_3)\left(\frac{2k^4n^2+\eta^4(-k^2+n^2)+\eta^2(k^4-2k^2n^2)}{\eta^2(\eta^2+n^2)^2}\right.\right. \\
&\left.\left.+i\frac{kn(\eta^2-2k^2)\sqrt{\eta^2-k^2}\sqrt{\eta^2+n^2}}{\eta^2(\eta^2+n^2)^2}\right)\right]k_{z'}^2 \\
&+\left[\frac{2\sqrt{3}kn\sqrt{\eta^2-k^2}}{\eta^4(\eta^2+n^2)^2}\gamma_2\Big((\eta^2-2k^2)\sqrt{\eta^2+n^2}(\eta^2+2n^2)\right. \\
&+i(4nk(\eta^2-2k^2)\sqrt{\eta^2-k^2})\Big) \\
&+\frac{2\sqrt{3}}{\eta^4(\eta^2+n^2)^2}\gamma_3\big(kn\sqrt{\eta^2+n^2}\sqrt{\eta^2-k^2}(\eta^2+2n^2)(-\eta^2+2k^2) \\
&\left.+i(\eta^8+2\eta^6n^2+\eta^4n^2(n^2-4k^2)+4\eta^2n^2k^2(k^2-n^2)+4k^4n^4)\big)\right]k_{x'}k_{y'} \\
&+\left[\frac{2\sqrt{3}k\eta}{\eta^4(\eta^2+n^2)^2}\gamma_2\Big((\eta^2-2k^2)\sqrt{\eta^2+n^2}(\eta^2+2n^2)\right. \\
&+i(4nk(\eta^2+n^2)\sqrt{\eta^2-k^2})\Big) \\
&+\frac{2\sqrt{3}k\eta}{\eta^4(\eta^2+n^2)^2}\gamma_3\Big(-\sqrt{\eta^2+n^2}\sqrt{\eta^2-k^2}(\eta^2-2k^2)(\eta^2+2n^2) \\
&\left.+i4k(-\eta^2+k^2)(\eta^2+n^2)\Big)\right]k_{y'}k_{z'} \\
&+\left[\frac{2n\eta}{\sqrt{3}\eta^4(\eta^2+n^2)^2}\gamma_2\Big((3\eta^2(\eta^4+2\eta^2k^2-2k^2)-3(\eta^2-2k^2)^2n^2)\right. \\
&-i(6nk(\eta^2-2k^2)\sqrt{\eta^2-k^2}\sqrt{\eta^2+n^2})\Big) \\
&+\frac{2\sqrt{3}n\eta}{\eta^4(\eta^2+n^2)^2}\gamma_3\Big(\eta^6-4k^4n^2+\eta^4(2k^2-n^2)-2\eta^2(k^4-2k^2n^2) \\
&\left.+i2kn(-\eta^2+2k^2)\sqrt{\eta^2-k^2}\sqrt{\eta^2+n^2}\Big)\right]k_{x'}k_{z'}
\end{aligned}
\tag{3.322}
$$

$$
\begin{aligned}
-S = &\left[\frac{2\sqrt{3}n}{\eta^3(\eta^2+n^2)^2}(\gamma_2-\gamma_3)(\eta^6-\eta^4n^2+2\eta^2k^2n^2-2k^4n^2\right.\\
&\left.-ikn(\eta^2-2k^2)\sqrt{\eta^2-k^2}\sqrt{\eta^2+n^2})\right]k_{x'}^2\\
&+\left[\frac{2\sqrt{3}k}{\eta^3(\eta^2+n^2)}(\gamma_2-\gamma_3)(2k(-\eta^2+k^2)\right.\\
&\left.+i(\eta^2-2k^2)\sqrt{\eta^2-k^2}\sqrt{\eta^2+n^2})\right]k_{y'}^2\\
&+\left[\frac{2\sqrt{3}}{\eta(\eta^2+n^2)^2}(\gamma_2-\gamma_3)(n(-\eta^4-2k^2+\eta^2(2k^2+n^2))\right.\\
&\left.-ik(\eta^2-2k^2)\sqrt{\eta^2-k^2}\sqrt{\eta^2+n^2})\right]k_{z'}^2\\
&+\left[\frac{4\sqrt{3}km}{\eta^3(\eta^2+n^2)^2}(\gamma_2-\gamma_3)\Big(n(\eta^2-2k^2)\sqrt{\eta^2-k^2}\sqrt{\eta^2+n^2}\right.\\
&\left.+i2k(\eta^2-k^2)(\eta^2+n^2)\Big)\right]k_{x'}k_{y'}\\
&-\left[\frac{2\sqrt{3}km}{\eta^2(\eta^2+n^2)^2}\Big(+2(\gamma_2-\gamma_3)k(\eta^2-2k^2)\sqrt{\eta^2-k^2}\sqrt{\eta^2+n^2}\right.\\
&\left.+i(\eta^2+n^2)[\eta^4\gamma_3-4(\gamma_2-\gamma_3)k^4+4\eta^2(\gamma_2-\gamma_3)+\eta^2\gamma_3n^2]\Big)\right]k_{y'}k_{z'}\\
&+\left[\frac{2\sqrt{3}}{\eta^2(\eta^2+n^2)^2}((\gamma_2-\gamma_3)(2\eta^4n^2+4k^4n^2-4\eta^2k^2n^2)+\gamma_3(\eta^6+\eta^2n^4)\right.\\
&\left.+\gamma_2 2\eta^4n^2+i2(\gamma_2-\gamma_3)kn(\eta^2-2k^2)\sqrt{\eta^2-k^2}\sqrt{\eta^2+n^2})\right]k_{y'}k_{z'},
\end{aligned}
\tag{3.323}
$$

where $\eta^2 = h^2 + n^2$. Here, we can write the crystalline surface in terms of the azimuthal angle ϕ and the polar angle θ as follows:

$$
\tan\phi = \frac{k}{h} \quad \text{and} \quad \tan\theta = \frac{\sqrt{h^2+k^2}}{h}. \tag{3.324}
$$

Next, we consider the specific case of the $(1ln)$ surface where the angle ϕ is fixed to be $\pi/4$.

3.6.3 (11n)-Oriented Zinc Blende Hamiltonian

The relation between the coordinate systems for vectors and tensors is given by Equation (3.311). Our object is to obtain P', Q', R', and S' in (x',y',z') coordinates from the Hamiltonian in Equation (3.319), which can be done by substituting Equation (3.311) into Equation (3.317). Here, we consider the case of quantum wells with $(11n)$ orientations with $\theta = \arctan(\sqrt{2}/n)$ and $\phi = \pi/4$. Then, from Equation (3.320) to (3.323),

$$
\begin{aligned}
P_{k'} &= \frac{\hbar^2}{2m_o}\gamma_1\left[k_x'^2 + k_y'^2 + k_z'^2\right],\\
Q_{k'} &= \frac{\hbar^2}{2m_o}\gamma_1\left\{\left[\frac{(n^2-4)(n^2-1)}{(n^2+2)^2}\gamma_2 + \frac{9n^2}{(n^2+2)^2}\gamma_3\right]k_x'^2\right.\\
&\quad + \left[\frac{(n^2+2)(n^2-1)}{(n^2+2)^2}\gamma_2 + \frac{3}{(n^2+2)}\gamma_3\right]k_y'^2\\
&\quad \left. + \left[-2\frac{(n^2-1)^2}{(n^2+2)^2}\gamma_2 - \frac{6+12n^2}{(n^2+2)^2}\gamma_3\right]k_z'^2\right\},\\
R_{k'} &= \frac{\hbar^2}{2m_o}\left\{\sqrt{3}\left[\frac{(n^2-4)}{(n^2+2)^2}\gamma_2 - \frac{5n^2+n^4}{(n^2+2)^2}\gamma_3\right]k_x'^2\right.\\
&\quad + \sqrt{3}\left[\frac{1}{(n^2+2)}\gamma_2 + \frac{1+n^2}{(n^2+2)^2}\gamma_3\right]k_y'^2\\
&\quad + 2\sqrt{3}\left[-\frac{(n^2-1)}{(n^2+2)^2}(\gamma_2-\gamma_3)\right]k_z'^2 + i2\sqrt{3}\frac{n^2\gamma_2+2\gamma_3}{(n^2+2)}k_x'k_y' \qquad (3.325)\\
&\quad \left. + 2\sqrt{6}\left[i\frac{n}{(n^2+2)}(\gamma_2-\gamma_3)\right]k_y'k_z' + 6\sqrt{6}\left[\frac{n}{(n^2+2)^2}(\gamma_2-\gamma_3)\right]k_x'k_z'\right\},\\
-S_{k'} &= \frac{\hbar^2}{2m_o}\left\{\left[\sqrt{6}\frac{n(4-n^2)}{(n^2+2)^2}(\gamma_2-\gamma_3)\right]k_x'^2 + \left[\sqrt{6}\frac{n}{(n^2+2)}(\gamma_3-\gamma_2)\right]k_y'^2\right.\\
&\quad + \left[2\sqrt{6}\frac{n(n^2-1)}{(n^2+2)}(\gamma_2-\gamma_3)\right]k_z'^2 + i2\sqrt{6}\frac{n}{(n^2+2)}(\gamma_2-\gamma_3)k_x'k_y'\\
&\quad + i2\sqrt{3}\left[\frac{2\gamma_2+n^2\gamma_3}{(n^2+2)}\right]k_y'k_z'\\
&\quad \left. + \sqrt{3}\left[-\frac{12n^2}{(n^2+2)^2}\gamma_2 - 2\frac{(4-2n^2+n^4)}{(n^2+2)^2}\gamma_3\right]k_x'k_z'\right\},
\end{aligned}
$$

$$
\begin{aligned}
P_{\varepsilon'} &= -a_v(\varepsilon'_{xx}+\varepsilon'_{yy}+\varepsilon'_{zz}),\\
Q_{\varepsilon'} &= -\frac{\begin{array}{c} b(n^2-1)(-\Gamma+3\sqrt{2}\varepsilon'_{xz}n+n^2\Gamma)\\ +d\sqrt{3}[\Gamma+\sqrt{2}\varepsilon'_{xz}n+n^2(2\Gamma-\sqrt{2}\varepsilon'_{xz}n)]\end{array}}{(2+n^2)^2},\\
R_{\varepsilon'} &= -\frac{b\sqrt{3}(-\Gamma+3\sqrt{2}\varepsilon'_{xz}n+n^2\Gamma)+d[\Gamma-3\sqrt{2}\varepsilon'_{xz}n-n^2\Gamma)}{(2+n^2)^2},\\
-S_{\varepsilon'} &= -\frac{\begin{array}{c} bn\sqrt{3}(\sqrt{2}\Gamma-6n\varepsilon'_{xz}-\sqrt{2}n^2\Gamma)\\ +d[\sqrt{2}n(n^2-1)\Gamma+\varepsilon'_{xz}(-4+2n^2-n^4)\end{array}}{(2+n^2)^2},
\end{aligned}
\tag{3.326}
$$

where $\Gamma = \varepsilon_{\|} - \varepsilon'_{zz}$, $\varepsilon_{\|} = (a_s^c - a_e^c)/a_e^c$. The biaxial strain components for a general crystal orientation are determined from the condition that the layer is grown pseudomorphically and these strain coefficients should minimize the strain energy of the layer simultaneously, as discussed in Section 3.6.5.

It is straightforward to obtain the 6×6 Hamiltonian of the (11n)-oriented zinc blende crystal by using components P', Q', R', and S' in (x', y', z') coordinates. That is, the 6×6 Hamiltonian for the valence-band structure in (x', y', z') coordinates can be written as

$$
H' = -\begin{pmatrix}
P'+Q' & -S' & R' & 0 & -S'/\sqrt{2} & \sqrt{2}R' \\
-S'^* & P'-Q' & 0 & R' & -\sqrt{2}Q' & \sqrt{3/2}S' \\
R'^* & 0 & P'-Q' & S' & \sqrt{3/2}S'^* & \sqrt{2}Q' \\
0 & R'^* & S'^* & P'+Q' & -\sqrt{2}R'^* & -S'^*/\sqrt{2} \\
-S'^*/\sqrt{2} & -\sqrt{2}Q' & \sqrt{3/2}S' & -\sqrt{2}R' & P'+\Delta & 0 \\
\sqrt{2}R'^* & \sqrt{3/2}S'^* & \sqrt{2}Q' & -S'/\sqrt{2} & 0 & P'+\Delta
\end{pmatrix}
\begin{matrix} |u_1\rangle' \\ |u_2\rangle' \\ |u_3\rangle' \\ |u_4\rangle', \\ |u_5\rangle' \\ |u_6\rangle', \end{matrix}
\tag{3.327}
$$

where

$$P' = P_{k'} + P_{\varepsilon'}, Q' = Q_{k'} + Q_{\varepsilon'}, R' = R_{k'} + R_{\varepsilon'}, S' = S_{k'} + S_{\varepsilon'}.$$

Here, the basis set used in (x', y', z') coordinates consists of the following basis functions:

$$
\begin{aligned}
|u_1\rangle' &= \left|\frac{3}{2},\frac{3}{2}\right\rangle' = -\frac{1}{\sqrt{2}}|(X'+iY')\uparrow'\rangle \\
|u_2\rangle' &= \left|\frac{3}{2},\frac{1}{2}\right\rangle' = \frac{1}{\sqrt{6}}|-(X'+iY')\downarrow' +2Z'\uparrow'\rangle \\
|u_3\rangle' &= \left|\frac{3}{2},-\frac{1}{2}\right\rangle' = \frac{1}{\sqrt{6}}|(X'-iY')\uparrow' +2Z'\downarrow'\rangle \\
|u_4\rangle' &= \left|\frac{3}{2},-\frac{3}{2}\right\rangle' = \frac{1}{\sqrt{2}}|(X'-iY')\downarrow'\rangle \\
|u_5\rangle' &= \left|\frac{1}{2},\frac{1}{2}\right\rangle' = \frac{1}{\sqrt{3}}|(X'+iY')\downarrow' +Z'\uparrow'\rangle \\
|u_6\rangle' &= \left|\frac{1}{2},-\frac{1}{2}\right\rangle' = \frac{1}{\sqrt{3}}|(X'-iY')\uparrow' -Z'\downarrow'\rangle.
\end{aligned}
\tag{3.328}
$$

3.6.4 Interband Optical-Matrix Elements in a (11n)-Oriented Zinc Blende Quantum Well

The hole wave function in a quantum well can be written as

$$
\Psi_m(z;k_t) = \frac{e^{i\mathbf{k}_t\cdot\mathbf{r}_t}}{\sqrt{A}}\sum_{\nu=1}^{6} g_m'^{(\nu)}(z;k_t)|u_\nu\rangle', \tag{3.329}
$$

where $g_m'^{(\nu)}$ ($\nu = 1,2,3,4,5,$ and 6) is the wave function for the mth subband in (x',y',z') coordinates, $|u_\nu\rangle'$ is given by Equation (3.328), and $g_m'^{(i)}$ follows the normalization rules

$$
\sum_{\nu=1}^{6}\int dz\left|g_m'^{(\nu)}(z;k_t)\right|^2 = 1. \tag{3.330}
$$

Since we solve one Schrödinger equation for electrons and a 6×6 Hamiltonian for holes, we need only the optical matrix elements for two cases: spin up and spin down. The polarization-dependent interband momentum matrix elements are written as TE-polarization ($\hat{\mathbf{e}}' = \cos\phi'\hat{\mathbf{x}}' + \sin\phi'\hat{\mathbf{y}}'$):

$$
\begin{aligned}
\left|\hat{\mathbf{e}}'\cdot\mathbf{M}'^{\uparrow}\right|^2 = \Bigg|&\cos\phi'\left\{-\frac{1}{\sqrt{2}}P_x\langle g_m'^{(1)}|\phi_l\rangle+\frac{1}{\sqrt{6}}P_x\langle g_m'^{(3)}|\phi_l\rangle+\frac{1}{\sqrt{3}}P_x\langle g_m'^{(6)}|\phi_l\rangle\right\}\\
&+\sin\phi'\left\{-i\frac{1}{\sqrt{2}}P_x\langle g_m'^{(1)}|\phi_l\rangle-i\frac{1}{\sqrt{6}}P_x\langle g_m'^{(3)}|\phi_l\rangle\right.\\
&\left.-i\frac{1}{\sqrt{3}}P_x\langle g_m'^{(6)}|\phi_l\rangle\right\}\Bigg|^2
\end{aligned}
\tag{3.331}
$$

$$
\begin{aligned}
\left|\hat{\mathbf{e}}'\cdot\mathbf{M}'^{\downarrow}\right|^2 = \Bigg|&\cos\phi'\left\{-\frac{1}{\sqrt{6}}P_x\langle g_m'^{(2)}|\phi_l\rangle+\frac{1}{\sqrt{2}}P_x\langle g_m'^{(4)}|\phi_l\rangle+\frac{1}{\sqrt{3}}P_x\langle g_m'^{(5)}|\phi_l\rangle\right\}\\
&+\sin\phi'\left\{-i\frac{1}{\sqrt{6}}P_x\langle g_m'^{(2)}|\phi_l\rangle-i\frac{1}{\sqrt{2}}P_x\langle g_m'^{(4)}|\phi_l\rangle\right.\\
&\left.+i\frac{1}{\sqrt{3}}P_x\langle g_m'^{(5)}|\phi_l\rangle\right\}\Bigg|^2
\end{aligned}
\tag{3.332}
$$

TM-polarization ($\hat{\mathbf{e}}'=\hat{\mathbf{z}}'$):

$$\left|\hat{\mathbf{e}}'\cdot\mathbf{M}'^{\uparrow}\right|^2=\left|-\frac{2}{\sqrt{6}}P_z\langle g_m'^{(2)}|\phi_l\rangle+\frac{1}{\sqrt{3}}P_z\langle g_m'^{(5)}|\phi_l\rangle\right|^2 \tag{3.333}$$

$$\left|\hat{\mathbf{e}}'\cdot\mathbf{M}'^{\downarrow}\right|^2=\left|\frac{2}{\sqrt{6}}P_x\langle g_m'^{(3)}|\phi_l\rangle-\frac{1}{\sqrt{3}}P_x\langle g_m'^{(6)}|\phi_l\rangle\right|^2. \tag{3.334}$$

3.6.5 Strain Tensors on Zinc Blende Semiconductors for a General Crystal Orientation

3.6.5.1 Derivation from Hooke's Law In general, the in-plane strain tensor components are known to be $\varepsilon'_{11}=\varepsilon'_{22}=\varepsilon'_{\parallel}$ for biaxial tensile or compressive strain, and $\varepsilon'_{12}=\varepsilon'_{21}=0$ due to no in-plane shear. The other three independent stain tensor elements $(\varepsilon'_{13}=\varepsilon'_{23}=\varepsilon'_{33})$ remain to be determined. The substrate applies a uniform in-plane stress to the film. This is the only external stress applied to the film. Thus, $T'_{33}=T'_{23}=T'_{13}=0$. On the other hand, T'_{11}, T'_{22}, and T'_{12} should be determined. Also, the stress and strain tensor components are related to Hooke's law within the linear elastic domain:

$$T'_{\alpha\beta}+\sum_{i,j=1}^{3}C'_{\alpha\beta ij}\varepsilon'_{ij}=0 \tag{3.335}$$

for $\alpha,\beta=1,2,3$. For $(\alpha\beta)=(33)$, (23), and (13), $T'_{\alpha\beta}=0$, as given above, and Hooke's law gives

$$\sum_{i,j=1}^{3} C'_{\alpha\beta ij}\varepsilon'_{ij} = 0, (\alpha\beta) = (33),(23),(13). \tag{3.336}$$

Using the symmetry of $C'_{\alpha\beta ij}$ and ε'_{ij},

$$\begin{aligned}
&C'_{\alpha\beta 11}\varepsilon'_{11} + C'_{\alpha\beta 22}\varepsilon'_{22} + C'_{\alpha\beta 33}\varepsilon'_{33} + 2C'_{\alpha\beta 12}\varepsilon'_{12} + 2C'_{\alpha\beta 31}\varepsilon'_{31} + 2C'_{\alpha\beta 23}\varepsilon'_{23} = 0,\\
&C'_{\alpha\beta 33}\varepsilon'_{33} + 2C'_{\alpha\beta 31}\varepsilon'_{31} + 2C'_{\alpha\beta 23}\varepsilon'_{23} = -(C'_{\alpha\beta 11} + C'_{\alpha\beta 22})\varepsilon'_{\parallel},\\
&(\alpha\beta) = (33),(23),(13).
\end{aligned} \tag{3.337}$$

This can be written as a matrix equation:

$$\begin{pmatrix} C'_{3333} & C'_{3323} & C'_{3331} \\ C'_{2333} & C'_{2323} & C'_{2331} \\ C'_{3133} & C'_{3123} & C'_{3131} \end{pmatrix} \begin{pmatrix} \varepsilon'_{33}/2 \\ \varepsilon'_{23} \\ \varepsilon'_{31} \end{pmatrix} = -\frac{\varepsilon'_{\parallel}}{2} \begin{pmatrix} C'_{3311} + C'_{3322} \\ C'_{2311} + C'_{2322} \\ C'_{3111} + C'_{3122} \end{pmatrix}. \tag{3.338}$$

The elastic stiffness tensor of a material in the cubic crystal class $O_h(m3m)$ has only three distinct components, which are c_{11}, c_{12}, and c_{44}, in the standard contracted index notation. That is, the first two and the last two subscripts in c_{ijkl} are replaced by a single subscript running from 1 to 6 as follows:

$$\begin{array}{lcccccc} \text{tensor notation} & 11 & 22 & 33 & 23,32 & 31,13 & 12,21 \\ \text{matrix notation} & 1 & 2 & 3 & 4 & 5 & 6 \end{array}. \tag{3.339}$$

Then, using the relation

$$C'_{\gamma\delta kl} = \sum_{\alpha,\beta,i,j=1}^{3} U_{\alpha\gamma}U_{\beta\delta}U_{ik}U_{jl}C_{\alpha\beta ij}, \tag{3.340}$$

with $c_{11} = c_{22} = c_{33}, c_{12} = c_{23} = c_{13}$, and $c_{44} = c_{55} = c_{66}$ for a cubic crystal, we obtain

$$\begin{aligned}
C'_{\gamma\delta kl} = {}& c_{11}U_{1\gamma}U_{1\delta}U_{1k}U_{1l} + c_{22}U_{2\gamma}U_{2\delta}U_{2k}U_{2l} + c_{33}U_{3\gamma}U_{3\delta}U_{3k}U_{3l}\\
&+ c_{12}(U_{1\gamma}U_{1\delta}U_{2k}U_{2l} + U_{2\gamma}U_{2\delta}U_{1k}U_{1l})\\
&+ c_{23}(U_{2\gamma}U_{2\delta}U_{3k}U_{3l} + U_{3\gamma}U_{3\delta}U_{2k}U_{2l})\\
&+ c_{13}(U_{1\gamma}U_{1\delta}U_{3k}U_{3l} + U_{3\gamma}U_{3\delta}U_{1k}U_{1l})\\
&+ c_{44}(U_{2\gamma}U_{3\delta}U_{2k}U_{3l} + U_{2\gamma}U_{3\delta}U_{3k}U_{2l}\\
&+ U_{3\gamma}U_{2\delta}U_{2k}U_{3l} + U_{3\gamma}U_{2\delta}U_{3k}U_{2l})\\
&+ c_{55}(U_{3\gamma}U_{1\delta}U_{3k}U_{1l} + U_{3\gamma}U_{1\delta}U_{1k}U_{3l}\\
&+ U_{1\gamma}U_{3\delta}U_{3k}U_{1l} + U_{1\gamma}U_{3\delta}U_{1k}U_{3l})\\
&+ c_{66}(U_{1\gamma}U_{2\delta}U_{1k}U_{2l} + U_{1\gamma}U_{2\delta}U_{2k}U_{1l}\\
&+ U_{2\gamma}U_{1\delta}U_{1k}U_{2l} + U_{2\gamma}U_{1\delta}U_{2k}U_{1l})
\end{aligned} \tag{3.341}$$

$$
\begin{aligned}
C'_{\gamma\delta kl} &= c_{11}\sum_{\alpha=1}^{3} U_{\alpha\gamma}U_{\alpha\delta}U_{\alpha k}U_{\alpha l} \\
&+ c_{12}\sum_{\beta=2}^{3}\sum_{\alpha=1}^{\beta-1}(U_{\alpha\gamma}U_{\alpha\delta}U_{\beta k}U_{\beta l} + U_{\beta\gamma}U_{\beta\delta}U_{\alpha k}U_{\alpha l}) \\
&+ c_{44}\sum_{\beta=2}^{3}\sum_{\alpha=1}^{\beta-1}(U_{\alpha\gamma}U_{\beta\delta} + U_{\beta\gamma}U_{\alpha\delta})(U_{\alpha k}U_{\beta l} + U_{\beta k}U_{\alpha l})
\end{aligned} \tag{3.342}
$$

Having calculated the various elements $C'_{\gamma\delta kl}$, Equation (3.338) can be inverted numerically to determine ε'_{33}, ε'_{23}, and ε'_{31}. Up to this point, the discussion is perfectly general and applicable to cases of substrates of any arbitrary orientation. We can obtain ε_{ij} from Equation (3.338) and, in turn, obtain ε'_{ij} from the relation Equation (3.311). For example, the components of the strain tensor for $(11n)$ surfaces are given by

$$
\begin{aligned}
\varepsilon'_{xx} &= \varepsilon'_{yy} = \varepsilon_{\|}, \\
\varepsilon'_{xy} &= \varepsilon'_{yz} = 0, \\
\varepsilon'_{xz} &= \frac{K_1}{\sqrt{2}K_2}\varepsilon_{\|}, \\
\varepsilon'_{zz} &= \frac{K_3}{K_2}\varepsilon_{\|},
\end{aligned} \tag{3.343}
$$

where

$$
\begin{aligned}
K_1 &= (C_{11} + 2C_{12})(-C_{11} + C_{12} + 2C_{44})n(n^2 - 1), \\
K_2 &= 2C_{11}C_{44} + 2C_{12}C_{44} + 4C_{44}^2 + (C_{11}^2 + C_{11}C_{12} - 2C_{12}^2 + 2C_{11}C_{44} \\
&\quad - 4C_{12}C_{44} + C_{11}C_{44}n^2)n^2, \\
K_3 &= -2[C_{11}C_{44} + 3C_{12}C_{44} - 2C_{44}^2 + (C_{11}^2 + C_{11}C_{12} - 2C_{12}^2 - 2C_{11}C_{44} \\
&\quad + C_{12}C_{44}n^2)n^2].
\end{aligned} \tag{3.344}
$$

3.6.5.2 Derivation from the Condition Grown Pseudomorphically The strain coefficients in the (x, y, z) coordinates for a general crystal orientation can be also determined from the condition that the layer is grown pseudomorphically, and these strain coefficients should minimize the strain energy of the layer simultaneously. We define the unit vectors $\hat{\mathbf{x}}'$, $\hat{\mathbf{y}}'$, and $\hat{\mathbf{z}}'$ along the $\mathbf{x}'$-, $\mathbf{y}'$-, and $\mathbf{z}'$-axes, and they are related to unit vectors $\hat{\mathbf{x}}$, $\hat{\mathbf{y}}$, and $\hat{\mathbf{z}}$ along the $\mathbf{x}$-, $\mathbf{y}$-, and $\mathbf{z}$-axes through the rotation matrix Equation (3.310).

The zinc blende primitive translational vectors are

$$
\begin{aligned}
\boldsymbol{\alpha}_i &= \frac{a_i}{2}(\hat{\mathbf{x}}+\hat{\mathbf{y}}),\\
\boldsymbol{\beta}_i &= \frac{a_i}{2}(\hat{\mathbf{y}}+\hat{\mathbf{z}}),\\
\boldsymbol{\gamma}_i &= \frac{a_i}{2}(\hat{\mathbf{x}}+\hat{\mathbf{z}}),
\end{aligned}
\tag{3.345}
$$

where a is the lattice constant and i labels the epilayer (e) and the substrate (s). When the crystal is translated, the primitive translational vectors become

$$
\begin{aligned}
\boldsymbol{\alpha}_i' &= \frac{a_i}{2}(\mathbf{x}_i+\mathbf{y}_i),\\
\boldsymbol{\beta}_i' &= \frac{a_i}{2}(\mathbf{y}_i+\mathbf{z}_i),\\
\boldsymbol{\gamma}_i' &= \frac{a_i}{2}(\mathbf{x}_i+\mathbf{z}_i),
\end{aligned}
\tag{3.346}
$$

where

$$
\begin{aligned}
\mathbf{x}_i &= (1+\varepsilon_{xx}^i)\hat{\mathbf{x}}+\varepsilon_{xy}^i\hat{\mathbf{y}}+\varepsilon_{xz}^i\hat{\mathbf{z}},\\
\mathbf{y}_i &= \varepsilon_{yx}^i\hat{\mathbf{x}}+(1+\varepsilon_{yy}^i)\hat{\mathbf{y}}+\varepsilon_{yz}^i\hat{\mathbf{z}},\\
\mathbf{z}_i &= \varepsilon_{zx}^i\hat{\mathbf{x}}+\varepsilon_{zy}^i\hat{\mathbf{y}}+(1+\varepsilon_{zz}^i)\hat{\mathbf{z}}.
\end{aligned}
\tag{3.347}
$$

We defined the set of orthonormal vectors $\{\hat{\mathbf{N}}_l\}$ so that $\hat{\mathbf{N}}_1$ and $\hat{\mathbf{N}}_2$ lie in the superlattice plane, and $\hat{\mathbf{N}}_3$ is orthogonal to this plane as follows:

$$
\begin{aligned}
\hat{\mathbf{N}}_1 &= \cos\theta\cos\phi\hat{i}+\cos\theta\sin\phi\hat{j}-\sin\theta\hat{k},\\
\hat{\mathbf{N}}_1 &= -\sin\theta\hat{i}+\cos\phi\hat{j},\\
\hat{\mathbf{N}}_1 &= \sin\theta\cos\phi\hat{i}+\sin\theta\sin\phi\hat{j}+\cos\theta\hat{k}.
\end{aligned}
\tag{3.348}
$$

When the first atomic layers are deposited on the substrate, these layers will be strained to match the substrate, and a pseudomorphic (or coherent) interface will be formed. Thus, the condition for a pseudomorphic interface means that the projections of the strain-distorted primitive translational vectors of the epilayer and the substrate on the growth plane should be equal:

$$\begin{aligned}
\boldsymbol{\alpha}'_e \cdot \hat{\mathbf{N}}_1 &= \boldsymbol{\alpha}'_s \cdot \hat{\mathbf{N}}_1, \\
\boldsymbol{\alpha}'_e \cdot \hat{\mathbf{N}}_2 &= \boldsymbol{\alpha}'_s \cdot \hat{\mathbf{N}}_2, \\
\boldsymbol{\beta}'_e \cdot \hat{\mathbf{N}}_1 &= \boldsymbol{\beta}'_s \cdot \hat{\mathbf{N}}_1, \\
\boldsymbol{\beta}'_e \cdot \hat{\mathbf{N}}_2 &= \boldsymbol{\beta}'_s \cdot \hat{\mathbf{N}}_2, \\
\boldsymbol{\gamma}'_e \cdot \hat{\mathbf{N}}_1 &= \boldsymbol{\gamma}'_s \cdot \hat{\mathbf{N}}_1, \\
\boldsymbol{\gamma}'_e \cdot \hat{\mathbf{N}}_2 &= \boldsymbol{\gamma}'_s \cdot \hat{\mathbf{N}}_2.
\end{aligned} \tag{3.349}$$

The expression of the strain component for general crystal orientation is complicated, and thus we will explain the case of the quantum well with the (111) orientation as an example. Then, angles θ and ϕ are given as $\theta = \arctan(\sqrt{2})$ and $\phi = \pi/4$. Equation (3.349) is easily calculated with the relations of $\varepsilon_{xx} = \varepsilon_{yy} = \varepsilon_{zz}$, $\varepsilon_{xy} = \varepsilon_{yz} = \varepsilon_{zx}$. Then, conditions of $\alpha_{e'} \cdot \hat{\mathbf{N}}_1 = \alpha_{s'} \cdot \hat{\mathbf{N}}_1$ and $\alpha_{e'} \cdot \hat{\mathbf{N}}_2 = \alpha_{s'} \cdot \hat{\mathbf{N}}_2$ give the following relation

$$\begin{aligned}
a_e(1+\varepsilon^e_{xx}+\varepsilon^e_{xy})/\sqrt{6} - a_e\varepsilon^e_{xy}\sqrt{2/3} &= a_s(1+\varepsilon^s_{xx}+\varepsilon^s_{xy})/\sqrt{6} - a_s\varepsilon^s_{xy}\sqrt{2/3} \\
-a_e(1+\varepsilon^e_{xx}+\varepsilon^e_{xy})/(2\sqrt{2}) + a_e\varepsilon^e_{xy}/\sqrt{2} &= -a_s(1+\varepsilon^s_{xx}+\varepsilon^s_{xy})/(2\sqrt{2}) + a_s\varepsilon^s_{xy}/\sqrt{2},
\end{aligned} \tag{3.350}$$

with analogous conditions for β'_i and γ'_i. These constraint equations are linearized in the lattice constant difference $a_e - a_s$. Only five of the constraint equations are linearly independent because the fourth and the fifth equations in Equation (3.349) give the same form. Four unknowns ε^e_{xx}, ε^e_{xy}, ε^s_{xx}, and ε^s_{xy} are determined by the condition of the minimization of the strain energy density, in addition to constraint equations (Eq. (3.350)).

The strain energy density of a zinc blende structure material is

$$\begin{aligned}
U_i = &\frac{1}{2}C^i_{11}[(\varepsilon^i_{xx})^2 + (\varepsilon^i_{yy})^2 + (\varepsilon^i_{zz})^2] \\
&+ 2C^i_{44}[(\varepsilon^i_{xy})^2 + (\varepsilon^i_{yz})^2 + (\varepsilon^i_{xz})^2] \\
&+ C^i_{12}[(\varepsilon^i_{yy})(\varepsilon^i_{zz}) + (\varepsilon^i_{xx})(\varepsilon^i_{zz}) + (\varepsilon^i_{xx})(\varepsilon^i_{yy})],
\end{aligned} \tag{3.351}$$

where C_{ij}s are the elastic constants. For a stained-layer superlattice made up of two zinc blende materials, the strain energy density is

$$U_{es} = \frac{U_e h_e + U_s h_s}{h_e + h_s}, \tag{3.352}$$

where $h_e(h_s)$ is the layer thickness of material $e(s)$ in the superlattice. In a case of the single quantum well, we put h_s to be infinite in a final result.

For (111) crystal orientation, for example, U_i $(i = e, s)$ is given by

$$U_i = (3C_{11}^i/2 + 3C_{12}^i)\varepsilon_{xx}^{i\ 2} + 6C_{44}^i\varepsilon_{xy}^{i\ 2}. \tag{3.353}$$

To obtain ε_{xy}^e in terms of ε_{xx}^e, first, we replace the ε_{xy}^e in U_{es} with the expression obtained from Equation (3.350). Then, we differentiate U_{es} with respect to ε_{xx}^e and obtain

$$\frac{\partial U_{es}}{\partial \varepsilon_{xx}^e} = \left(\frac{\partial U_e}{\partial \varepsilon_{xx}^e} + \frac{\partial U_s}{\partial \varepsilon_{xx}^s}\right) \Big/ (h_e + h_s) = 0. \tag{3.354}$$

Thus,

$$\frac{\partial U_e}{\partial \varepsilon_{xx}^e} = (3C_{11}^e/2 + 3C_{12}^e)2\varepsilon_{xx}^e + 6C_{44}^e\varepsilon_{xy}^e\frac{\partial \varepsilon_{xy}^e}{\partial \varepsilon_{xx}^e} = 0, \tag{3.355}$$

with $\partial\varepsilon_{xy}^e/\partial\varepsilon_{xx}^e = 1/2$ from Equation (3.350). Finally, we obtain the expression

$$\varepsilon_{xy}^e = -\frac{(C_{11}^e + 2C_{12}^e)}{4C_{44}^e}(\varepsilon_{xx}^e) \equiv -A^e\varepsilon_{xx}^e, \tag{3.356}$$

where $A^e = (C_{11}^e + 2C_{12}^e)/(4C_{44}^e)$. Similarly, we can obtain

$$\varepsilon_{xy}^s = -A^s\varepsilon_{xx}^s, \tag{3.357}$$

where $A^s = (C_{11}^s + 2C_{12}^s)/(4C_{44}^s)$.

Next, we replace the ε_{xy}^s in U_{es} with the expression obtained from Equation (3.350), and differentiate U_{es} with respect to ε_{xx}^e. Then,

$$\frac{\partial U_{es}}{\partial \varepsilon_{xx}^e} = \left(\frac{\partial U_e}{\partial \varepsilon_{xx}^e} + \frac{\partial U_s}{\partial \varepsilon_{xx}^s}\right) \Big/ (h_e + h_s) = 0, \tag{3.358}$$

where

$$\frac{\partial U_e}{\partial \varepsilon_{xx}^e} = \left(\frac{3}{2}C_{11}^l + 3C_{12}^e\right)2(\varepsilon_x x^e)h^e + \left(\frac{3}{2}C_{11}^s + 3C_{12}^s\right)2(\varepsilon_{xx}^s)\frac{\partial \varepsilon_{xy}^s}{\partial \varepsilon_{xx}^e}h^s, \tag{3.359}$$

and $\partial\varepsilon_{xy}^s/\partial\varepsilon_{xx}^e = a^e/a^s$ from Equation (3.350). Thus,

$$\varepsilon_{xx}^{s} = -\frac{(C_{11}^{e} + 2C_{12}^{e})}{(C_{11}^{s} + 2C_{12}^{s})}(\varepsilon_{xx}^{e})\frac{h^{e}}{h^{s}}\frac{a^{s}}{a^{e}} \equiv -B\frac{h^{e}}{h^{s}}\frac{a^{s}}{a^{e}}\varepsilon_{xx}^{e}, \tag{3.360}$$

where $B = (C_{11}^{e} + 2C_{12}^{e})/(C_{11}^{s} + 2C_{12}^{s})$.

The ε_{xx}^{e} is obtained by substituting Equations (3.356), (3.357), and (3.360) into Equation (3.350) as follows:

$$\varepsilon_{xx}^{e} = -\frac{1}{(1+A^{e}) + (1+A^{s})B(h_{e}/h_{s})(a^{s}/a^{e})^{2}}\left(\frac{a^{s}}{a^{e}} - 1\right). \tag{3.361}$$

In the case of a single quantum well, taking h_s to be infinite, we obtain

$$\varepsilon_{xx}^{e} = -\frac{4C_{44}^{e}}{4C_{44}^{e} + C_{11}^{e} + 2C_{12}^{e}} \tag{3.362}$$

The strain components for ε'_{ij} can be obtained using the rotational matrix and the obtained ε_{ij}. Similarly, we can obtain the expressions ε_{ij} for a general crystal orientation (hkl).

3.6.6 Piezoelectric Fields in the (11n)-Oriented Zinc Blende Hamiltonian

Off-diagonal strains in zinc blende structure semiconductors induce a polarization given by

$$P_{i}^{S} = 2e_{14}\varepsilon_{jk}, \tag{3.363}$$

where P^{S} is the induced polarization, e_{14} is the piezoelectric constant, and $i, j, k = x, y, z$. Here, i, j, k are in cyclic order. A strained layer with a (001) growth direction has only diagonal strains. Thus, the (001)-oriented layer will not have strain-induced piezoelectric polarization fields. In addition, in the cubic gallium nitride (GaN) crystal structure (c-GaN), spontaneous polarization does not exist due to the higher crystal symmetry. However, strained layers with any other growth direction will have piezoelectric polarization fields. The piezoelectric polarization along the growth direction is important, because the electric field in the quantum well originates from polarization charges at heterojunction interfaces. The electric field induced perpendicular to the growth direction due to the strain can be calculated as

$$E_{\perp} = -\frac{\mathbf{P}^{\mathrm{S}} \cdot \hat{\mathbf{u}}}{\varepsilon}, \tag{3.364}$$

where ε is the dielectric constant, and the unit vector along the growth direction is given by $\hat{\mathbf{u}} = \sqrt{1/(n^2+2)}(\hat{\mathbf{i}} + \hat{\mathbf{j}} + n\hat{\mathbf{k}})$. For a (11n) growth direction,

$$E_{\perp} = -\frac{2e_{14}}{\sqrt{n^2+2}}(n^2\varepsilon_{xy} + \varepsilon_{yz} + \varepsilon_{zx}), \tag{3.365}$$

where

$$\begin{aligned}
\varepsilon_{xy} &= \frac{1}{(n^2+2)}[-\varepsilon'_{xx} + \sqrt{2}n\varepsilon'_{xz} + 2n\varepsilon'_{zz}], \\
\varepsilon_{yz} &= \frac{1}{2(n^2+2)}[-2n\varepsilon'_{xx} - \sqrt{2}(2-n^2)\varepsilon'_{xz} + 2n\varepsilon'_{zz}], \\
\varepsilon_{zx} &= \varepsilon_{yz}.
\end{aligned} \tag{3.366}$$

3.7 CRYSTAL ORIENTATION EFFECTS ON A WURTZITE HAMILTONIAN

3.7.1 Strain Tensors on Wurtzite Semiconductors

We consider only the θ dependence of physical quantities in the following for simplicity [12]. However, we can easily extend the results to a general case with $\phi \neq 0$ for special cases, such as nonpolar planes. Figure 3.2 shows (a) a configuration of the coordinate systems (x', y', z') in $(hkil)$-oriented crystals, and (b) a coordinate system in a wurtzite primitive cell and nonpolar a- and m-planes with the growth direction parallel to the c-axis. The z-axis corresponds to the c-axis (0001), and the growth axis (defined as the z'-axis) is normal to the quantum well plane $(hkil)$. The coordinate system (x, y, z) denotes the primary crystallographic axes. The relation between the coordinate systems for vectors and tensors is given by Equation (3.311). The strain coefficients in the (x, y, z) coordinates for a general crystal orientation are determined from the condition that the layer is grown pseudomorphically, and these strain coefficients should minimize the strain energy of the layer simultaneously. We define the unit vectors $\hat{\mathbf{x}}', \hat{\mathbf{y}}'$, and $\hat{\mathbf{z}}'$ along the $\mathbf{x}'$-, $\mathbf{y}'$-, and $\mathbf{z}'$-axes, and they are related to unit vectors $\hat{\mathbf{x}}, \hat{\mathbf{y}}$, and $\hat{\mathbf{z}}$ along the $\mathbf{x}$-, $\mathbf{y}$-, and $\mathbf{z}$-axes through the rotation matrix Equation (3.310). The hexagonal primitive translational vectors are

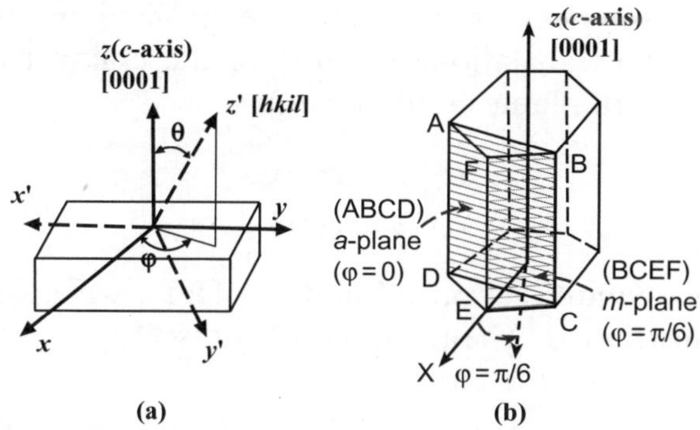

Figure 3.2. (a) Configuration of the coordinate systems (x', y', z') in $(hkil)$-oriented crystals. The growth axis or the z'-axis is normal to the substrate surface $(hkil)$, and the coordinate system (x, y, z) denotes the primary crystallographic axes. (b) The coordinate system in a wurtzite primitive cell and nonpolar a- and m-planes with the growth direction parallel to the c-axis. $\theta = \pi/2$ with $\phi = 0$ (a-plane) corresponds to the $z' = [11\bar{2}0]$ growth direction, and $\theta = \pi/2$ with $\phi = \pi/6$ (m-plane) corresponds to the $z' = [10\bar{1}0]$ growth direction.

$$\begin{aligned}\boldsymbol{\alpha}_i &= a_i\hat{\mathbf{x}},\\ \boldsymbol{\beta}_i &= -\frac{a_i}{2}\hat{\mathbf{x}} + \frac{\sqrt{3}a_i}{2}\hat{\mathbf{y}},\\ \boldsymbol{\gamma}_i &= c_i\hat{\mathbf{z}},\end{aligned} \tag{3.367}$$

where a is the lattice constant, and i labels the epilayer (e) and the substrate (s). When the crystal is translated, the primitive translational vectors become

$$\begin{aligned}\boldsymbol{\alpha}_i'' &= a_i\hat{\mathbf{x}}'',\\ \boldsymbol{\beta}_i'' &= -\frac{a_i}{2}\hat{\mathbf{x}}'' + \frac{\sqrt{3}a_i}{2}\hat{\mathbf{y}}'',\\ \boldsymbol{\gamma}_i'' &= c_i\hat{\mathbf{z}}'',\end{aligned} \tag{3.368}$$

where

$$\begin{aligned}\hat{\mathbf{x}}'' &= (1+\varepsilon_{xx})\hat{\mathbf{x}} + \varepsilon_{xy}\hat{\mathbf{y}} + \varepsilon_{xz}\hat{\mathbf{z}},\\ \hat{\mathbf{y}}'' &= \varepsilon_{yx}\hat{\mathbf{x}} + (1+\varepsilon_{yy})\hat{\mathbf{y}} + \varepsilon_{yz}\hat{\mathbf{z}},\\ \hat{\mathbf{z}}'' &= \varepsilon_{zx}\hat{\mathbf{x}} + \varepsilon_{zy}\hat{\mathbf{y}} + (1+\varepsilon_{zz})\hat{\mathbf{z}}.\end{aligned} \tag{3.369}$$

When the first atomic layers are deposited on the substrate, these layers will be strained to match the substrate, and a pseudomorphic (or coherent) interface will be formed. Thus, the condition for a pseudomorphic interface means that the projections of the strain-distorted primitive translational vectors of the epilayer and the substrate on the growth plane should be equal:

$$\begin{aligned} \boldsymbol{\alpha}_e''\cdot\hat{\mathbf{x}}' &= \boldsymbol{\alpha}_s''\cdot\hat{\mathbf{x}}', \\ \boldsymbol{\alpha}_e''\cdot\hat{\mathbf{y}}' &= \boldsymbol{\alpha}_s''\cdot\hat{\mathbf{y}}', \end{aligned} \tag{3.370}$$

with analogous conditions on β'' and γ''. Here, we assume that the substrate is unstrained. Then, the constraints (Eq. (3.370)) yield the following relations for the strain tensors:

$$\begin{aligned} \varepsilon_{xx} &= \varepsilon_{xx}^{(0)} + \varepsilon_{xz}\frac{\sin\theta}{\cos\theta}, \\ \varepsilon_{yy} &= \varepsilon_{xx}^{(0)}, \\ \varepsilon_{zz} &= \varepsilon_{xz}\frac{\cos\theta}{\sin\theta} + \varepsilon_{zz}^{(0)}, \\ \varepsilon_{xy} &= \varepsilon_{yz} = 0, \end{aligned} \tag{3.371}$$

where $\varepsilon_{xx}^{(0)} = (a_s - a_e)/a_e$ and $\varepsilon_{zz}^{(0)} = (c_s - c_e)/c_e$. Under the engineering notation, the strain energy density is given by

$$\begin{aligned} W = \frac{1}{2}\big[&C_{11}\varepsilon_{xx}^2 + C_{11}\varepsilon_{yy}^2 + C_{33}\varepsilon_{zz}^2 + 2C_{12}\varepsilon_{xx}\varepsilon_{yy} \\ &+ 2C_{13}(\varepsilon_{xx}\varepsilon_{zz} + \varepsilon_{yy}\varepsilon_{zz}) + 4C_{44}\varepsilon_{xz}^2\big]. \end{aligned} \tag{3.372}$$

Using the above relations, the strain energy can be expressed through only one strain component, ε_{xz}, which can be found by minimizing the strain energy with respect to the variable ε_{xz}. This procedure gives the following expression for ε_{xz}

$$\varepsilon_{xz} = -\frac{\begin{aligned}&\big[(c_{11} + c_{12} + c_{13}\varepsilon_{zz}^{(0)}/\varepsilon_{xx}^{(0)})\sin^2\theta \\ &+ (2c_{13} + c_{33}\varepsilon_{zz}^{(0)}/\varepsilon_{xx}^{(0)})\cos^2\theta\big]\varepsilon_{xx}^{(0)}\cos\theta\sin\theta\end{aligned}}{c_{11}\sin^4\theta + 2(c_{13} + 2c_{44})\sin^2\theta\cos^2\theta + c_{33}\cos^4\theta}. \tag{3.373}$$

Hence, we obtain a 6×6 Hamiltonian in the (x', y', z') coordinates by substituting the transformation relation for the vector k in Equation (3.311), and the strain coefficients for a general crystal orientation into the Hamiltonian for the (0001) orientation.

3.7.2 Valence Band Edges of Bulk Wurtzite Semiconductors

At the band edges, $k_x' = k_y' = k_z' = 0$, the Hamiltonian in the (x', y', z') coordinates can be simplified

$$H'(\mathbf{k}'=0,\varepsilon)=\begin{pmatrix} F' & -K'^* & -H'^* & 0 & 0 & 0 \\ -K' & G' & H' & 0 & 0 & \Delta \\ -H' & H'^* & \lambda' & 0 & \Delta & 0 \\ 0 & 0 & 0 & F' & -K' & H' \\ 0 & 0 & \Delta & -K'^* & G' & -H'^* \\ 0 & \Delta & 0 & H'^* & -H' & \lambda' \end{pmatrix}\begin{matrix} |U_1\rangle' \\ |U_2\rangle' \\ |U_3\rangle' \\ |U_4\rangle' \\ |U_5\rangle' \\ |U_6\rangle' \end{matrix} \tag{3.374}$$

where

$$\begin{aligned} F' &= \Delta_1+\Delta_2+\lambda'_\varepsilon+\theta'_\varepsilon, \\ G' &= \Delta_1-\Delta_2+\lambda'_\varepsilon+\theta'_\varepsilon, \\ K' &= D_5(\varepsilon_{xx}-\varepsilon_{yy}), \\ H' &= D_6\varepsilon_{xz}, \\ \lambda'_\varepsilon &= D_1\varepsilon_{zz}+D_2(\varepsilon_{xx}+\varepsilon_{yy}), \\ \theta'_\varepsilon &= D_3\varepsilon_{zz}+D_4(\varepsilon_{xx}+\varepsilon_{yy}), \\ \Delta &= \sqrt{2}\Delta_3, \end{aligned} \tag{3.375}$$

and the strain coefficients ε_{ij} for a general crystal orientation θ are given by Equations (3.371) and (3.373). Note that K' and H' are not zero because $\varepsilon_{xx}\neq\varepsilon_{yy}$, and $\varepsilon_{xz}\neq 0$ in the general coordinates. Also, all elements are constant. This Hamiltonian can be block diagonalized. The transformation matrix is defined as

$$T=\begin{pmatrix} \alpha^* & 0 & 0 & \alpha & 0 & 0 \\ 0 & \beta & 0 & 0 & \beta^* & 0 \\ 0 & 0 & \beta^* & 0 & 0 & \beta \\ \alpha^* & 0 & 0 & -\alpha & 0 & 0 \\ 0 & \beta & 0 & 0 & -\beta^* & 0 \\ 0 & 0 & -\beta^* & 0 & 0 & \beta \end{pmatrix}, \tag{3.376}$$

where

$$\alpha=\frac{1}{\sqrt{2}}e^{i(3\pi/4)},\beta=\frac{1}{\sqrt{2}}e^{i(\pi/4)}. \tag{3.377}$$

The block-diagonalized Hamiltonian, $H''=T^*H'T^t$, is given by

$$H'' = \begin{pmatrix} H'^U & 0 \\ 0 & H'^L \end{pmatrix}, \tag{3.378}$$

where the upper 3 × 3 Hamiltonian is

$$H'^U = \begin{pmatrix} F' & K' & -iH' \\ K' & G' & \Delta - iH' \\ iH' & \Delta + iH' & \lambda' \end{pmatrix} \tag{3.379}$$

and the lower 3 × 3 Hamiltonian is

$$H'^L = \begin{pmatrix} F' & K' & iH' \\ K' & G' & \Delta + iH' \\ -iH' & \Delta - iH' & \lambda' \end{pmatrix}. \tag{3.380}$$

The lower Hamiltonian is the complex conjugate of the upper Hamiltonian. Therefore, they have exactly the same eigenvalues, since the eigenenergies are real. The piezoelectric field in bulk crystal does not change the in-plane dispersion curves. The wave functions of the lower Hamiltonian are the complex conjugates of the corresponding wave functions of the upper Hamiltonian. The eigenequation for the valence band energy E' at the band edge is given by

$$\begin{pmatrix} F' - E' & K' & -iH' \\ K' & G' - E' & \Delta - iH' \\ iH' & \Delta + iH' & \lambda' - E' \end{pmatrix} \begin{pmatrix} g_1' \\ g_2' \\ g_3' \end{pmatrix} = 0. \tag{3.381}$$

The eigenequation is a third-order polynomial equation, which can always be solved analytically. These three analytical solutions can be expressed as

$$\begin{aligned} E_1' &= 2(S_1 + S_2) - \frac{C_2}{3}, \\ E_2' &= -\frac{1}{2}(S_1 + S_2) - \frac{C_2}{3} + i\frac{\sqrt{3}}{2}(S_1 - S_2), \\ E_3' &= -\frac{1}{2}(S_1 + S_2) - \frac{C_2}{3} - i\frac{\sqrt{3}}{2}(S_1 - S_2), \end{aligned} \tag{3.382}$$

where

$$
\begin{aligned}
q &= \frac{1}{3}C_1 - \frac{1}{9}C_2^2, \\
r &= \frac{1}{6}(C_1C_2 - 3C_0) - \frac{C_2^3}{27}, \\
S_1 &= [r + \sqrt{q^3 + r^2}]^{1/3}, \\
S_2 &= [r - \sqrt{q^3 + r^2}]^{1/3},
\end{aligned}
\tag{3.383}
$$

and

$$
\begin{aligned}
C_0 &= -(F'G'\lambda' + 2K'H'^2 - H'^2G' - K'^2\lambda' - F'\Delta^2 - F'H'^2), \\
C_1 &= G'\lambda' + F'\lambda' + F'G' - \Delta^2 - 2H'^2 - K'^2, \\
C_2 &= -(G' + F' + \lambda).
\end{aligned}
\tag{3.384}
$$

The envelope functions corresponding to the three eigenvalues are determined by

$$
\begin{pmatrix} g_1' \\ g_2' \\ g_3' \end{pmatrix} = \frac{1}{D} \begin{pmatrix} D_1 \\ D_2 \\ D_3 \end{pmatrix} \tag{3.385}
$$

where

$$
\begin{aligned}
D_1 &= iH'(G' - E') + (\Delta - iH')K', \\
D_2 &= iH'K'(F' - E')(\Delta - iH'), \\
D_3 &= (F' - E')(G' - E') - K'^2,
\end{aligned}
\tag{3.386}
$$

and

$$
D = \sqrt{|D_1|^2 + |D_2|^2 + |D_3|^2}. \tag{3.387}
$$

The complete Bloch wave functions for the valence band are given by

$$
\Psi'^{v}(\mathbf{r}', \mathbf{k}' = 0) = \frac{1}{\sqrt{V}} e^{i\mathbf{k}'\cdot\mathbf{r}'} \sum_{i=1}^{3} g_i'(\mathbf{k}')|i\rangle', \tag{3.388}
$$

where

$$
\begin{aligned}
|1\rangle' &= \alpha^*|U_1\rangle' + \alpha|U_4\rangle', \\
|2\rangle' &= \beta|U_2\rangle' + \beta^*|U_5\rangle', \\
|3\rangle' &= \beta^*|U_3\rangle' + \beta|U_6\rangle',
\end{aligned}
\tag{3.389}
$$

and we can obtain similar expressions for the lower Hamiltonian (Eq. (3.380)).

3.7.3 Polarization-Dependent Interband Optical-Matrix Elements of Bulk Wurtzite Semiconductors

The optical momentum matrix elements in the (x', y', z') coordinates are given by

$$|\hat{\mathbf{e}}' \cdot \mathbf{M}'^{\eta}|^2 = |\langle \Psi'^{c\eta} | \hat{\mathbf{e}}' \cdot \mathbf{p}' | \Psi'^{v} \rangle|^2. \tag{3.390}$$

Using the expressions for the band-edge wave functions of the conduction and valence bands, we obtain the polarization-dependent interband momentum matrix elements:

TE-polarization ($\hat{\mathbf{e}}' = \hat{\mathbf{x}}'$ or $\hat{\mathbf{y}}'$):

$$\begin{aligned} |M'^{\eta}_{x'}|^2 &= \Big| -\frac{1}{\sqrt{2}} g'_1 \alpha^* \cos\theta \langle S|p_x|X\rangle \\ &\quad + \frac{1}{\sqrt{2}} g'_2 \beta \cos\theta \langle S|p_x|X\rangle - g'_3 \beta^* \sin\theta \langle S|p_z|Z\rangle \Big|^2 \\ &\quad \textit{for } \sigma = U \\ &= \Big| -\frac{1}{\sqrt{2}} g'_4 \alpha^* \cos\theta \langle S|p_x|X\rangle \\ &\quad + \frac{1}{\sqrt{2}} g'_5 \beta \cos\theta \langle S|p_x|X\rangle + g'_6 \beta^* \sin\theta \langle S|p_z|Z\rangle \Big|^2 \\ &\quad \textit{for } \sigma = L \end{aligned} \tag{3.391}$$

TM-polarization ($\hat{\mathbf{e}}' = \hat{\mathbf{z}}'$):

$$\begin{aligned} |M'^{\eta}_{z'}|^2 &= \Big| -\frac{1}{\sqrt{2}} g'_1 \alpha^* \sin\theta \langle S|p_x|X\rangle \\ &\quad + \frac{1}{\sqrt{2}} g'_2 \beta \sin\theta \langle S|p_x|X\rangle + g'_3 \beta^* \cos\theta \langle S|p_z|Z\rangle \Big|^2 \\ &\quad \textit{for } \sigma = U \\ &= \Big| -\frac{1}{\sqrt{2}} g'_4 \alpha^* \sin\theta \langle S|p_x|X\rangle \\ &\quad + \frac{1}{\sqrt{2}} g'_5 \beta \sin\theta \langle S|p_x|X\rangle - g'_6 \beta^* \cos\theta \langle S|p_z|Z\rangle \Big|^2 \\ &\quad \textit{for } \sigma = L. \end{aligned} \tag{3.392}$$

The above analytical results include the crystal orientation dependence, and the band-edge matrix elements are as follows:

$$P_x = P_y = \langle S|p_x|X\rangle = \langle S|p_y|Y\rangle = \frac{m_o}{\hbar}P_2, P_z = \langle S|p_z|Z\rangle = \frac{m_o}{\hbar}P_1,$$

$$P_1^2 = \frac{\hbar^2}{2m_o}\left(\frac{m_o}{m_e^z}-1\right)\frac{(E_g+\Delta_1+\Delta_2)(E_g+2\Delta_2)-2\Delta_3^2}{E_g+2\Delta_2}, \tag{3.393}$$

$$P_2^2 = \frac{\hbar^2}{2m_o}\left(\frac{m_o}{m_e^t}-1\right)\frac{E_g[(E_g+\Delta_1+\Delta_2)(E_g+2\Delta_2)-2\Delta_3^2]}{(E_g+\Delta_1+\Delta_2)(E_g+\Delta_2)-\Delta_3^2}.$$

3.7.4 Piezoelectric Fields and Spontaneous Polarizations in a Quantum Well Structure

If a stress τ_{jk} is applied to GaN crystals, there is an induced piezoelectric polarization with a magnitude proportional to the applied stress τ_{jk}:

$$P_i = d_{ijk}\tau_{jk}, \tag{3.394}$$

where d_{ijk} are the piezoelectric moduli or piezoelectric tensor elements. The stress τ_{ij} is related to the strain by

$$\tau_{ij} = c_{ijkl}\varepsilon_{kl}, \tag{3.395}$$

where c_{ijkl} are the stiffness constants of the wurtzite GaN crystal. Here, it is convenient to replace the tensor notation with the engineering notation for d_{ijk} and c_{ijkl} using their symmetry properties. That is, the second and third subscripts in d_{ijk}, and the first two and the last two subscripts in c_{ijkl}, are replaced by a single subscript running from 1 to 6 as follows:

$$\begin{pmatrix} \text{tensor notation} & 11 & 22 & 33 & 23,32 & 31,13 & 12,21 \\ \text{matrix notation} & 1 & 2 & 3 & 4 & 5 & 6 \end{pmatrix}. \tag{3.396}$$

Then, the piezoelectric polarization in the (x, y, z) coordinates for a general crystal orientation is given by

$$\begin{pmatrix} P_x \\ P_y \\ P_z \end{pmatrix} = \begin{pmatrix} 0 & 0 & 0 & 0 & d_{15} & 0 \\ 0 & 0 & 0 & d_{15} & 0 & 0 \\ d_{31} & d_{31} & d_{33} & 0 & 0 & 0 \end{pmatrix} \begin{pmatrix} c_{11} & c_{12} & c_{13} & 0 & 0 & 0 \\ c_{12} & c_{11} & c_{13} & 0 & 0 & 0 \\ c_{13} & c_{13} & c_{33} & 0 & 0 & 0 \\ 0 & 0 & 0 & c_{44} & 0 & 0 \\ 0 & 0 & 0 & 0 & c_{44} & 0 \\ 0 & 0 & 0 & 0 & 0 & c_{66} \end{pmatrix} \begin{pmatrix} \varepsilon_{xx} \\ \varepsilon_{yy} \\ \varepsilon_{zz} \\ 0 \\ 2\varepsilon_{xz} \\ 0 \end{pmatrix}. \tag{3.397}$$

We then obtain the polarization components along the x-, y-, and z-axes:

$$\begin{aligned}
P_x &= 2d_{15}c_{44}\varepsilon_{xz}, \\
P_y &= 0, \\
P_z &= [d_{31}(c_{11}+c_{12})+d_{33}c_{13}](\varepsilon_{xx}+\varepsilon_{yy})+[2d_{31}c_{13}+d_{33}c_{33}]\varepsilon_{zz}.
\end{aligned} \tag{3.398}$$

The normal polarization with respect to the growth plane (along the growth direction) can be obtained by taking the scalar product between the polarization vector **P** in the (x, y, z) coordinates and the unit vector $\hat{\mathbf{z}}'$:

$$P_{z'} = \mathbf{P}\cdot\hat{\mathbf{z}}' = P_x \sin\theta + P_z \cos\theta. \tag{3.399}$$

Similarly, we can obtain the parallel polarization on the growth plane:

$$\begin{aligned}
P_{x'} &= \mathbf{P}\cdot\hat{\mathbf{x}}' = P_x \cos\theta - P_z \sin\theta, \\
P_{y'} &= \mathbf{P}\cdot\hat{\mathbf{y}}' = 0.
\end{aligned} \tag{3.400}$$

The strain-induced polarization P_i will lead to an electric field E_i, given by

$$D_{i'} = \varepsilon_{ij}^{p\prime} E_{j'} + P_{i'}, \tag{3.401}$$

where $D_{i'}$ is the electric displacement, and $\varepsilon_{ij}^{p\prime}$ is the permittivity tensor for the epilayer given by

$$\varepsilon' = U\varepsilon^p U^{-1} \tag{3.402}$$

and

$$\varepsilon^p = \begin{pmatrix} \varepsilon_t & 0 & 0 \\ 0 & \varepsilon_t & 0 \\ 0 & 0 & \varepsilon_z \end{pmatrix}. \tag{3.403}$$

Here, ε_t and ε_z are permittivities perpendicular and along the c-axes for the (0001) crystal orientation, respectively. Then, the electric displacement $D_{i'}$ is given by

$$\begin{aligned}
D_{x'} &= (\varepsilon_t \cos^2\theta + \varepsilon_z \sin^2\theta)E_{x'} + (\varepsilon_t - \varepsilon_z)\sin\theta\cos\theta E_{z'} + P_{x'}, \\
D_{y'} &= \varepsilon_t E_{y'} + P_{y'}, \\
D_{z'} &= (\varepsilon_t - \varepsilon_z)\sin\theta\cos\theta E_{x'} + (\varepsilon_t \sin^2\theta + \varepsilon_z \cos^2\theta)E_{z'} + P_{z'}.
\end{aligned} \tag{3.404}$$

If there are no external charges, and $\varepsilon_t \simeq \varepsilon_z$, $D_{z'}$ vanishes, and the electric field reduces to

$$E_{z'} \simeq -\frac{P_{z'}}{\varepsilon_t}. \tag{3.405}$$

The electric fields in the well and barrier due to the spontaneous (SP) and piezoelectric (PZ) polarization along the (0001)-direction for (0001)-oriented wurtzite structure can be estimated from the periodic boundary condition for a superlattice structure as follows:

$$\begin{aligned} F_{z'} &= \frac{(F_{SP}^b + F^b - F_{SP}^w - F_{PZ}^w)L^b}{L^b\varepsilon^w + L^w\varepsilon^b}, \\ F_{z'}^b &= -\frac{L_w}{L_b} F_{z'}^w, \end{aligned} \tag{3.406}$$

where superscripts w and b mean the well and barrier, respectively, and L and ε are the layer thickness and the static dielectric constant, respectively. The strain-induced piezoelectric field F_{PZ}^i $(i = w, b)$ is given by

$$F_{PZ}^i = \frac{2d_{31}^i}{\varepsilon^i}\left(C_{11}^i + C_{12}^i - \frac{2C_{13}^{i\ 2}}{C_{33}^i}\right)\varepsilon_{xx}^{\prime i(3)}, \tag{3.407}$$

where d_{31} is the piezoelectric constant.

The spontaneous polarization P_{SP} along the growth direction in the arbitrary crystal orientation is estimated by the relation $P_{SP} = P_{SP}^{(0001)}\cos\theta$, since the spontaneous polarization in wurtzite (0001) GaN-based quantum well structures exists along the (0001) axis. The strain-induced piezoelectric polarization normal with respect to the growth plane (along the growth direction) in an arbitrary crystal orientation can be expressed as

$$P_{PZ} = P_x \sin\theta + P_z \cos\theta. \tag{3.408}$$

3.7.5 Valence Band Structure of a Quantum Well and Self-Consistent Calculations with the Screening Effect

The constant energy contour of the (0001) orientation is isotropic. On the other hand, the valence band structure of the quantum well structures with general crystal orientations shows anisotropy in the quantum well plane unlike the (0001)-oriented structure. Thus, the valence band structure and the momentum matrix element should be calculated as a function of the angle Φ in the quantum well plane. That is, the angle is related to $k_{\|}$ by $k'_x = k_{\|}\cos\Phi$ and $k'_y = k_{\|}\sin\Phi$. Here, k'_x and k'_y are the in-plane wave vectors in the (x', y', z') coordinate system. Please note that, for simplicity, we omit the angle Φ dependence in the notation of physical functions such as the wave function and the momentum matrix element.

For a quantum well structure of a width L_w, the valence band structure is calculated by using 6×6 Hamiltonian for an arbitrary crystal orientation, which is obtained by using Equations (3.285) and (3.311). The envelope functions $\{g_m^{(\nu)}\}$ satisfy

$$\sum_{\nu'}\left[H_{\nu\nu'}\left(k'_x, k'_y, -i\frac{\partial}{\partial z'}\right) + (V_\nu(z') + qE'_z z')\delta_{\nu\nu'}\right] g_m^{(\nu)}(k'_x, k'_y, z')$$
$$= E'_m(k'_x, k'_y) g_m^{(\nu)}(k'_x, k'_y, z'), \tag{3.409}$$

where $\nu, \nu' = 1, 2, 3, 4, 5, 6$, and E'_z is given in Equation (3.405). These coupled differential equations can be solved using a finite-difference method.

The self-consistent band structures and wave functions are obtained by iteratively solving the Schrödinger equation for electrons, the block-diagonalized Hamiltonian for holes, and Poisson's equation. The total potential profiles for the electrons and the holes are

$$\begin{aligned} V_c(z) &= V_{cw}(z) - |e|\phi(z), \\ V_\nu(z) &= V_{vw}(z) - |e|\phi(z), \end{aligned} \tag{3.410}$$

where $V_{cw}(z)$ and $V_{vw}(z)$ are the square-well potential plus the potential due to the internal field without the screening effect for the conduction band and the valence band, respectively, and $\phi(z)$ is the screening potential induced by the charged carriers and satisfies the Poisson equation

$$\frac{d}{dz}\left(\varepsilon(z)\frac{d}{dz}\right)\phi(z) = -|e|[p(z) - n(z)], \tag{3.411}$$

where $\varepsilon(z)$ is the dielectric constant.

The electron and the hole concentrations, $p(z)$ and $n(z)$, are related to the wave functions of the nth conduction subband and the mth valence subband by

$$n(z) = \frac{kTm_e}{\pi\hbar^2}\sum_n |f_n(z)|^2 \ln\left(1 + e^{[E_{fc} - E_{cn}(0)]/kT}\right) \tag{3.412}$$

and

$$p(z)=\sum_{m}\int dk_{\parallel}\frac{k_{\parallel}}{2\pi}\sum_{\nu}\left|g_{mk_{\parallel}}^{(\nu)}(z)\right|^{2}\left(\frac{1}{1+e^{[E_{fv}-E_{vm}(k_{\parallel})]/kT}}\right), \tag{3.413}$$

where m_e is the effective mass of electrons, $\hbar$ is Planck's constant divided by 2π, n and m are the quantized subband indices for the conduction and the valence bands, E_{fc} and E_{fv} are the quasi-Fermi levels of the electrons and the holes, respectively, $E_{cn}(0)$ is the quantized energy level of the electrons, $E_{vm}(k_{\parallel})$ is the energy for the mth subband in the valence band, $k_{\parallel}$ is the in-plane wave vector, ν refers to the new bases for the Hamiltonian, and $f_n(z)$ and $g_{mk_{\parallel}}^{(\nu)}(z)$ are the envelope functions in the conduction and the valence bands, respectively. The potential $\phi(z)$ is obtained by integration:

$$\phi(z)=-\int_{-L/2}^{z}E(z')dz', \tag{3.414}$$

where

$$E(z)=\int_{-L/2}^{z}\frac{1}{\varepsilon(z)}\rho(z')dz'. \tag{3.415}$$

Note that the origin of the coordinate is chosen at the center of the well, and $2L$ is the total length containing the well and the barrier.

The procedures for the self-consistent calculations consist of the following steps: (1) Start with the potential profiles V_c and V_v with $\phi^{(0)}(z)=0$ in Equation (3.410); (2) solve the Schrödinger equation (for electrons) and the 6×6 Hamiltonian (for holes) with the potential profiles $\phi^{(n-1)}(z)$ in step (1) to obtain band structures and wave functions; (3) for a given carrier density, obtain the Fermi-energies by using the band structures and the charge distribution by using the wave functions; (4) solve Poisson's equation to find $\phi^{(n)}(z)$; (5) check if $\phi^{(n)}(z)$ converges to $\phi^{(n-1)}(z)$. If not, set $\phi^{(n)}(z)=w\phi^{(n)}(z)+(1-w)\phi^{(n-1)}(z)$, $n=n+1$; then, return to step (2). If yes, the band structures and the wave functions obtained with $\phi^{(n-1)}(z)$ are solutions. An adjustable parameter $w\,(0<w<1)$ is typically set to 0.5 at low carrier densities. With increasing carrier densities, a smaller value of w is needed for rapid convergence. The procedure for the self-consistent calculation is shown in Figure 3.3.

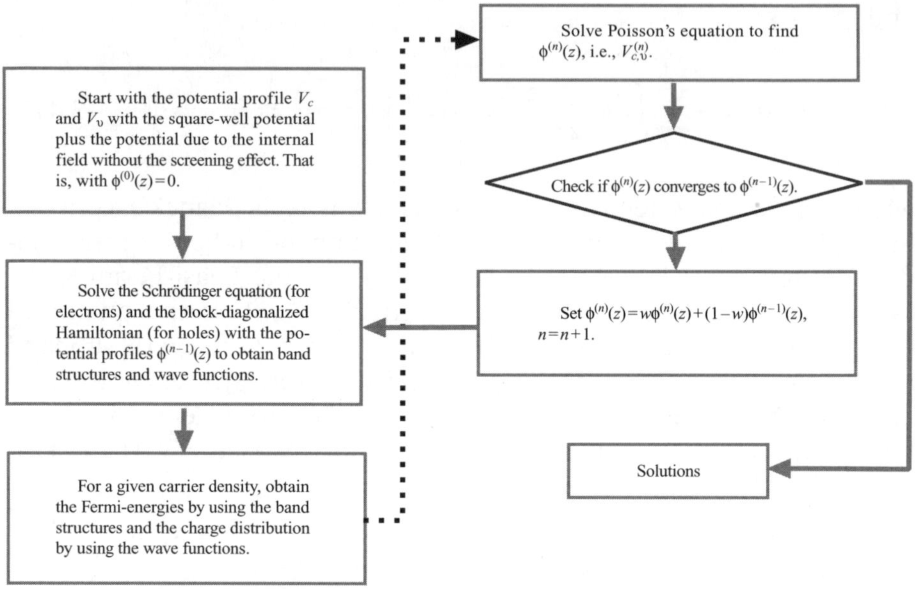

Figure 3.3. Procedure for the self-consistent calculation.

3.7.6 Polarization-Dependent Interband Optical-Matrix Elements of Wurtzite Quantum Wells

The optical momentum matrix elements in the (x', y', z') coordinates for a general crystal orientation are given by Equation (3.390). Since we solve one Schrödinger equation for electrons and a 6×6 Hamiltonian for holes, we need only the optical matrix elements for two cases: spin up and spin down. On the other hand, for a block-diagonalized 3×3 Hamiltonian, we needed the optical matrix elements for four cases: spin up and spin down for upper block, spin up and spin down for lower Hamiltonian. The polarization-dependent interband momentum matrix elements for a wurtizte quantum well structure can be written as TE-polarization ($\hat{\mathbf{e}}' = \cos\phi'\hat{\mathbf{x}}' + \sin\phi'\hat{\mathbf{y}}'$):

$$\begin{aligned}|\hat{\mathbf{e}}'\cdot\mathbf{M}'^{\uparrow}|^2 = \Bigg|&\cos\phi'\left\{-\frac{1}{\sqrt{2}}\cos\theta P_x\langle g_m'^{(1)}|\phi_l\rangle\right.\\&\left.+\frac{1}{\sqrt{2}}\cos\theta P_x\langle g_m'^{(2)}|\phi_l\rangle-\sin\theta P_z\langle g_m'^{(3)}|\phi_l\rangle\right\}\\&+\sin\phi'\left\{-i\frac{1}{\sqrt{2}}P_x\langle g_m'^{(1)}|\phi_l\rangle-i\frac{1}{\sqrt{2}}P_x\langle g_m'^{(2)}|\phi_l\rangle\right\}\Bigg|^2\end{aligned} \tag{3.416}$$

$$
\begin{aligned}
\left|\hat{\mathbf{e}}'\cdot\mathbf{M}'^{\downarrow}\right|^2 = \Bigg| & \cos\phi'\left\{\frac{1}{\sqrt{2}}\cos\theta P_x\left\langle g_m'^{(4)}\middle|\phi_l\right\rangle\right. \\
& \left.-\frac{1}{\sqrt{2}}\cos\theta P_x\left\langle g_m'^{(5)}\middle|\phi_l\right\rangle-\sin\theta P_z\left\langle g_m'^{(6)}\middle|\phi_l\right\rangle\right\} \\
& +\sin\phi'\left\{-i\frac{1}{\sqrt{2}}P_x\left\langle g_m'^{(4)}\middle|\phi_l\right\rangle-i\frac{1}{\sqrt{2}}P_x\left\langle g_m'^{(5)}\middle|\phi_l\right\rangle\right\}\Bigg|^2
\end{aligned}
\tag{3.417}
$$

TM-polarization ($\hat{\mathbf{e}}'=\hat{\mathbf{z}}'$):

$$
\begin{aligned}
\left|\hat{\mathbf{e}}'\cdot\mathbf{M}'^{\uparrow}\right|^2 = \Bigg| & -\frac{1}{\sqrt{2}}\sin\theta P_x\left\langle g_m'^{(1)}\middle|\phi_l\right\rangle+\frac{1}{\sqrt{2}}\sin\theta P_x\left\langle g_m'^{(2)}\middle|\phi_l\right\rangle \\
& +\cos\theta P_z\left\langle g_m'^{(3)}\middle|\phi_l\right\rangle\Bigg|^2
\end{aligned}
\tag{3.418}
$$

$$
\begin{aligned}
\left|\hat{\mathbf{e}}'\cdot\mathbf{M}'^{\downarrow}\right|^2 = \Bigg| & \frac{1}{\sqrt{2}}\sin\theta P_x\left\langle g_m'^{(4)}\middle|\phi_l\right\rangle-\frac{1}{\sqrt{2}}\sin\theta P_x\left\langle g_m'^{(5)}\middle|\phi_l\right\rangle \\
& +\cos\theta P_z\left\langle g_m'^{(6)}\middle|\phi_l\right\rangle\Bigg|^2,
\end{aligned}
\tag{3.419}
$$

where $g_m'^{(\nu)}$ ($\nu=1,2,3,4,5,$ and 6) is the wave function for the mth subband in (x',y',z') coordinates. Also, we can obtain the optical matrix element for the (0001) crystal orientation by substituting $\theta=0$ in Equation (3.419): TE-polarization ($\hat{\mathbf{e}}=\hat{\mathbf{x}}$ or $\hat{\mathbf{y}}$):

$$
\left|\hat{\mathbf{e}}\cdot\mathbf{M}^{\uparrow}\right|^2=\left|-\frac{1}{\sqrt{2}}P_x\left\langle g_m^{(1)}\middle|\phi_l\right\rangle+\frac{1}{\sqrt{2}}P_x\left\langle g_m^{(2)}\middle|\phi_l\right\rangle\right|^2 \tag{3.420}
$$

$$
\left|\hat{\mathbf{e}}\cdot\mathbf{M}^{\downarrow}\right|^2=\left|\frac{1}{\sqrt{2}}P_x\left\langle g_m'^{(4)}\middle|\phi_l\right\rangle-\frac{1}{\sqrt{2}}P_x\left\langle g_m'^{(5)}\middle|\phi_l\right\rangle\right|^2 \tag{3.421}
$$

TM-polarization ($\hat{\mathbf{e}}=\hat{\mathbf{z}}$):

$$
\left|\hat{\mathbf{e}}\cdot\mathbf{M}^{\uparrow}\right|^2=\left|P_z\left\langle g_m'^{(3)}\middle|\phi_l\right\rangle\right|^2 \tag{3.422}
$$

$$
\left|\hat{\mathbf{e}}\cdot\mathbf{M}^{\downarrow}\right|^2=\left|P_z\left\langle g_m'^{(6)}\middle|\phi_l\right\rangle\right|^2. \tag{3.423}
$$

3.7.7 Band Structure of Nonpolar *a* and *m*-Planes

3.7.7.1 {10$\bar{1}$0} m-Plane Hamiltonian The m-plane Hamiltonian with a set of faces {10$\bar{1}$0} can be obtained by substituting $\phi=\pi/6$ and $\theta=\pi/2$ into Equation (3.311) [13].

$$H(\mathbf{k}',\varepsilon')=\begin{pmatrix} F' & -K'^{*} & -H'^{*} & 0 & 0 & 0 \\ -K' & G' & H' & 0 & 0 & \Delta \\ -H' & H'^{*} & \lambda' & 0 & \Delta & 0 \\ 0 & 0 & 0 & F' & -K' & H' \\ 0 & 0 & \Delta & -K'^{*} & G' & -H'^{*} \\ 0 & \Delta & 0 & H'^{*} & -H' & \lambda' \end{pmatrix}\begin{matrix}|U_1\rangle' \\ |U_2\rangle' \\ |U_3\rangle' \\ |U_4\rangle' \\ |U_5\rangle' \\ |U_6\rangle',\end{matrix} \tag{3.424}$$

where

$$\begin{aligned}
F' &= \Delta_1 + \Delta_2 + \lambda' + \theta', \\
G' &= \Delta_1 - \Delta_2 + \lambda' + \theta', \\
\lambda' &= \frac{\hbar^2}{2m_o}[A_1 k_x'^2 + A_2(k_y'^2 + k_z'^2)] + D_1\varepsilon_{zz}^{(0)} + D_2\left(2\varepsilon_{xx}^{(0)} + \frac{2}{\sqrt{3}}S\right), \\
\theta' &= \frac{\hbar^2}{2m_o}[A_3 k_x'^2 + A_4(k_y'^2 + k_z'^2)] + D_3\varepsilon_{zz}^{(0)} + D_4\left(2\varepsilon_{xx}^{(0)} + \frac{2}{\sqrt{3}}S\right), \\
K' &= \frac{\hbar^2}{2m_o}\left[-A_5\frac{1+i\sqrt{3}}{2}k_x'^2 + A_5(-\sqrt{3}+i)k_y'k_z' \right. \\
&\quad \left. + A_5\frac{1+i\sqrt{3}}{2}k_z'^2 + \frac{D_5}{\sqrt{3}}(1+3i)S\right], \\
H' &= \frac{\hbar^2}{2m_o}\left[A_6\frac{1-i\sqrt{3}}{2}k_x'k_y' - A_6\frac{\sqrt{3}-i}{2}k_x'k_z'\right], \\
S &= -\sqrt{3}\,\frac{(C_{11}+3C_{12})(\varepsilon_{xx}^{(0)}+1) + C_{12}(-3+\varepsilon_{xx}^{(0)}) + C_{11}(-1+3\varepsilon_{xx}^{(0)}) + 4C_{13}\varepsilon_{zz}^{(0)}}{5C_{11}+3C_{12}}.
\end{aligned} \tag{3.425}$$

Note that the old bases are used in Equation (3.424), which were defined in Equation (3.279). This use of old bases is convenient since P_x, P_y, and P_z in the old bases are known quantities, as defined in Equation (3.374). The optical momentum matrix elements for the *m*-plane are given by

$$|\hat{\mathbf{e}}'\cdot\mathbf{M}'^{\eta}|^2 = |\langle\Psi_l'^{c\eta}|\hat{\mathbf{e}}'\cdot\mathbf{p}'|\Psi_m'^{v}\rangle|^2, \tag{3.426}$$

where Ψ'^c and Ψ'^v are wave functions for the conduction and the valence bands, respectively, and $\eta = \uparrow$ and $\downarrow$ for both electron spins. The polarization-dependent interband momentum matrix elements can be written as

TE-polarization $\hat{\mathbf{e}}' = \cos\phi'\hat{\mathbf{x}}' + \sin\phi'\hat{\mathbf{y}}'$:

$$\begin{aligned}\left|\hat{\mathbf{e}}'\cdot\mathbf{M}'^{\uparrow}\right|^2 = \Big| &-\cos\phi' P_z \langle g_m'^{(3)}|\phi_l\rangle + \sin\phi' \frac{P_x}{2\sqrt{2}}\{(1-\sqrt{3}i)\langle g_m'^{(1)}|\phi_l\rangle \\ &-(1+\sqrt{3}i)\langle g_m'^{(2)}|\phi_l\rangle\}\Big|^2\end{aligned} \tag{3.427}$$

$$\begin{aligned}\left|\hat{\mathbf{e}}'\cdot\mathbf{M}'^{\downarrow}\right|^2 = \Big| &-\cos\phi' P_z \langle g_m'^{(6)}|\phi_l\rangle - \sin\phi' \frac{P_x}{2\sqrt{2}}\{(1+i\sqrt{3})\langle g_m'^{(4)}|\phi_l\rangle \\ &-(1-i\sqrt{3})\langle g_m'^{(5)}|\phi_l\rangle\}\Big|^2\end{aligned} \tag{3.428}$$

TM-polarization $\hat{\mathbf{e}}' = \hat{\mathbf{z}}'$:

$$\left|\hat{\mathbf{e}}'\cdot\mathbf{M}'^{\uparrow}\right|^2 = \left|-\frac{P_x}{2\sqrt{2}}\{(\sqrt{3}+i)\langle g_m'^{(1)}|\phi_l\rangle - (\sqrt{3}-i)\langle g_m'^{(2)}|\phi_l\rangle\}\right|^2 \tag{3.429}$$

$$\left|\hat{\mathbf{e}}'\cdot\mathbf{M}'^{\downarrow}\right|^2 = \left|\frac{P_x}{2\sqrt{2}}\{(\sqrt{3}-i)\langle g_m'^{(4)}|\phi_l\rangle - (\sqrt{3}+i)\langle g_m'^{(5)}|\phi_l\rangle\}\right|^2. \tag{3.430}$$

3.7.7.2 {11$\bar{2}$0} a-Plane Hamiltonian The a-plane Hamiltonian with a set of faces {11$\bar{2}$0} can be obtained by substituting $\phi = 0$ and $\theta = \pi/2$ into Equation (3.311). The matrix elements in Equation (3.424) are given by

$$\begin{aligned}
F' &= \Delta_1 + \Delta_2 + \lambda' + \theta', \\
G' &= \Delta_1 - \Delta_2 + \lambda' + \theta', \\
\lambda' &= \frac{\hbar^2}{2m_o}[A_1 k_x'^2 + A_2(k_y'^2 + k_z'^2)] + D_1\varepsilon_{zz}^{(0)} + D_2\left(2\varepsilon_{xx}^{(0)} - \frac{C_{12}}{C_{11}}\varepsilon_{xx}^{(0)} - \frac{C_{13}}{C_{11}}\varepsilon_{zz}^{(0)}\right), \\
\theta' &= \frac{\hbar^2}{2m_o}[A_3 k_x'^2 + A_4(k_y'^2 + k_z'^2)] + D_3\varepsilon_{zz}^{(0)} + D_4\left(2\varepsilon_{xx}^{(0)} - \frac{C_{12}}{C_{11}}\varepsilon_{xx}^{(0)} - \frac{C_{13}}{C_{11}}\varepsilon_{zz}^{(0)}\right), \\
K' &= \frac{\hbar^2}{2m_o}[-A_5 k_y'^2 + 2iA_5 k_y' k_z' + A_5 k_z'^2] + D_5\left(-\frac{C_{12}}{C_{11}}\varepsilon_{xx}^{(0)} - \frac{C_{13}}{C_{11}}\varepsilon_{zz}^{(0)} - \varepsilon_{xx}^{(0)}\right), \\
H' &= \frac{\hbar^2}{2m_o}[-iA_6 k_x' k_y' - A_6 k_x' k_z'].
\end{aligned} \tag{3.431}$$

The polarization-dependent interband momentum matrix elements for an a-plane are given by

TE-polarization $\hat{\mathbf{e}}' = \cos\phi'\hat{\mathbf{x}}' + \sin\phi'\hat{\mathbf{y}}'$:

$$|\hat{\mathbf{e}}' \cdot \mathbf{M}'^{\uparrow}|^2 = \left| -P_z \cos\phi' \langle g_m'^{(3)} | \phi_l \rangle + \sin\phi' \frac{P_x}{\sqrt{2}} \{ -i \langle g_m'^{(1)} | \phi_l \rangle - i \langle g_m'^{(2)} | \phi_l \rangle \} \right|^2 \tag{3.432}$$

$$|\hat{\mathbf{e}}' \cdot \mathbf{M}'^{\downarrow}|^2 = \left| -P_z \cos\phi' \langle g_m'^{(6)} | \phi_l \rangle + \sin\phi' \frac{P_x}{\sqrt{2}} \{ -i \langle g_m'^{(4)} | \phi_l \rangle - i \langle g_m'^{(5)} | \phi_l \rangle \} \right|^2 \tag{3.433}$$

TM-polarization $\hat{\mathbf{e}}' = \hat{\mathbf{z}}'$:

$$|\hat{\mathbf{e}}' \cdot \mathbf{M}'^{\uparrow}|^2 = \left| -\frac{1}{\sqrt{2}} P_x \langle g_m'^{(1)} | \phi_l \rangle + \frac{1}{\sqrt{2}} P_x \langle g_m'^{(2)} | \phi_l \rangle \right|^2 \tag{3.434}$$

$$|\hat{\mathbf{e}}' \cdot \mathbf{M}'^{\downarrow}|^2 = \left| \frac{1}{\sqrt{2}} P_x \langle g_m'^{(4)} \phi_l \rangle - \frac{1}{\sqrt{2}} P_x \langle g_m'^{(5} | \phi_l \rangle \right|^2. \tag{3.435}$$

PROBLEMS

1. Derive D_{YY}, D_{ZZ}, D_{XY}, D_{YZ}, and D_{ZX} in Section 3.2.3.
2. Derive H_0^{c1l1} and H_0^{c1s1} in Hamiltonian Equation (3.210) of Section 3.3.2.
3. Show that a (001)-oriented Hamiltonian with wurtzite bases is given by Equation (3.283) in Section 3.5.2.
4. Derive Equation (3.134) for a degenerate band without spin–orbit coupling.
5. Derive the expression Equation (3.373) for ε_{xz} by minimizing the strain energy Equation (3.372) with respect to the variable ε_{xz} using Equation (3.371) in Section 3.7.1.
6. Derive three analytical solutions Equation (3.382) for 3×3 Hamiltonian Equation (3.381).
7. Derive the block-diagonalized Hamiltonian expression Equation (3.293) using the transformation matrix Equation (3.291).
8. Derive matrix form Equation (3.185) for Equations (3.178) and (3.179) using the matrix elements in Equations (3.181), (3.183), and (3.184).

REFERENCES

[1] S. Datta, *Quantum Phenomena*. Modular Series on Solid State Devices, Vol. 8, New York: Addison-Wesley, 1989.

[2] J. M. Luttinger and W. Kohn, "Motion of electrons and holes in perturbed periodic fields," *Phy. Rev.*, 97, p. 869, 1955.

[3] S. L. Chuang, *Physics of Optoelectronic Devices*. New York: Wiley, 1995.

[4] J. F. Nye, *Physical Properties of Crystals*. Oxford: Clarendon, 1989.

[5] D. Ann, S. J. Yoon, S. L. Chuang, and C.-S. Chang, "Theory of optical gain in strained-layer quantum wells within the 6×6 Luttinger-Kohn model," *J. Appl. Phys.*, 78, p. 15, 1995.

[6] G. L. Bir and G. E. Pikus, *Symmetry and Strain-Induced Effects in Semiconductors*. New York: Wiley, 1974.

[7] S. L. Chuang and C. S. Chang, "k·p method for strained wurtzite semiconductors," *Phys. Rev. B*, 54, p. 2491, 1996.

[8] S. H. Park and S. L. Chuang, "Piezoelectric effects on electrical and optical properties of wurtzite GaN/AlGaN quantum well lasers," *J. Appl. Phys.*, 87, p. 353, 2000.

[9] S. H. Park, "Optical anisotropy in (110)-oriented zinc-blende GaN/AlGaN quantum wells," *Semicond. Sci. Technol.*, 23, p. 1, 2008.

[10] Y. Kajikawa, "Optical anisotropy of (111)-oriented strained quantum-wells calculated with the effect of the spin-orbit split-off band," *J. Appl. Phys.*, 86, p. 5663, 1999.

[11] D. L. Smith and C. Mailhiot, "Theory of semiconductor superlattice electronic structure," *Rev. Mod. Phys.*, 62, p. 173, 1990.

[12] S. H. Park arid S. L. Chuang, "Crystal-orientation effects on the piezoelectric field and electronic properties of strained wurtzite semiconductors," *Phys. Rev. B*, 59, p. 4725, 1999.

[13] S. H. Park, D. Ahn, and S. L. Chuang, "Electronic and optical properties of a- and m-plane wurtzite InGaN/GaN quantum wells," *IEEE J. Quantum Electron.*, 43, p. 1175, 2007.

PART II
Modern Applications

4 Quantum Information Science

4.1 QUANTUM BITS AND TENSOR PRODUCTS

Quantum information science is an interdisciplinary subject that includes physics, mathematics, chemistry, electrical engineering, and computer science, and has the potential to revolutionize many areas of science and technology. It exploits fundamentally new modes of computation and information processing, because it is based on the laws of quantum mechanics instead of those of classical physics. It holds the promise of immense computing power beyond the capabilities of any classical computer, and it is directly linked to emerging quantum technologies, such as quantum teleportation. The key element of quantum information science is a quantum system with two states that can represent "0" and "1," the basic building blocks of a digital logic circuit called "qubit." A qubit is a unit vector in a two-dimensional Hilbert space, whose basis vectors (computational basis) are defined as [1]

$$|0\rangle = \begin{pmatrix} 1 \\ 0 \end{pmatrix} \text{and}\, |1\rangle = \begin{pmatrix} 0 \\ 1 \end{pmatrix}. \tag{4.1}$$

In order to implement any useful quantum information processing unit, we need to deal with many qubits at a time, and the appropriate model is a tensor product of qubits. Specifically, if we have n qubits, each with a given computational basis in a two dimensional Hilbert space H_i, then the tensor product is a 2^n-dimensional Hilbert space with a computational basis consisting of 2^n vectors

$$|i_1 i_2 \cdots i_n\rangle = |i_1\rangle \otimes |i_2\rangle \otimes \cdots \otimes |i_n\rangle, \tag{4.2}$$

where the indices i_k take on the values 0 and 1. In the following, we give a more detailed description of a tensor product. Let A be an $m \times n$ matrix, and B be a $p \times q$ matrix.

Engineering Quantum Mechanics, First Edition. Doyeol Ahn, Seoung-Hwan Park.

Then

$$A \otimes B = \begin{pmatrix} a_{11}B & a_{12}B & \cdots & a_{1n}B \\ a_{21}B & a_{22}B & \cdots & a_{2n}B \\ & \cdots & & \\ a_{m1}B & a_{m2}B & \cdots & a_{mn}B \end{pmatrix} \tag{4.3}$$

is an $mp \times nq$ matrix called the tensor product of A and B. The tensor product of vectors $|\psi\rangle = \begin{pmatrix} a \\ b \end{pmatrix}$ and $|\varphi\rangle = \begin{pmatrix} c \\ d \end{pmatrix}$ is then given by

$$|\psi\rangle \otimes |\varphi\rangle = \begin{pmatrix} a|\varphi\rangle \\ b|\varphi\rangle \end{pmatrix} = \begin{pmatrix} ac \\ ad \\ bc \\ bd \end{pmatrix}. \tag{4.4}$$

The tensor product $|\psi\rangle \otimes |\varphi\rangle$ is often denoted as $|\psi\rangle|\varphi\rangle$ or $|\psi\varphi\rangle$ when there is no cause for confusion. If C is an $n \times r$ matrix, and D is a $q \times s$ matrix, then one can show that

$$(A \otimes B)(C \otimes D) = (AC) \otimes (BD). \tag{4.5}$$

Moreover, we obtain

$$(A \otimes B)^{\dagger} = A^{\dagger} \otimes B^{\dagger} \tag{4.6}$$

and

$$(A \otimes B)^{-1} = A^{-1} \otimes B^{-1}. \tag{4.7}$$

When A_k and $|\psi_k\rangle$ are the linear operator and the vector in the Hilbert space H_k, respectively, then Equation (4.5) implies that

$$(A_1 \otimes \cdots \otimes A_n)(|\psi_1\rangle \otimes \cdots \otimes |\psi_n\rangle) = (A_1|\psi_1\rangle) \otimes \cdots \otimes (A_n|\psi_n\rangle). \tag{4.8}$$

Let $A = A_1 \otimes \cdots \otimes A_n$ and $|\psi\rangle = |\psi_1\rangle \otimes \cdots \otimes |\psi_n\rangle$, then

$$\langle\psi|A|\psi\rangle = \langle\psi_1|A_1|\psi_1\rangle \cdots \langle\psi_n|A_n|\psi_n\rangle. \tag{4.9}$$

4.2 QUANTUM ENTANGLEMENT

Suppose we have n qubit states. An entangled state is defined as the superposition of tensor product states, which is not necessarily decomposable into a tensor product form. A general entangled state is represented as [1,2]

$$|\psi\rangle = \sum_{i_k=0,1} a_{i_1 i_2 \cdots i_n} |i_1\rangle \otimes |i_2\rangle \otimes \cdots \otimes |i_n\rangle. \tag{4.10}$$

Entangled states play very important roles in such aspects of quantum information science as quantum computing and quantum teleportation. Consider a system of two qubits. The basis for the two-qubit system is $\{|00\rangle, |01\rangle, |10\rangle, |11\rangle\}$. More generally, the basis for a system of n qubits could be written as $\{|c_{n-1}c_{n-2}\cdots c_0\rangle\}$, where $c_{n-1}, c_{n-2}, \cdots, c_0 \in \{0,1\}$. We can also represent this basis in terms of the decimal system. In decimal system, we denote

$$|x\rangle = |c_{n-1}c_{n-2}\cdots c_0\rangle, \tag{4.11}$$

where

$$x = c_{n-1}2^{n-1} + c_{n-2}2^{n-2} + \cdots + c_0 \tag{4.12}$$

Using this decimal notation, the basis for the two-qubit system can be written as $\{|0\rangle, |1\rangle, |2\rangle, |3\rangle\}$. The basic principles of quantum computation can be described easily using the entangled state with decimal notation. Assuming that we have the system of n qubits from which we form following $2n$ qubit state:

$$|\Psi_x\rangle = |x\rangle \otimes |0\rangle. \tag{4.13}$$

Let U_f be a unitary operator, which leaves the first n qubit, but changes the second n qubit as follows:

$$U_f(|x\rangle \otimes |0\rangle) = |x\rangle \otimes |f(x)\rangle. \tag{4.14}$$

Now, we have the new state formed by the superposition of the first n qubits given by

$$|\Phi\rangle = \sum_{x=0,1,\cdots 2^{n-1}} a_x |x\rangle \otimes |0\rangle. \tag{4.15}$$

If we apply the unitary operator U_f on $|\Phi\rangle$, we obtain

$$
\begin{aligned}
U_f|\Phi\rangle &= \sum_x a_x U_f(|x\rangle \otimes |0\rangle) \\
&= \sum_x a_x |x\rangle \otimes |f(x)\rangle.
\end{aligned} \tag{4.16}
$$

The above result suggests that 2^n different computations can be done at once as the result of unitary operation. Imagine that each computation takes 1 second. When n is equal to 32, one needs to do four billion computations, which takes approximately 32 years using the conventional computer. On the other hand, if one has a quantum processor, the same computation can be done in one second. That's a billionfold increase in computation speed!

In quantum information science, the unitary operators are often called quantum gates. The most elementary gates are single-qubit gates and two-qubit gates. General single qubit gate can be built from the following Pauli operators X, Y, Z, and the identity operator I:

$$
\begin{aligned}
I &= \begin{pmatrix} 1 & 0 \\ 0 & 1 \end{pmatrix}, \\
X = \sigma_x &= \begin{pmatrix} 0 & 1 \\ 1 & 0 \end{pmatrix}, \\
Y = \sigma_y &= \begin{pmatrix} 0 & -i \\ i & 0 \end{pmatrix}, \\
Z = \sigma_z &= \begin{pmatrix} 1 & 0 \\ 0 & -1 \end{pmatrix}.
\end{aligned} \tag{4.17}
$$

The operation I is the identity transformation, while X is the negation (NOT), Z the phase shift, and $Y = iXZ$ the combination of both X and Z.

The CNOT (controlled-NOT) gate is the most important two-qubit gate in quantum computation. The gate flips the second qubit (the target qubit) when the first qubit (the control qubit) is $|1\rangle$, while leaving the second qubit unchanged when the first qubit is $|0\rangle$. Let $\{|00\rangle, |01\rangle, |10\rangle, |11\rangle\}$ be the basis for the two-qubit system. The action of CNOT gate, whose matrix expression will be written as U_{CNOT}, is

$$
U_{\text{CNOT}} : |00\rangle \mapsto |00\rangle,\ |01\rangle \mapsto |01\rangle,\ |10\rangle \mapsto |11\rangle,\ |11\rangle \mapsto |10\rangle. \tag{4.18}
$$

The matrix expression of U_{CNOT} is given by

$$U_{\text{CNOT}} = \begin{pmatrix} 1 & 0 & 0 & 0 \\ 0 & 1 & 0 & 0 \\ 0 & 0 & 0 & 1 \\ 0 & 0 & 1 & 0 \end{pmatrix} \tag{4.19}$$
$$= |0\rangle\langle 0| \otimes I + |1\rangle\langle 1| \otimes X.$$

Here, the outer product $|u\rangle\langle v|$ between two vectors

$$|u\rangle = \begin{pmatrix} u_1 \\ u_2 \\ \vdots \\ u_n \end{pmatrix} \quad \text{and} \quad |v\rangle = \begin{pmatrix} v_1 \\ v_2 \\ \vdots \\ v_m \end{pmatrix}$$

is defined as

$$|u\rangle\langle v| = \begin{pmatrix} u_1v_1 & u_1v_2 & \cdots & u_1v_m \\ u_2v_1 & u_2v_2 & \cdots & u_2v_m \\ & \cdots & & \\ u_nv_1 & u_nv_2 & \cdots & u_nv_m \end{pmatrix}. \tag{4.20}$$

Another important two-qubit gate is the SWAP gate, which swaps the first and second qubits:

$$U_{\text{SWAP}} |\psi\rangle \otimes |\phi\rangle = |\phi\rangle \otimes |\psi\rangle, \tag{4.21}$$

or

$$U_{\text{SWAP}} = |\Phi^+\rangle\langle\Phi^+| + |\Phi^-\rangle\langle\Phi^-| + |\Psi^+\rangle\langle\Psi^+| + e^{i\pi}|\Psi^-\rangle\langle\Psi^-|, \tag{4.22}$$

in terms of the Bell states

$$|\Phi^\pm\rangle = \frac{1}{\sqrt{2}}(|00\rangle \pm |11\rangle) \quad \text{and} \quad |\Psi^\pm\rangle = \frac{1}{\sqrt{2}}(|01\rangle \pm |10\rangle). \tag{4.23}$$

Recently, it was shown that an arbitrary two-qubit operation can be implemented using only three $(\text{SWAP})^\alpha$ gates and six single-qubit operations. The $(\text{SWAP})^\alpha$ gate is defined as follows

$$U_{(\mathrm{SWAP})^\alpha} = |\Phi^+\rangle\langle\Phi^+| + |\Phi^-\rangle\langle\Phi^-| + |\Psi^+\rangle\langle\Psi^+| + e^{i\alpha\pi}|\Psi^-\rangle\langle\Psi^-|$$
$$= \begin{pmatrix} 1 & 0 & 0 & 0 \\ 0 & \frac{1+e^{i\alpha\pi}}{2} & \frac{1+e^{i\alpha\pi}}{2} & 0 \\ 0 & \frac{1+e^{i\alpha\pi}}{2} & \frac{1+e^{i\alpha\pi}}{2} & 0 \\ 0 & 0 & 0 & 1 \end{pmatrix}, \tag{4.24}$$

where the exponent α is a control parameter. The general two-qubit operation can be implemented when the Hamiltonian for the two-qubit system is of the form given below:

$$H = h_x\sigma_x \otimes \sigma_x + h_y\sigma_y \otimes \sigma_y + h_z\sigma_z \otimes \sigma_z, \tag{4.25}$$

where $\pi/4 \geq h_x \geq h_y \geq h_z \geq 0$. In this case, H can be written as

$$\begin{aligned} H &= \lambda_{00}|\Phi^+\rangle\langle\Phi^+| + \lambda_{01}|\Psi^+\rangle\langle\Psi^+| + \lambda_{10}|\Phi^-\rangle\langle\Phi^-| + \lambda_{11}|\Psi^-\rangle\langle\Psi^-|, \\ \lambda_{00} &= h_x - h_y + h_z, \quad \lambda_{01} = h_x + h_y - h_z, \\ \lambda_{10} &= -h_x + h_y + h_z, \quad \lambda_{11} = -h_x - h_y - h_z. \end{aligned} \tag{4.26}$$

4.3 QUANTUM TELEPORTATION

Quantum teleportation is perhaps the most unique feature of quantum information science, in which an unknown quantum state of a qubit is transmitted to the recipient who can reproduce exactly the same state as the original qubit state. It is important to note that the qubit itself is not transmitted, but the information required to reproduce the quantum state is sent to the recipient. The original quantum state is destroyed in the process. Assume that Alice and Bob share an entangled state, for example,

$$|\Phi^+\rangle_{AB} = \frac{1}{\sqrt{2}}(|00\rangle_{AB} + |11\rangle_{AB}) \tag{4.27}$$

and Alice wants to teleport an unknown quantum state belonging to Charlie

$$|\phi\rangle_C = a|0\rangle_C + b|1\rangle_C \tag{4.28}$$

to Bob. Then the total state is given by

$$
\begin{aligned}
|\Psi\rangle &= |\phi\rangle_C \otimes |\Phi^+\rangle_{AB} \\
&= \left(a|0\rangle_C + b|1\rangle_C \right) \otimes \frac{1}{\sqrt{2}} \left(|00\rangle_{AB} + |11\rangle_{AB} \right) \\
&= \frac{1}{\sqrt{2}} (a|000\rangle_{CAB} + a|011\rangle_{CAB} + b|100\rangle_{CAB} + b|111\rangle_{CAB}).
\end{aligned}
\tag{4.29}
$$

Now Alice has the first two qubits while Bob has the third qubit. First, Alice does the Bell state measurement of her two qubits. In quantum mechanics, by taking a measurement of a system, we project the state vector to one of the basis vectors that the measurement defines. In order to do this, we define the measurement operator M_m such that the probability $p(m)$ of obtaining the outcome m in the state $|\psi\rangle$ is [1]

$$
p(m) = \langle \psi | M_m M_m | \psi \rangle, \tag{4.30}
$$

and the state immediately after the measurement is

$$
|m\rangle = \frac{M_m |\psi\rangle}{\sqrt{p(m)}}. \tag{4.31}
$$

The measurement operators satisfy the completeness relation,

$$
\sum_m M_m^\dagger M_m = I. \tag{4.32}
$$

Before Alice performs her measurement, it is convenient to rewrite the Equation (4.29) into [1,2]

$$
\begin{aligned}
|\Psi\rangle &= \frac{1}{\sqrt{2}} (a|000\rangle_{CAB} + a|011\rangle_{CAB} + b|100\rangle_{CAB} + b|111\rangle_{CAB}) \\
&= \frac{1}{\sqrt{2}} \left(\frac{a}{\sqrt{2}} \{ |\Phi^+\rangle_{CA} + |\Phi^-\rangle_{CA} \} \otimes |0\rangle_B + \frac{a}{\sqrt{2}} \{ |\Psi^+\rangle_{CA} + |\Psi^-\rangle_{CA} \} \otimes |1\rangle_B \right. \\
&\quad \left. + \frac{b}{\sqrt{2}} \{ |\Psi^+\rangle_{CA} - |\Psi^-\rangle_{CA} \} \otimes |0\rangle_B + \frac{b}{\sqrt{2}} \{ |\Phi^+\rangle_{CA} - |\Phi^-\rangle_{CA} \} \otimes |1\rangle_B \right) \\
&= \frac{1}{2} (|\Phi^+\rangle_{CA} \otimes \{ a|0\rangle_B + b|1\rangle_B \} + |\Phi^-\rangle_{CA} \otimes \{ a|0\rangle_B - b|1\rangle_B \} \\
&\quad + |\Psi^+\rangle_{CA} \otimes \{ a|1\rangle_B + b|0\rangle_B \} + |\Psi^-\rangle_{CA} \otimes \{ a|1\rangle_B - b|0\rangle_B \}).
\end{aligned}
\tag{4.33}
$$

If Alice measures the two qubits in her hand, she will obtain one of the states $|\Phi^+\rangle, |\Phi^-\rangle, |\Psi^+\rangle$, or $|\Psi^-\rangle$ with equal probability ¼. Bob's qubit immediately collapses to $a|0\rangle + b|1\rangle$, $a|0\rangle - b|1\rangle$, $a|1\rangle + b|0\rangle$, or $a|1\rangle - b|0\rangle$, respectively, depending on the result of Alice's measurement. Alice then sends her results to Bob using two classical bits through a classical communication channel, such as a telephone. Once Bob learns the result of Alice's measurement, he knows which qubit state is in his hand. Bob reconstructs the initial state $|\phi\rangle$ by applying unitary operators I, Z, X, or Y, respectively, depending on Alice's result. As an example, suppose Alice lets Bob know that her measurement result was $|\Phi^-\rangle$. Then Bob applies an operator Z on his state to reconstruct $|\phi\rangle$ as follows:

$$Z(a|0\rangle - b|1\rangle) = a|0\rangle + b|1\rangle = |\phi\rangle. \tag{4.34}$$

4.4 EVOLUTION OF THE QUANTUM STATE: QUANTUM INFORMATION PROCESSING

In previous sections, we have seen that quantum entanglement plays a very important role in aspects of quantum information science such as quantum computing and quantum teleportation. Considerable progress has been made in developing a general quantum theory of information processing, analogous to classical information theory founded by Shannon 60 years ago. In this section, we would like to understand how the quantum state evolves through the transmission process [1,2]. Suppose R and Q are two quantum systems, and Q is described by a Hilbert space H_Q of finite dimension d. Initially, the joint system RQ is prepared in an entangled state $|\Psi^{RQ}\rangle$, whose vector is uniquely specified. Such a state is called a pure state. On the other hand, it may happen that a quantum system under consideration is in the state $|\Psi_i\rangle$ with a probability p_i so that we cannot say definitely which state the system is in. Such a system is said to be in a mixed state. The system R is dynamically isolated and zero internal Hamiltonian, while the system Q undergoes evolution that possibly involves interaction with the environment E. The evolution of Q may represent a coding, transmission, and decoding process via some quantum mechanical channel or gate for the quantum information in Q. The final state of RQ is possibly mixed, and is described by the density operator $\rho^{RQ'}$. In order to characterize the process, we first define the fidelity of the quantum mechanical process by

$$F_e = \langle \Psi^{RQ} | \rho^{RQ'} | \Psi^{RQ} \rangle, \tag{4.35}$$

which is the probability that the final state $\rho^{RQ'}$ would pass a test checking whether it agreed with the initial state $|\Psi^{RQ}\rangle$. F_e measures how successfully the quantum process preserves the entanglement of Q with the reference system R. Imagine that the system Q is prepared in an initial state ρ^Q, and then subjected to some dynamical process, after which the state becomes $\rho^{Q'}$. The dynamical process is described by a map ε^Q, so that the evolution is

$$\rho^Q \mapsto \rho^{Q'} = \varepsilon^Q(\rho^Q). \tag{4.36}$$

In the most general case, the map ε^Q must be a trace-preserving, completely positive linear map. In other words, we have the following [1,2]:

(1) ε^Q must be linear in the density operators. That is, if $\rho^Q = p_1\rho_1^Q + p_2\rho_2^Q$, then

$$\begin{aligned}\varepsilon^Q(\rho^Q) &= \varepsilon^Q(p_1\rho_1^Q + p_2\rho_2^Q) \\ &= p_1\varepsilon^Q(\rho_1^Q) + p_2\varepsilon^Q(\rho_2^Q).\end{aligned} \tag{4.37}$$

A probabilistic mixture of inputs to ε^Q leads to a probabilistic mixture of outputs. Here, ε^Q is a superoperator that acts on the space of linear operators, such as density operators on H_Q.

(2) ε^Q must be trace preserving, so that

$$Tr(\rho^Q) = Tr(\rho^{Q'}) = 1. \tag{4.38}$$

(3) ε^Q must be positive. We use the term "positive" to refer to operators that are Hermitian and have no negative eigenvalues. This means that if ρ^Q is positive, then $\rho^{Q'} = \varepsilon^Q(\rho^Q)$ must be positive. These properties mean that the superoperator ε^Q takes normalized density operators to normalized density operators in a reasonable way.

(4) ε^Q must be completely positive. This means that if we extend the evolution superoperator ε^Q to an evolution superoperator for the compound system RQ, yielding $I^R \otimes \varepsilon^Q$, where I^R is the identity superoperator on R states. This means adjoining a system R that has trivial dynamics (no state of R is changed), and which does not interact with Q. If, for all such trivial extensions, the resulting superoperator $I_R \otimes \varepsilon^Q$ is positive, the superoperator ε^Q is said to be "completely positive."

A completely positive map is not only a reasonable map from density operators to density operators for Q, but also, it is extensible in a trivial way to a reasonable map from density operators to density operators on any larger system RQ. Since we cannot exclude *a priori* that our system Q is in fact initially entangled with some distant isolated system R, any acceptable ε^Q must satisfy this condition.

Completely positive, trace-preserving linear maps include all unitary evolutions of the state

$$\rho^{Q'} = U^Q \rho^Q U^{Q\dagger}. \tag{4.39}$$

They also include unitary evolutions involving interactions with an external system. Suppose we consider an environment system that is initially in the pure state $|0^E\rangle$. Then we could have

$$\varepsilon^Q(\rho^Q) = Tr_E\left(U^{QE}\left(\rho^Q \otimes |0^E\rangle\langle 0^E|\right)U^{QE\dagger}\right), \tag{4.40}$$

where U^{QE} is the unitary evolution on the joint system QE. This map is also trace preserving and completely positive. If we can write a superoperator ε^Q as a unitary evolution on an extended system QE, followed by a partial trace over E, we say that we have a "unitary representation" of the superoperator. Such a representation is not unique, since many different unitary operators U^{QE} will lead to the same ε^Q.

There is another useful representation for completely positive maps that employs only operators on H_Q. Let A_μ^Q be a collection of such operators indexed by μ. Then the map ε^Q given by

$$\varepsilon^Q(\rho^Q) = \sum_\mu A_\mu^Q \rho^Q A_\mu^Q \tag{4.41}$$

is a completely positive map. If, in addition, A_μ^Q operators satisfy

$$\sum_\mu A_\mu^{Q\dagger} A_\mu^Q = I^Q, \tag{4.42}$$

then the map is also trace preserving. Such a representation for ε^Q in terms of operators A_μ^Q is called an "operator representation" for ε^Q. A single ε^Q will admit many different operator–sum representations. In fact, the "unitary representation" and the "operator–sum representation" are related to each other. Suppose that $\rho^Q = |\phi^Q\rangle\langle\phi^Q|$, and let $\{|\mu^E\rangle\}$ be a complete orthonormal set of states of E.

Then

$$\begin{aligned}\varepsilon^Q(\rho^Q) &= Tr_E\{U^{QE}(|\phi^Q\rangle\langle\phi^Q|\otimes|0^E\rangle\langle 0^E|)U^{QE\dagger}\}\\ &= \sum_\mu \langle\mu^E|U^{QE}(|\phi^Q\rangle\langle\phi^Q|\otimes|0^E\rangle\langle 0^E|)U^{QE\dagger}|\mu^E\rangle\\ &= \sum_\mu \langle\mu^E|U^{QE}(|\phi^Q\rangle\otimes|0^E\rangle)(\langle\phi^Q|\otimes\langle 0^E|)U^{QE\dagger}|\mu^E\rangle\\ &= \sum_\mu A^Q_\mu|\phi^Q\rangle\langle\phi^Q|A^{Q\dagger}_\mu,\end{aligned} \tag{4.43}$$

if we define the operator A^Q_μ by

$$A^Q_\mu|\phi^Q\rangle = \langle\mu^E|U^{QE}(|\phi^Q\rangle\otimes|0^E\rangle). \tag{4.44}$$

It can be shown that every trace-preserving, completely positive linear map ε^Q has a unitary representation and an operator–sum representation. These are the class of allowed evolutions of a quantum system.

4.5 A MEASURE OF INFORMATION

In classical communication systems involving transmission and processing of data, the information or data and the channel are appropriately represented by probabilistic models. Suppose the nonnegative numbers $p_1, p_2, \cdots$ sum to unity and thus form a probability distribution. The Shannon entropy $H(\vec{p})$ of this probability distribution is [1,2]

$$H(\vec{p}) = -\sum_k p_k \log p_k,\ \vec{p} = (p_1, p_2, \cdots). \tag{4.45}$$

We specify the base of our logarithms to be 2. If $\vec{p}$ forms the probability for some random variable X, so that $p(x_k) = p_k$ for various values x_k of a random variable X, then we often write this entropy as $H(X)$.

The Shannon entropy $H(X)$ is the fundamental measure in classical information theory, and it represents the average number of binary digits (bits) required to represent the value of X. It can be thought of as a measure of the uncertainty in the value of X expressed by the probability distribution. We can use it to define various information theoretic quantities, such as the conditional entropy

$$H(X \mid Y) = \sum_k p(y_k)H(X \mid y_k) = -\sum_{j,k} p(x_j, y_k) \log p(x_j \mid y_k) \tag{4.46}$$

for a joint distribution $p(x_j, y_k)$ and the conditional probability $p(x_j \mid y_k)$ over values of two variables X and Y. $H(X \mid Y)$ is interpreted as the average information (over x_j and y_k required to specify x_j after y_k is known. A particularly important quantity is the mutual information $I(X:Y)$ between two random variables X and Y;

$$I(X:Y) = H(X) - H(X \mid Y), \tag{4.47}$$

which is the average amount that the uncertainty about X decreases when the value of Y is known. If X represents the input of a communication channel, and Y represents the output, then $I(X:Y)$ represents the amount of information conveyed by the channel. It turns out that

$$I(X:Y) = I(Y:X). \tag{4.48}$$

The quantum mechanical definition of entropy was first given by von Neumann. Suppose ρ^Q is the density operator representing a mixed state of Q. Then the entropy is

$$S(\rho^Q) = -Tr\left(\rho^Q \log \rho^Q\right). \tag{4.49}$$

If $\lambda_1, \lambda_2, \cdots$ are the eigenvalues of ρ^Q, then

$$S(\rho^Q) = H(\bar{\lambda}), \bar{\lambda} = (\lambda_1, \lambda_2, \cdots). \tag{4.50}$$

The von Neumann entropy is the average number of two-level quantum systems (or qubits) needed to faithfully represent one of the pure states of an ensemble described by ρ^Q.

4.6 QUANTUM BLACK HOLES

Quantum entanglement is the most valuable resource in quantum information science. It is found that quantum entanglement is universally found not only in microscopic quantum systems, but also in cosmological objects such as black holes. A black hole is a singularity formed when a celestial body collapses beyond a certain radius called the event horizon. More than 30 years ago, Hawking discovered that a black hole can be regarded a quantum mechanical object that emits quantum particles called Hawking radiation. This remarkable phenomenon has been called the Hawking effect ever since the discovery. Before the Hawking effect was discovered, it was generally believed that a black hole absorbs everything surround it and nothing comes out. Hawking effects implies that a black hole also emits radiation and eventually evaporates. The question of whether the information consumed by a black hole is destroyed and lost forever, or might be recovered from the Hawking radiation that is emitted as the black hole evaporates, is one of the fundamental problems in modern physics. It requires a clear understanding of phenomena ranging from gravity to information theory. Hawking's semiclassical argument predicts that the process of black hole formation and evaporation is not unitary [3,4]. The vacuum state under the strong gravity of the black hole is composed of infalling and outgoing particle states inside and outside the event horizon. To an observer outside the black hole, the unitarity is lost because he would not be able to make any measurement in the interior of the event horizon, and as a result he would be forced to make an average over the states in H_{in} that corresponds to the Hilbert space inside of the horizon to obtain the density operator in H_{out}, the Hilbert space outside the black hole. On the other hand, there is some evidence in string theory that the formation and evaporation of black hole is a unitary process [4]. Nonetheless, the Hawking effect, discovered nearly 30 years ago, is generally accepted as very credible, and considered as would be an essential ingredient of the yet unknown correct theory of quantum gravity.

4.6.1 Microcanonical Model of a Black Hole Evaporation

Horowitz and Maldacena (HM) proposed a final-state boundary condition (FBC) in microcanonical form to reconcile the unitarity of the black hole evaporation with Hawking's semiclassical reasoning [5]. The essence of the HM proposal is to impose a unique final boundary condition at the black hole singularity, such that no information is absorbed by the singularity. The final boundary state is a maximally entangled state of collapsing matter and infalling Hawking radiation. When a black hole evaporates, particles are created in entangled pairs, with one falling into the black hole, and the other radiated to infinity. The projection of the final boundary state at the black hole singularity collapses the state into one associated with the collapsing matter, and transfers the information to the outgoing Hawking radiation. The HM model is further refined by including the unitary interactions between the collapsing matter and infalling Hawking radiation, and a random purification of the final boundary state. The black hole evaporation process as seen by the observer outside the black hole is a unitary process, which looks like a quantum teleportation process without the classical information transmitted. This indicates that nonlocal physics would be required to transmit the information outside the black hole, and inside and outside the Hilbert spaces do not have independent existence. Then one could raise the following question: "In quantum theory, which is more fundamental, unitarity or locality?" What locality requires is that there be no influence on an object due to any action taken in a region that is at spacelike separation with respect to the object. If one forces the quantum gravity be unitary to observers outside the event horizon, then nonlocality should be an essential feature of the theory.

In the HM model, the boundary state outside the event horizon is assumed to be the Unruh vacuum state [4]. As a matter of fact, Hawking's original discovery can be regarded as imposing a boundary condition at the event horizon. We would like to denote it as Hawking boundary condition (HBC), in contrast with the FBC proposed by HM (Fig. 4.1). HBC dictates that the quantum states inside and outside

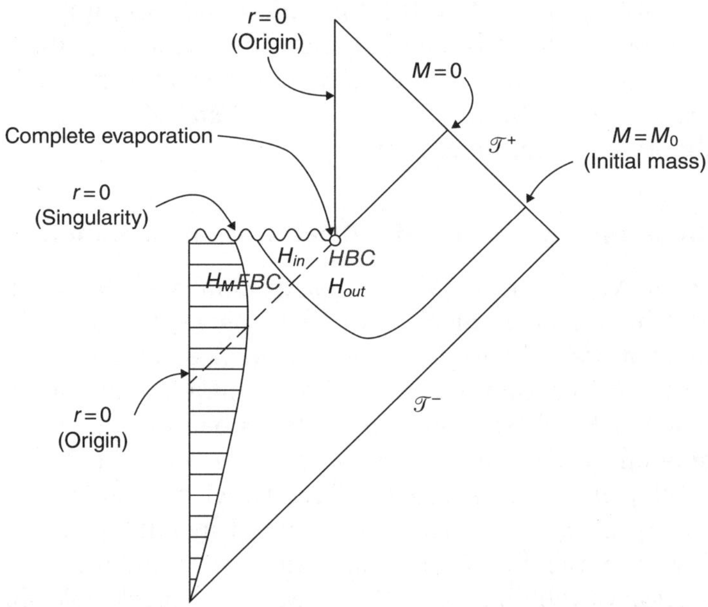

Figure 4.1. Penrose diagram of for the black hole formation and evaporation processes. A semiclassical theory that predicts the unitary evolution of black hole formation and evaporation has two boundary conditions: (1) Hawking boundary conditions (HBC) at the event horizon with quantum states in H_{in} and H_{out} maximally entangled, and (2) the final-state boundary condition (FBC) inside the black hole for the quantum states of collapsing matter in H_M and the infalling Hawking radiation in H_{in} maximally entangled. J_+ and J_- are future and past null infinity respectively. (This figure is reproduced from D. Ahn, *Phys. Rev. D*, 74, 084010 [2006].)

the event horizon of the black hole are maximally entangled. The significance of the HM proposal is that the black hole formation and evaporation process can be put into a unified picture by combining HBC together with FBC. Moreover, the process can be unitary as predicted by string theory.

It is shown [6] that the black hole evaporation process will be affected by the boundary condition outside the event horizon. Boundary states outside the event horizon affect the final state projection because the quantum states inside and outside the event horizon are entangled by HBC.

One of the critical assumptions in the HM proposal is that the internal quantum state of the black hole can be represented by maximally entangled states of collapsing matter and infalling Hawking radiation. The HM model is also based on the simplified microcanonical form, and it would be an interesting question how the FBC will look in a more general case, such as a Schwarzschild black hole.

Here, it is shown that the internal stationary state of the Schwarzschild black hole can be represented by a maximally entangled two-mode squeezed state of collapsing matter and infalling Hawking radiation. The outgoing Hawking radiation is obtained by the final state projection on the total wave function, which looks like a quantum teleportation process without the classical information transmitted. The black hole evaporation process as seen by the observer outside the black hole is a unitary process, but nonlocality is required to transmit the information outside the black hole. It is also shown that the final state projection by the evaporation process is affected by HBC at the event horizon, which clearly violates the locality principle.

The purpose of this section is to demonstrate that the FBC of Horowitz and Maldacena necessarily implies a breakdown of locality in whatever quantum theory of gravity one might construct that incorporates this proposal. We proceed by (1) reviewing the result that the original vacuum outside of a black hole evolves into a maximally entangled two-mode squeezed state on H_{in} and H_M, and then (2) showing that the interior state of the black hole is a maximally entangled two-mode squeezed state on H_{in} and H_M, where H_M is the Hilbert space for the collapsing matter. The FBC is then applied to the latter state, and the outgoing radiation is obtained by projection of this onto the former state. However, if one took an excited state outside of the black hole as the HBC, then it too could be written as a maximally entangled two-mode squeezed state on H_{in} and H_{out}—but the state of outgoing radiation that one would obtain via the FBC is orthogonal to what one obtains from the vacuum state (Eq. (4.30)). Hence, one obtains the result that the final outgoing particle state for black hole evaporation is dependent on the HBC. Since the interior and the exterior regions of the event horizon are causally disconnected, this is contrary to expectations based on locality. We interpret this as indicating that nonlocality is required to transmit information outside a black hole.

We first review the original HM proposal briefly. We assume that the quantum state of the collapsing matter belongs to a Hilbert space H_M with dimension N and $|\psi\rangle_M$ be the initial quantum state of the collapsing matter. The Hilbert space of fluctuations on the background spacetime for black hole formation and evaporation is separated into H_{in} and H_{out}, which contain quantum states localized inside and outside the event horizon, respectively. In the HM proposal, HBC is assumed to be the Unruh vacuum state $|\Phi_0\rangle_{in\otimes out}$ belonging to $H_{in} \otimes H_{out}$ in microcanonical form:

$$|\Phi_0\rangle_{in\otimes out} = \frac{1}{\sqrt{N}} \sum_i |i\rangle_{in} \otimes |i\rangle_{out}, \tag{4.51}$$

where $\{|i\langle_{in}\}$ and $\{|i\langle_{out}\}$ are orthonormal bases for H_{in} and H_{out}, respectively. The FBC imposed at the singularity requires a maximally entangled quantum state in $|\psi\rangle_M$, which is called the final boundary state, and is given by

$$_{M\otimes in}\langle\Psi| = \frac{1}{\sqrt{N}} \sum_l {}_M\langle l| \otimes_{in} \langle l|(S \otimes I), \tag{4.52}$$

where S is a random unitary transformation. The initial matter state $|\psi\rangle_M$ evolves into a state in $H_M \otimes H_{in} \otimes H_{out}$ under HBC, which is given by $|\Psi_0\rangle_{M\otimes in\otimes out} = |\psi\rangle_M \otimes |\Phi_0\rangle_{in\otimes out}$. Then, the transformation from the quantum state of collapsing matter to the state of outgoing Hawking radiation is given by the following final state projection

$$|\phi_0\rangle_{out} = {}_{M\otimes in}\langle\Psi||\Psi_0\rangle_{M\otimes in\otimes out} = \sum_i {}_M\langle i|S|\psi\rangle_M |i\rangle_{out}, \tag{4.53}$$

where the right side of Equation (4.53) is properly normalized. Let us assume that the orthonormal bases $\{|i\rangle_{out}\}$ and $\{|l\rangle_M\}$ are related by the unitary transformation T'. The quantum state of the collapsing matter is transferred to the state of the outgoing Hawking radiation with fidelity

$$f_0 = |{}_{out}\langle\phi_0|T'|\psi\rangle_M|^2. \tag{4.54}$$

We can also regard T' as a tunneling Hamiltonian, and the evaporation rate will be proportional to $(2\pi/\hbar)f_0$.

4.6.2 Elementary Relativity

Before we investigate the quantum state of a black hole, it is necessary to have some working knowledge of special and general relativity, which we are going to provide in this section. Special relativity is based on the experimental fact that the speed of light $c = 3\times10^8 m/\text{sec}$) is the same for all inertial observers. This fact leads to some rather surprising conclusions. Newtonian intuition about the absolute nature of time, the concept of simultaneity, and other familiar ideas must be revised. In comparing the coordinates of events, two inertial observers, called Lorentz observers, find that the appropriate coordinate transformations mix space and time.

In special relativity, events are characterized by the value of four coordinates: a time coordinate t, and three spatial coordinates x, y, and z. It is convenient to collect these four numbers in the form (ct, x, y, z), where the time coordinate is scaled by the speed of light, so that all coordinates have units of length. We use indices to relabel the space and time coordinates as follows [7]:

$$x^\mu = (x^0, x^1, x^2, x^3) = (ct, x, y, z). \tag{4.55}$$

Here, the superscript μ takes four values: 0, 1, 2, and 3. The x^μ are spacetime coordinates. Consider a Lorentz frame S in which two events are represented by the coordinates x^μ and $x^\mu + \Delta x^\mu$. Consider now a second Lorentz frame S', in which the same two events are described by the coordinates x'^μ and $x'^\mu + \Delta x'^\mu$, respectively. In general, not only are the coordinates x^μ and x'^μ different, so too are the coordinate differences Δx^μ and $\Delta x'^\mu$. On the other hand, both observers will agree on the value of the interval Δs^2. The interval is defined by

$$-\Delta s^2 \equiv -(\Delta x^0)^2 + (\Delta x^1)^2 + (\Delta x^2)^2 + (\Delta x^3)^2. \tag{4.56}$$

Note the minus sign in front of $(\Delta x^0)^2$, as opposed to the plus sign appearing before the spacelike differences $(\Delta x^i)^2$ $(i = 1,2,3)$. This sign encodes the fundamental difference between time and space coordinates. The agreement on the value of the interval is expressed as

$$\begin{aligned} &-(\Delta x^0)^2 + (\Delta x^1)^2 + (\Delta x^2)^2 + (\Delta x^3)^2 \\ &= -(\Delta x'^0)^2 + (\Delta x'^1)^2 + (\Delta x'^2)^2 + (\Delta x'^3)^2, \end{aligned} \tag{4.57}$$

or, in brief:

$$\Delta s^2 = \Delta s'^2. \tag{4.58}$$

The minus sign in the left-hand side of Equation (4.56) implies that $\Delta s^2 > 0$ for events that are timelike separated. Timelike separated events are events for which

$$(\Delta x^0)^2 > (\Delta x^1)^2 + (\Delta x^2)^2 + (\Delta x^3)^2. \tag{4.59}$$

The history of a particle is represented in spacetime as a curve, the world line of the particle. Any two events on the world line of a particle are timelike separated, because no particle can move faster than light, and therefore the distance the light would have travelled in the time interval that separates the events must be larger than the space separation between events. This is what Equation (4.59) implies. Events connected by the world line of a photon are said to be lightlike separated. For such a pair of events, we have $\Delta s^2 = 0$, because in this case, the two sides of Equation (4.59) are identical: The spatial separation between the events coincides with the distance that light would have travelled in the time that separates the events. Two events for which $\Delta s^2 < 0$ are said to be spacelike separated. Events that are simultaneous in a Lorentz frame but occur at different points in that same frame are spacelike separated. It is because Δs^2 can be negative that it is not written as $(\Delta s)^2$. For timelike separated events, however, we do define

$$\Delta s \equiv \sqrt{\Delta s^2} \text{ if } \Delta s^2 > 0. \tag{4.60}$$

Many times, it is useful to consider events that are infinitesimally close to each other. Small coordinate differences are needed to define velocities, and are also useful in general relativity. Infinitesimal coordinate differences are written as dx^μ, and the associated invariant interval is written as ds^2. Following Equation (4.56), we have

$$-ds^2 \equiv -(dx^0)^2 + (dx^1)^2 + (dx^2)^2 + (dx^3)^2. \tag{4.61}$$

The equality of intervals is the statement

$$ds^2 = ds'^2.$$

A very useful notation can be motivated by trying to simplify the expression for the invariant ds^2. To do this, we introduce symbols that carry subscripts instead of superscripts. Let us define

$$dx_0 \equiv -dx^0, dx_1 \equiv dx^1, dx_2 \equiv dx^2, dx_3 = dx^3. \tag{4.62}$$

The only significant change is the introduction of a minus sign for the zeroth component. All together, we write

$$dx_\mu = (dx_0, dx_1, dx_2, dx_3) \equiv (-dx^0, dx^1, dx^2, dx^3). \tag{4.63}$$

Now we can rewrite ds^2 in terms of dx^μ and dx_μ:

$$\begin{aligned} -ds^2 &\equiv -(dx^0)^2 + (dx^1)^2 + (dx^2)^2 + (dx^3)^2 \\ &= dx_0 dx^0 + dx_1 dx^1 + dx_2 dx^2 + dx_3 dx^3 \\ &= \sum_{\mu=0}^{3} dx_\mu dx^\mu. \end{aligned} \tag{4.64}$$

In the following, we will use Einstein's summation convention. In this convention, indices repeated in a single term are to be summed over the appropriate set of values. Using this summation convention, we rewrite Equation (4.64) as

$$-ds^2 = dx_\mu dx^\mu. \tag{4.65}$$

For infinitesimal timelike intervals, we define the quantity

$$ds = \sqrt{ds^2} \text{ if } ds^2 > 0. \tag{4.66}$$

We can also express the interval ds^2 using the Minkowski metric tensor $\eta_{\mu\nu}$, such that

$$-ds^2 = \eta_{\mu\nu} dx^\mu dx^\nu, \tag{4.67}$$

where $\eta_{00} = -1$, $\eta_{11} = \eta_{22} = \eta_{33} = 1$, and all the other components vanish. In matrix form, the Minkowski metric is written as:

$$\eta_{\mu\nu} = \begin{pmatrix} -1 & 0 & 0 & 0 \\ 0 & 1 & 0 & 0 \\ 0 & 0 & 1 & 0 \\ 0 & 0 & 0 & 1 \end{pmatrix}. \tag{4.68}$$

In this equation, which follows a common index identification of two-index objects with matrices, we think of μ, the first index in η, as the row index, and ν, the second index in η, as the column index. The Minkowski metric can be used to "lower indices." Indeed, Equation (4.62) can be rewritten as

$$dx_\mu = \eta_{\mu\nu} dx^\nu . \tag{4.69}$$

If we are handed a set of quantities b^μ, we can always define

$$b_\mu \equiv \eta_{\mu\nu} b^\nu . \tag{4.70}$$

Given objects a^μ and b^μ, the relativistic scalar product $a \cdot b$ is defined as

$$a \cdot b = a^\mu b_\mu = \eta_{\mu\nu} = -a^0 b^0 + a^1 b^1 + a^2 b^2 + a^3 b^3 = a_\mu b^\mu . \tag{4.71}$$

It is convenient to introduce the inverse of $\eta_{\mu\nu}$, written conveniently as $\eta^{\mu\nu}$, whose matrix form is:

$$\eta^{\mu\nu} = \begin{pmatrix} -1 & 0 & 0 & 0 \\ 0 & 1 & 0 & 0 \\ 0 & 0 & 1 & 0 \\ 0 & 0 & 0 & 1 \end{pmatrix} . \tag{4.72}$$

In index notation the inverse property is written as

$$\eta^{\nu\rho} \eta_{\rho\mu} = \delta^\nu_\mu , \tag{4.73}$$

where the Kronecker delta δ^ν_μ is defined by

$$\delta^\nu_\mu = \begin{cases} 1 \text{ if } \mu = \nu \\ 0 \text{ if } \mu \neq \nu \end{cases} . \tag{4.74}$$

Einstein's theory of general relativity is a theory of gravitation. In this elegant theory, the dynamical variables encode the geometry of spacetime. When gravitational fields are sufficiently weak and velocities are small, Newtonian gravitation is accurate enough, and one needs not work with the more complex machinery of general relativity. The spacetime of special relativity, Minkowski spacetime, is the arena for physics in the absence of gravitational fields. The geometrical properties of Minkowski spacetime are encoded by the metric formula Equation (4.67), which gives the invariant interval separating two nearby events.

Minkowski space is said to be a flat space. In the presence of a gravitational field, the metric becomes dynamical. We then write

$$-ds^2 = g_{\mu\nu}(x)dx^\mu dx^\nu, \tag{4.75}$$

where the constant metric $\eta_{\mu\nu}$ is replaced by the metric $g_{\mu\nu}(x)$. If there is a gravitational field, the metric is in general a nontrivial function of the spacetime coordinates. The metric $g_{\mu\nu}(x)$ is defined to be symmetric

$$g_{\mu\nu}(x) = g_{\nu\mu}(x). \tag{4.76}$$

It is also customary to define $g^{\mu\nu}(x)$ as the inverse of the $g_{\mu\nu}(x)$:

$$g^{\mu\rho}(x)g_{\rho\nu}(x) = \delta^\mu_\nu. \tag{4.77}$$

Because the metric is now a function of spacetime, the derivatives involving spacetime coordinates are modified in general relativity. Especially, the divergence of a given dynamical quantity V^μ is modified as

$$\frac{\partial}{\partial x^\mu}V^\mu \mapsto \frac{1}{\sqrt{-g}}\frac{\partial}{\partial x^\mu}\left(\sqrt{-g}V^\mu\right) = \frac{1}{\sqrt{-g}}\frac{\partial}{\partial x^\mu}\left(\sqrt{-g}g^{\mu\nu}V_\nu\right) \tag{4.78}$$

where a scalar density g is defined by

$$g = \det(g_{\mu\nu}). \tag{4.79}$$

On the other hand, the gradient of a scalar function takes the same form in general relativity. As a result, the wave Equation in general relativity is written as [8]

$$\frac{\partial}{\partial x^\mu}\frac{\partial}{\partial x_\mu}\phi(x) \mapsto (-g)^{1/2}\frac{\partial}{\partial x^\mu}\left[g^{\mu\nu}(-g)^{1/2}\frac{\partial}{\partial x^\nu}\right]\phi(x) = 0. \tag{4.80}$$

4.6.3 Hawking Radiation from a Schwarzschild Black Hole

In this section, we first study the derivation of Hawking radiation and extend the results to the field of a collapsing matter inside the event horizon. The stationary Schwarzschild black hole is represented by the metric [4,8]

$$ds^2 = -\left(1-\frac{2M}{r}\right)dt^2 + \frac{dr^2}{1-\frac{2M}{r}} + r^2\left(d\theta^2 + \sin^2\theta d\varphi^2\right), \tag{4.81}$$

where M is the mass of the black hole. At $r = 2M$, the Schwarzschild spacetime has an event horizon. The general coordinate is $x^\mu = (t, r, \theta, \varphi)$, with the metric tensor given by

$$g_{tt} = -\left(1-\frac{2M}{r}\right),\ g_{rr} = \frac{1}{1-\frac{2M}{r}},\ g_{\theta\theta} = r^2,\ g_{\varphi\varphi} = r^2\sin^2\theta. \tag{4.82}$$

The massless scalar field satisfies the wave equation

$$(-g)^{1/2}\frac{\partial}{\partial x^\mu}\left[g^{\mu\nu}(-g)^{1/2}\frac{\partial}{\partial x^\nu}\right]\phi = 0, \tag{4.83}$$

and the positive frequency normal mode solution is given by

$$\phi_{\omega lm} = (2\pi|\omega|)^{-1/2}e^{-i\omega t}f_{\omega l}(r)Y_{lm}(\theta,\varphi), \tag{4.84}$$

where $f_{\omega l}(r)$ satisfies

$$\frac{\partial^2 f_{\omega l}}{\partial r^{*2}} + \omega^2 f_{\omega l} - \left(1-\frac{2M}{r}\right)\left(\frac{l(l+1)}{r^2} + \frac{2M}{r^3}\right)f_{\omega l} = 0, \tag{4.85}$$

with $r^* = r + 2M\ln(r/2M - 1)$. If we denote the radiation part of the wave coming out of the past horizon of the black hole by $f_{\omega l}^-$, then it is given by

$$f_{\omega l}^-(r) \approx e^{i\omega r^*} + A_{\omega l}^- e^{-i\omega r^*}. \tag{4.86}$$

In Kruskal coordinate, the Schwarzschild metric becomes [4]

$$\begin{aligned} &ds^2 = -2M\frac{e^{-r/2M}}{r}d\bar{u}d\bar{v} + r^2 d\theta^2 + r^2\sin^2\theta d\varphi^2, \\ &\bar{u} = -4Me^{-u/4M},\ \bar{v} = 4Me^{v/4M}, \\ &u = t - r^*,\ v = t + r^*, \\ &r^* = r + 2M\ln(r/2M - 1). \end{aligned} \tag{4.87}$$

The Kruskal extension of Schwarzschild spacetime is shown in Figure 4.2. Since the Killing vector in Kruskal coordinates is given by $\partial/\partial\bar{u}$ on H^- (Fig. 4.2), the solution in Kruskal coordinates is given by

$$\bar{\phi}_{\varpi lm} = (2\pi|\varpi|)^{-1/2} e^{-i\varpi\bar{u}} Y_{lm}(\theta,\varphi). \tag{4.88}$$

On the other hand, the original positive frequency normal mode on H^- can be written by

$$\phi^-_{\omega lm} = (2\pi|\omega|)^{-1/2}(e^{-i\omega u} + A^-_{\omega l} e^{-i\omega v}) Y_{lm}(\theta,\varphi) \tag{4.89}$$

Using $e^{-i\omega u} = (|\bar{u}|/4M)^{i4M\omega}$ and $e^{-i\omega v} = (\bar{v}/4M)^{-i4M\omega}$, and the fact that $\bar{v} = 0$ on H^-, we obtain

$$\phi^-_{\omega lm} = (2\pi|\omega|)^{-1/2}(|\bar{u}|/4M)^{i4M\omega} Y_{lm}(\theta,\varphi). \tag{4.90}$$

Since $\bar{u} < 0$ in region I and $\bar{u} > 0$ in region II of the Figure 4.2, the wave coming out of the past horizon of the black hole on H^- can be written by

$$\phi^-_{\omega lm} = (e^{2\pi M\omega}\,{}_{out}\phi_{\omega lm} + e^{-2\pi M\omega}\,{}_{in}\phi_{\omega lm})/(2\sinh(4\pi M\omega))^{1/2}, \tag{4.91}$$

where ${}_{out}\phi_{\omega lm}$ vanishes inside the event horizon (region II), and ${}_{in}\phi_{\omega lm}$ vanishes in the exterior region of the black hole (region I). In Equation

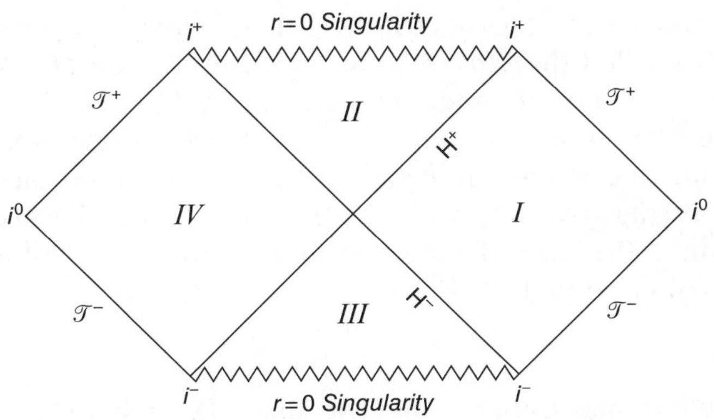

Figure 4.2. The Kruskal extention of the Schwarzschild spacetime. In region I, null asymptotes H_+ and H_- act as future and past event horizons, respectively. The boundary lines labelled J^+ and J^- are future and past null infinities, respectively, and i^o is the spacelike infinity. (This figure is reproduced from the paper by the author, D. Ahn, *Phys. Rev. D*, 74, 084010 [2006]).

(4.91), we have used the fact that $(-1)^{-i4M\omega} = e^{4\pi M\omega}$. The above definition of the positive frequency solution in terms of ${}_{out}\phi_{\omega lm}$ and ${}_{in}\phi_{\omega lm}$ leads to the Bogoliubov transformations for the particle creation and annihilation operators in Schwarzschild and Kruskal spacetimes (Appendix A):

$$\begin{aligned} a_{K,\omega lm} &= \cosh r_\omega b_{out,\omega lm} - \sinh r_\omega b^\dagger_{in,\omega lm}, \\ a^\dagger_{K,\omega lm} &= \cosh r_\omega b^\dagger_{out,\omega lm} - \sinh r_\omega b_{in,\omega lm}, \\ \tanh r_\omega &= e^{-4\pi M\omega}, \cosh r_\omega = \left(1 - e^{-8\pi M\omega}\right)^{-1/2}, \end{aligned} \tag{4.92}$$

where $a^\dagger_{K,\omega lm}$ and $a_{K,\omega lm}$ are the creation and annihilation operators, respectively, acting on the Kruskal vacuum in region I, $b^\dagger_{out,\omega lm}$ and $b_{out,\omega lm}$ are the creation and annihilation operators acting on the Schwarzschild vacuum of the exterior region of a black hole, and $b^\dagger_{in,\omega lm}$ and $b_{in,\omega lm}$ are the creation and annihilation operators acting on the Schwarzschild vacuum inside the event horizon. Then the ground state $|\Phi_o\rangle_{in\otimes out}$, which looks like the vacuum in the far past, is a maximally entangled two-mode squeezed state on $H_{in} \otimes H_{out}$ (Appendix B):

$$|\Phi_o\rangle_{in\otimes out} = \frac{1}{\cosh r_\omega} \sum_n e^{-4\pi M\omega n} |n\rangle_{in} \otimes |n\rangle_{out}, \tag{4.93}$$

where $\{|n\rangle_{in}\}$ and $\{|n\rangle_{out}\}$ are orthonormal bases (normal mode solutions) for H_{in} and H_{out}, respectively. Equation (4.93) shows that the original vacuum state evolves to a two-mode squeezed state $|\Phi_o\rangle_{in\otimes out}$, which is also called the Unruh state, which resides on $H_{in} \otimes H_{out}$. This state contains a flux of outgoing particles in H_{out}. For the observer outside the black hole, the unitarity is lost because he would not be able to do any measurement in H_{in}, and as a result, he would be forced to take an average over the states in H_{in} to obtain the density operator in H_{out}. Unlike the case of a microcanonical form, the Hilbert spaces are infinite dimensional.

4.6.4 Gravitational Collapse and the Black Hole State

Here, we show that the field inside the event horizon can also be decomposed into the collapsing matter field, and the advanced wave incoming from infinity has a similar form as the Hawking radiation, thus paving a road to the final boundary condition. The Penrose diagram of a collapsing star is shown in Figure 4.1. The region I is a fragmentation of Figure 4.2, including the region *II* (black hole). The collapsing shell metric in two dimensions is given by

$$ds^2 = \begin{cases} -d\tau^2 + dr^2, r < R(\tau) \\ -\left(1 - \frac{2M}{r}\right)dt^2 + \frac{dr^2}{1 - \frac{2M}{r}}, r > R(\tau), \end{cases} \tag{4.94}$$

with the shell radius $R(\tau)$ defined by

$$R(\tau) = \begin{cases} R_o, \tau < 0 \\ R_o - \nu\tau, \tau > 0. \end{cases} \tag{4.95}$$

We define the advanced and retarded null coordinates as

$$\begin{aligned} V &= \tau + r - R_o, \quad U = \tau - r + R_o, \\ v^* &= t + r - R_o^*, \quad u^* = t - r^* + R_o^*, \end{aligned} \tag{4.96}$$

with $R_o^* = R_o + 2M\ln(R_o/2M - 1)$. The null coordinates are chosen such that the shell begins to collapse at $U = V = u^* = v^* = 0$. After some mathematical manipulations, we obtain near the shell surface,

$$v^* \approx 4M\ln\left(1 - \frac{\nu V}{(1-\nu)(R_o - 2M)}\right), \tag{4.97}$$

and

$$u^* \approx -4M\ln\left(1 - \frac{\nu U}{(1+\nu)(R_o - 2M)}\right). \tag{4.98}$$

We now consider the massless scalar field inside the black hole, incoming from infinity, which is given by

$$\phi^+_{\omega lm} = (2\pi|\omega|)^{-1/2}(e^{-iwv} - A^+_{\omega l}e^{-i\omega u})Y_{lm}(\theta,\varphi), \tag{4.99}$$

where $A^+_{\omega l}$ is chosen such that the field vanishes at $r = 0$. The normal mode on H^+ becomes

$$\begin{aligned} \phi^+_{\omega lm} &= (2\pi|\omega|)^{-1/2} e^{i\omega R_o^*}\left|1 - \frac{\nu V}{(1-\nu)(R_o - 2M)}\right|^{-i4M\omega} Y_{lm}(\theta,\varphi) \\ &= \left(e^{2\pi M\omega}\,{}_M\phi_{\omega lm} + e^{-2\pi M\omega}\,{}_{in}\phi_{\omega lm}\right)/(2\sinh(4\pi M\omega))^{-1/2}, \end{aligned} \tag{4.100}$$

where ${}_{M}\phi_{\omega lm}$ is a mode that vanishes for outside the shell, $V > (1-\nu)(R_o - 2M)/\nu$ and ${}_{in}\phi_{\omega lm}$ is the solution vanishing inside the shell, $V < (1-\nu)(R_o - 2M)/\nu$.

The above definition of the positive frequency solution in terms of ${}_{M}\phi_{\omega lm}$ and ${}_{in}\phi_{\omega lm}$ leads to the Bogoliubov transformation for the particle creation and annihilation operators in Schwarzschild and Kruskal spacetimes, as in the case of the exterior region of the black hole:

$$\begin{aligned} c_{K,\omega lm} &= \cosh r_\omega b_{M,\omega lm} - \sinh r_\omega b^\dagger_{in,\omega lm}, \\ c^\dagger_{K,\omega lm} &= \cosh r_\omega b^\dagger_{M,\omega lm} - \sinh r_\omega b_{in,\omega lm}, \\ \tanh r_\omega &= e^{-4\pi M\omega}, \cosh r_\omega = \left(1 - e^{-8\pi M\omega}\right)^{-1/2}, \end{aligned} \tag{4.101}$$

where $c^\dagger_{K,\omega lm}$ and $c_{K,\omega lm}$ are the creation and annihilation operators, respectively, acting on the Kruskal vacuum in region II, $b^\dagger_{M,\omega lm}$ and $b_{M,\omega lm}$ are the creation and annihilation operators for the collapsing matter acting on the Schwarzschild vacuum, and $b^\dagger_{in,\omega lm}$ and $b_{in,\omega lm}$ are the creation and annihilation operators acting on the Schwarzschild vacuum inside the event horizon. Then the stationary state $|\Phi_o\rangle_{M\otimes in}$ inside the black hole is a maximally entangled two-mode squeezed state on $H_M \otimes H_{in}$

$$|\Phi_o\rangle_{M\otimes in} = \frac{1}{\cosh r_\omega} \sum_n e^{-4\pi M\omega n} |n\rangle_M \otimes |n\rangle_{in}, \tag{4.102}$$

where $\{|n\rangle_{in}\}$ and $\{|n\rangle_M\}$ are orthonormal bases (normal mode solutions) for H_{in} and H_M, respectively.

4.6.4 The Final State and Hawking Boundary Conditions

In the previous section, we have shown that the internal state for a fixed shell can be represented by the two-mode squeezed state of collapsing matter and infalling Hawking radiation. We now apply the final boundary condition at the singularity, which is given by the state $|\Psi\rangle_{M\otimes in}$:

$$\begin{aligned} {}_{M\otimes in}\langle\Psi| &= {}_{M\otimes in}\langle\Phi_o|(S\otimes I) \\ &= \frac{1}{\cosh r_\omega} \sum_n e^{-4\pi M\omega n} \left({}_M\langle n|S \right) \otimes {}_{in}\langle n|, \end{aligned} \tag{4.103}$$

where S is a random unitary transformation. The random unitary transformation S represents that the interior of the black hole is a turbulent place, and it is difficult to distinguish the two subsystems presumed to compose the interior of the black hole. The stationary state $|\Phi_o\rangle_{M\otimes in}$, a maximally entangled two-mode squeezed state of the infalling Hawking radiation and collapsing matter, derived by the semiclassical description, corresponds to the state far from the singularity.

As in the case of the microcanonical form, the initial matter state $|\psi\rangle_M \in H_M$, a pure state that will form the black hole, evolves into a state in $H_M \otimes H_{in} \otimes H_{out}$ under HBC, which is given by

$$\begin{aligned}|\Psi_0\rangle_{M\otimes in\otimes out} &= |\psi\rangle_M \otimes |\Phi_0\rangle_{in\otimes out} \\ &= |\psi\rangle_M \otimes \left(\frac{1}{\cosh r_\omega}\sum_n e^{-4\pi M\omega n}|n\rangle_{in}\otimes|n\rangle_{out}\right).\end{aligned} \tag{4.104}$$

The transformation from the quantum state of collapsing matter into the state of outgoing Hawking radiation when the black hole evaporates is given by the following final state projection

$$\begin{aligned}|\phi_0\rangle_{out} &= {}_{M\otimes in}\langle\Psi||\Psi_0\rangle_{M\otimes in\otimes out} \\ &= \frac{1}{\cosh^2 r_\omega}\sum_{n,m} e^{-4\pi M\omega(n+m)}\ {}_M\langle m|S|\psi\rangle_M\ {}_{in}\langle m||n\rangle_{in}\otimes|n\rangle_{out} \\ &= \frac{1}{\cosh^2 r_\omega}\sum_n e^{-8\pi M\omega n}\ {}_M\langle n|S|\psi\rangle_M |n\rangle_{out}.\end{aligned} \tag{4.105}$$

Equation (4.105) shows that the black hole evaporation process is a unitary process from a pure state $|\psi\rangle_M$ in H_M to another pure state $|\phi_0\rangle_{out}$ in H_{out}, which looks like a quantum teleportation process without the classical information transmitted. On the other hand, this indicates that nonlocal physics would be required to transmit the information outside the black hole, and inside and outside the Hilbert spaces do not have independent existence. To continue our discussion of the nonlocality, we consider the change of HBC on the final evaporation process. If quantum gravity is a local theory, then the final outgoing particle state at the black hole evaporation should be independent of the HBC, because the interior and the exterior regions of the event horizon are causally disconnected. In the following, we show that this is not the case. We assume the Kruskal excited state as HBC, which is given by (Appendix B):

$$
\begin{aligned}
|\Phi_1\rangle_{in\otimes out} &= a^{\dagger}_{K,\omega lm}|\Phi_o\rangle_{in\otimes out} \\
&= \left(\cosh r_\omega b^{\dagger}_{out,\omega lm} - \sinh r_\omega b_{in,\omega lm}\right)\left(\frac{1}{\cosh r_\omega}\sum_n e^{-4\pi M\omega n}|n\rangle_{in}\otimes|n\rangle_{out}\right) \\
&= \frac{1}{\cosh^2 r_\omega}\sum_n e^{-4\pi M\omega n}\sqrt{n+1}|n\rangle_{in}\otimes|n+1\rangle_{out}.
\end{aligned}
\tag{4.106}
$$

Then the initial matter state $|\psi\rangle_M$ evolves into a state in $H_M \otimes H_{in} \otimes H_{out}$ under the new HBC, which is given by

$$
\begin{aligned}
|\Psi_1\rangle_{M\otimes in\otimes out} &= |\psi\rangle_M \otimes |\Phi_1\rangle_{in\otimes out} \\
&= |\psi\rangle_M \otimes \left(\frac{1}{\cosh^2 r_\omega}\sum_n e^{-4\pi M\omega n}\sqrt{n+1}|n\rangle_{in}\otimes|n+1\rangle_{out}\right),
\end{aligned}
\tag{4.107}
$$

and the state of outgoing Hawking radiation when the black hole evaporates given by the final state projection is

$$
\begin{aligned}
|\phi_1\rangle_{out} &= {}_{M\otimes in}\langle\Psi||\Psi_1\rangle_{M\otimes in\otimes out} \\
&= \frac{1}{\cosh^3 r_\omega}\sum_{n,m} e^{-4\pi M\omega(n+m)}\,{}_M\langle m|S|\psi\rangle_M\,{}_{in}\langle m||n\rangle_{in}\otimes\sqrt{n+1}|n+1\rangle_{out} \\
&= \frac{1}{\cosh^3 r_\omega}\sum_n e^{-8\pi M\omega n}\,{}_M\langle n|S|\psi\rangle_M\sqrt{n+1}|n+1\rangle_{out}.
\end{aligned}
\tag{4.108}
$$

We would like to calculate the inner product between $|\phi_o\rangle_{out}$ and $|\phi_1\rangle_{out}$ to see how they are related to the final state projection:

$$
\begin{aligned}
&{}_{out}\langle\phi_1||\phi_o\rangle_{out} \\
&= \frac{1}{\cosh^5 r_\omega}\sum_{n,m}\sqrt{m+1}e^{-8\pi M\omega(n+m)}\,{}_M\langle\psi|S^{\dagger}|m\rangle_M\,{}_M\langle n|S|\psi\rangle_M\,{}_{out}\langle m+1||n\rangle_{out} \\
&= \frac{1}{\cosh^5 r_\omega}\sum_n\sqrt{n}e^{-8\pi M\omega(2n-1)}\,{}_M\langle\psi|S^{\dagger}|n-1\rangle_M\,{}_M\langle n|S|\psi\rangle_M \\
&= {}_M\langle\psi|\left(\frac{1}{\cosh^5 r_\omega}\sum_n\sqrt{n}e^{-8\pi M\omega(2n-1)}S^{\dagger}|n-1\rangle_M\,{}_M\langle n|S\right)|\psi\rangle_M \\
&= {}_M\langle\psi|\hat{Z}|\psi\rangle_M,
\end{aligned}
\tag{4.109}
$$

where an operator $\hat{Z}$ is defined by

$$\hat{Z} = \frac{1}{\cosh^5 r_\omega} \sum_n \sqrt{n} e^{-8\pi M\omega(2n-1)} S^\dagger |n-1\rangle_{M\ M}\langle n|S. \tag{4.110}$$

From an elementary quantum theory, the term ${}_M\langle\psi|\hat{Z}|\psi\rangle_M$ in Equation (4.84) is nothing but an expectation value of $\hat{Z}$ averaged over the initial matter state, which can also be obtained by the trace operation, which is given by

$$\begin{aligned}
{}_M\langle\psi|\hat{Z}|\psi\rangle_M &\cong tr(\hat{Z}) \\
&= tr\left(\frac{1}{\cosh^5 r_\omega} \sum_n \sqrt{n} e^{-8\pi M\omega(2n-1)} S^\dagger |n-1\rangle_{M\ M}\langle n|S\right) \\
&= tr\left(\frac{1}{\cosh^5 r_\omega} \sum_n \sqrt{n} e^{-8\pi M\omega(2n-1)} |n-1\rangle_{M\ M}\langle n|SS^\dagger\right) \\
&= tr\left(\frac{1}{\cosh^5 r_\omega} \sum_n \sqrt{n} e^{-8\pi M\omega(2n-1)} |n-1\rangle_{M\ M}\langle n|\right) \\
&= 0,
\end{aligned} \tag{4.111}$$

where we have used the fact that $tr(AB) = tr(BA)$, and S is unitary. Equation (4.111) shows that $|\phi_o\rangle_{out}$ and $|\phi_1\rangle_{out}$ are orthogonal states, and the final state projection by the evaporation process is definitely affected by the HBC, which clearly violates the locality principle.

We have shown that the internal stationary state of the Schwarzschild black hole can be represented by an entangled two-mode squeezed state of collapsing matter and infalling Hawking radiation. The final boundary condition at the singularity is then described by the random unitary transformation acting on the collapsing matter field. The outgoing Hawking radiation is obtained by the final state projection on the total wave function, which looks like a quantum teleportation process without the classical information transmitted. The black hole evaporation process as seen by the observer outside the black hole is now a unitary process, but nonlocal physics is required to transmit the information outside the black hole. It is also shown that the final state projection by the evaporation process is definitely affected by the quantum state outside the event horizon, which clearly violates the locality principle.

APPENDIX A: DERIVATION OF EQUATION (4.82)

Let the quantum fields corresponding to the wave functions in Kruskal and Schwarzschild spacetimes be denoted as

$$ {}_K\hat{\phi}_{\omega lm} = a_{K,\omega lm}\, {}_K u_{\omega lm} + a^{\dagger}_{K,\omega lm}\, {}_K u^{*}_{\omega lm}, \tag{4.A1} $$

and

$$ \hat{\phi}^{-}_{\omega lm} = \left(e^{2\pi M\omega}\, {}_{out}\hat{\phi}_{\omega lm} + e^{-2\pi M\omega}\, {}_{in}\hat{\phi}_{\omega lm}\right) / \left(2\sinh(4\pi M\omega)\right)^{1/2}, \tag{4.A2} $$

with

$$ \begin{aligned} {}_{out}\hat{\phi}_{\omega lm} &= b_{out,\omega lm}\, {}_{out}u_{\omega lm} + b^{\dagger}_{out,\omega lm}\, {}_{out}u^{*}_{\omega lm}, \\ {}_{in}\hat{\phi}_{\omega lm} &= b_{in,\omega lm}\, {}_{in}u_{\omega lm} + b^{\dagger}_{in,\omega lm}\, {}_{in}u^{*}_{\omega lm}. \end{aligned} \tag{4.A3} $$

Here, ${}_{\alpha}u_{\beta}$ is a modal function. Quantum fields given by Equations (4.A1) and (4.A2) should be equivalent on H^{-}, and, as a result, we obtain

$$ \begin{aligned} a_{K,\omega lm} = &\frac{e^{2\pi M\omega}}{\left(2\sinh(4\pi M\omega)\right)^{1/2}} b_{out,\omega lm}\left({}_K u_{\omega lm}, {}_{out}u_{\omega lm}\right) \\ &+ \frac{e^{2\pi M\omega}}{\left(2\sinh(4\pi M\omega)\right)^{1/2}} b^{\dagger}_{out,\omega lm}\left({}_K u_{\omega lm}, {}_{out}u^{*}_{\omega lm}\right) \\ &+ \frac{e^{-2\pi M\omega}}{\left(2\sinh(4\pi M\omega)\right)^{1/2}} b_{in,\omega lm}\left({}_K u_{\omega lm}, {}_{in}u_{\omega lm}\right) \\ &+ \frac{e^{-2\pi M\omega}}{\left(2\sinh(4\pi M\omega)\right)^{1/2}} b^{\dagger}_{in,\omega lm}\left({}_K u_{\omega lm}, {}_{in}u^{*}_{\omega lm}\right), \end{aligned} \tag{4.A4} $$

where (ϕ, ψ) is the Klein–Gordon inner product. Calculations of Klein–Gordon inner products are straightforward, and after some mathematical manipulations, we obtain

$$ \begin{aligned} a_{K,\omega lm} &= \frac{e^{2\pi M\omega}}{\left(2\sinh(4\pi M\omega)\right)^{1/2}} b_{out,\omega lm} - \frac{e^{-2\pi M\omega}}{\left(2\sinh(4\pi M\omega)\right)^{1/2}} b^{\dagger}_{in,\omega lm} \\ &= \cosh r_{\omega} b_{out,\omega lm} - \sinh r_{\omega} b^{\dagger}_{in,\omega lm}, \end{aligned} \tag{4.A5} $$

with $\cosh r_{\omega} = (1 - e^{-8\pi M\omega})^{-1/2}$.

APPENDIX B: DERIVATION OF EQUATIONS (4.93) AND (4.106)

Before we proceed, let us simplify the notation as follows:

$$a_{K,\omega lm} = a,\ b_{out,\omega lm} = b,\ b^{\dagger}_{in,\omega lm} = \bar{b}^{\dagger},\ r_{\omega} = r. \tag{4.B1}$$

Then, Equation (4.A5) becomes

$$a = \cosh r b - \sinh r \bar{b}^{\dagger}, \tag{4.B2}$$

From the condition $a|\Phi_o\rangle_{in\otimes out} = 0$, we have

$$(\cosh r b - \sinh r \bar{b}^{\dagger})|\Phi_o\rangle_{in\otimes out} = 0, \tag{4.B3}$$

or

$$(b - s\bar{b}^{\dagger})|\Phi_o\rangle_{in\otimes out} = 0,\ s = \tanh r. \tag{4.B4}$$

Now, we assume that the Kruskal vacuum $|\Phi_o\rangle_{in\otimes out}$ is related to the Schwarzschild vacuum $|0\rangle_S$ by

$$|\Phi_o\rangle_{in\otimes out} = F(b,\bar{b}^{\dagger})|0\rangle_S \tag{4.B5}$$

where F is some function to be determined later.

From $[b,b^{\dagger}] = 1$, we obtain

$$\left[b,(b^{\dagger})^m\right] = \frac{\partial}{\partial b^{\dagger}}(b^{\dagger})^m \quad \text{and} \quad [b,F] = \frac{\partial}{\partial b^{\dagger}}F.$$

Then, using Equations (4.B4) and (4.B5), we get the following differential Equation for F:

$$\frac{\partial F}{\partial b^{\dagger}} - s\bar{b}^{\dagger}F = 0, \tag{4.B6}$$

and the solution is given by

$$F \propto \exp(sb^{\dagger}\bar{b}^{\dagger}). \tag{4.B7}$$

By substituting Equation (4.B7) into Equation (4.B5), and by properly normalizing the state vector, we get

$$|\Phi_o\rangle_{in\otimes out} = (1-s^2)^{1/2}\sum_n s^n |n\rangle_{in} \otimes |n\rangle_{out}, \tag{4.B8}$$

which agrees with Equation (4.67).

In the following, we derive Equation (4.80). From $a^\dagger = \cosh r b^\dagger - \sinh r \bar{b}$, we have

$$\begin{aligned}
|\Phi_1\rangle_{in\otimes out} &= a^\dagger |\Phi_o\rangle_{in\otimes out} \\
&= \left(\cosh r b^\dagger - \sinh r \bar{b}\right)\left(1-s^2\right)^{1/2} \sum_n s^n |n\rangle_{in} \otimes |n\rangle_{out} \\
&= \left(1-s^2\right)^{1/2} \cosh r \sum_{n=0} s^n \sqrt{n+1}\,|n\rangle_{in} \otimes |n+1\rangle_{out} \\
&\quad - \left(1-s^2\right)^{1/2} \sinh r \sum_{n=1} s^n \sqrt{n}\,|n-1\rangle_{in} \otimes |n\rangle_{out} \\
&= \left(1-s^2\right)^{1/2} \cosh r \sum_{n=0} s^n \sqrt{n+1}\,|n\rangle_{in} \otimes |n+1\rangle_{out} \\
&\quad - \left(1-s^2\right)^{1/2} \cosh r \sum_{n=1} s^{n+2} \sqrt{n+1}\,|n\rangle_{in} \otimes |n+1\rangle_{out} \\
&= \left(1-s^2\right)^{1/2} \left(1-s^2\right) \cosh r \sum_{n=0} s^n \sqrt{n+1}\,|n\rangle_{in} \otimes |n+1\rangle_{out} \\
&= \frac{1}{\cosh^2 r} \sum_{n=0} \tanh^n r \sqrt{n+1}\,|n\rangle_{in} \otimes |n+1\rangle_{out},
\end{aligned}$$

where we have used the fact $\cosh r = (1-s^2)^{-1/2}$.

PROBLEMS

1. Given state vectors

$$|\psi_1\rangle = \frac{1}{\sqrt{2}}\begin{pmatrix} 1 \\ -1 \end{pmatrix} \quad \text{and} \quad |\psi_2\rangle = \frac{1}{\sqrt{5}}\begin{pmatrix} 2 \\ 1 \end{pmatrix},$$

calculate $|\psi_3\rangle = |\psi_1\rangle \otimes |\psi_2\rangle$.

2. Matrix representation of CNOT gate is given by

$$U_{\text{CNOT}} = \begin{pmatrix} 1 & 0 & 0 & 0 \\ 0 & 1 & 0 & 0 \\ 0 & 0 & 0 & 1 \\ 0 & 0 & 1 & 0 \end{pmatrix}.$$

For given computational basis

$$|0\rangle = \begin{pmatrix} 1 \\ 0 \end{pmatrix} \quad \text{and} \quad |1\rangle = \begin{pmatrix} 0 \\ 1 \end{pmatrix},$$

show that $U_{\text{CNOT}}(|1\rangle \otimes |0\rangle) = |1\rangle \otimes |1\rangle$ and $U_{\text{CNOT}}(|1\rangle \otimes |1\rangle) = |1\rangle \otimes |0\rangle$.

3. Let the purified quantum state $|\psi^{RQ}\rangle$ is given by $|\psi^{RQ}\rangle = \sum_k \sqrt{p_k}\,|k^R\rangle \otimes |\phi_k^Q\rangle$, where $\{|k^R\rangle\}$ is the orthonormal basis of R. Then for arbitrary operator X^Q, which operates on Q only show that $\langle \psi^{RQ}| I^R \otimes X^Q |\psi^{RQ}\rangle = \sum_k p_k \langle \phi_k^Q | X^Q | \phi_k^Q \rangle$.

REFERENCES

[1] M. A. Nielsen and I. L. Chuang, *Quantum Computation and Quantum Information*. New York: Cambridge University Press, 2000.

[2] M. Nakahara and T. Ohmi, *Quantum Computing*. London: CRC Press, 2008.

[3] R. M. Wald, *Quantum Field Theory in Curved Spacetime and Black Hole Thermodynamics*. Chicago, IL: University of Chicago Press, 1994.

[4] N. D. Birrell and P. C. W. Davies, *Quantum Fields in Curved Space*. New York: Cambridge University Press, 1982.

[5] G. T. Horowotz and J. Maldacena, *J. High Energy Phys.*, 2, p. 8, 2004.

[6] D. Ahn, *Phys. Rev. D*, 74, p. 84010, 2006.

[7] B. Zwiebach, *A First Course in String Theory*. Cambridge: Cambridge University Press, 2004.

[8] S. Weinberg, *Gravitation and Cosmology*. New York: John Wiley & Sons, 1972.

5 Modern Semiconductor Laser Theory

Optical gain is probably one of the most important basic properties of semiconductor lasers. A free carrier model with a phenomenological damping is often used in an optical gain calculation for simplicity. This model provides some useful insight to the basic properties of a semiconductor laser. However, we need to include a Coulomb interaction between charged carriers for a more realistic description of the semiconductor laser. For example, many-body effects have been known to be very important in gallium nitride (GaN)-based lasers. Since Coulomb interaction processes involve more than one carrier, these effects are often called many-body effects. Coulomb interaction includes attractive interaction between electrons and holes (interband) and repulsive interaction between same carriers (intraband), and is the dominant contributor to optical damping. The optical gain and the refractive index spectra are significantly affected by these Coulomb interactions. That is, the spectral position of the optical gain and the refractive index spectra are shifted through bandgap renormalization, and their shapes are modified through Coulomb correlation effects.

We need quantum mechanical many-body techniques to analyze these Coulomb interaction processes. The equations of motion derived from the many-body Hamiltonian couple expectation values of products of two particle operators to those of the product of four particle operators, which, in turn, are coupled to six operator expectation values, and so on. This results in an infinite hierarchy of coupled differential equations involving expectation values of products of higher numbers of field operators. Thus, we need to use truncation procedures to deal with this problem in reality. For example, the Hartree–Fock equation can be obtained by factorizing higher-order expectation values into products of expectation values of products of two particle operators. Improvements on Hartree–Fock equations are obtained by delaying the

Engineering Quantum Mechanics, First Edition. Doyeol Ahn, Seoung-Hwan Park.

factorization procedure to the next level of the hierarchy. Contributions beyond the Hartree–Fock approximation are often called collision or correlation contributions.

In the Hartree–Fock equations without collision contributions, important many-body effects such as bandgap renormalization and interband Coulomb enhancement appear. However, collision contributions give rise to carrier and polarization relaxation, as well as plasma screening. The effects of plasma screening are included phenomenologically by replacing the bare Coulomb potential V_q with the screened Coulomb potential V_{sq}. We can obtain the semiconductor Bloch equations in the screened Hartree–Fock approximation. Then, bandgap renormalization contains contributions from the Coulomb-hole (CH) self-energy, and from the screened-exchange (SX) shift in the screened Hartree–Fock approximation.

The collision contribution (e.g., $\partial p/\partial t\,|_{col}$ or $\partial n/\partial t\,|_{col}$) describes polarization decay (dephasing), and can be written as $\partial p/\partial t\,|_{col} \simeq -\gamma p$ in the simplest approximation. Then, using the rate equation approximation that the electric field intensity and the total carrier density vary little in the dipole lifetime $1/\gamma$, we obtain the expression for an optical gain with the Lorentzian line shape function. However, this model, which uses a Lorentzian line shape function with a constant intraband relaxation time, is insufficient in explaining the gain spectra of high-quality samples accurately. In particular, an anomalous absorption region below the bandgap and discrepancies between the transparency points and the Fermi-level separation appear in the gain spectra when a Lorentzian line shape is used. To overcome these problems, an optical gain model with many-body effects has been presented in the Markovian (Lorentzian) limit, where the carrier–carrier scattering is considered to obtain the line shape function. Another approach is to replace the Lorentzian line shape function with a sharper spectral function, that is, with a non-Markovian (Gaussian) line shape.

The effect of the Lorentzian line shape function with k-dependent intraband relaxation time is similar to that of the Gaussian line shape function in a sense that the inclusion of carrier–carrier scattering in the intraband relaxation time gives a steeper low energy tail than the Lorentzian line shape with a constant intraband relaxation time. The Gaussian line shape function is connected with memory effects in the system–reservoir interaction, while the Lorentzian line shape corresponds to instantaneous reaction to an external perturbation.

In this chapter, we derive the optical gain model with a non-Markovian (Gaussian) line shape. The optical gain and the line shape function under an external optical field are derived from time-convolutionless (TCL) quantum kinetic equations for electron–hole pairs near the band edge. These equations are the generalization of the semiconductor Bloch equations by incorporating the non-Markovian relaxation and the renormalization of the memory effects through the interference between the external driving field and the stochastic reservoir. Many-body effects such as band-gap renormalization and excitonic enhancement are included by taking into account the Coulomb interaction in the Hartree–Fock approximation. Next, we outline the numerical method to calculate the valence band structure using Hamiltonian derived in Chapter 3. In addition, some results for nanostructures such as a quantum well (QW), a quantum wire (QWR), and a quantum box are discussed.

5.1 DENSITY OPERATOR DESCRIPTION OF OPTICAL INTERACTIONS

In this section, we derive an equation of motion for a density operator of optical interactions. The eigenstates $\{|a_n\rangle\}$ of A ($A|a_i\rangle = a_i|a_i\rangle$) form a complete orthogonal basis for the Hilbert space of the states of the system, so an arbitrary state $|\psi\rangle$ can be represented as a linear combination of elements of the basis

$$|\psi(t)\rangle = \sum_n c_n|a_n\rangle, \tag{5.1}$$

where a_n are arbitrary complex numbers. For such a case, the expectation value is given by the sum of a product between a measurement of the observable and its probability. In essence, we are averaging over a very large (or even "infinite") number of measurements, each performed on a system whose state is precisely specified by $|\psi\rangle$. This large number of identically prepared systems is referred to as a *pure ensemble*. That is to say, the system can be expressed by only one state $|\psi\rangle$. On the other hand, there is the case where an ensemble is composed of distinct, precisely defined states $|\psi_s\rangle$, with the probability W_s of occurrence of the state s during measurement. This kind of ensemble is often referred to as a *mixed ensemble*. The ensemble average on the outcome of a measurement of the observable A on a mixed ensemble is

$$\langle A\rangle = \sum_s W_s\langle\psi_s|A|\psi_s\rangle. \tag{5.2}$$

Now we take a closer look at how the ensemble average is calculated in Equation (5.2). These states $\{|\psi_s\rangle\}$ may not be orthogonal, and their number may exceed the dimensionality of the system's Hilbert space—the only requirement is that they are distinct. First, we pick an arbitrary orthonormal basis $|\psi\rangle$ for the system's Hilbert space, and then we can write the ensemble average for a physical observable A as

$$\begin{aligned}\langle A\rangle &= \sum_s W_s \langle\psi_s|A|\psi_s\rangle \\ &= \sum_s W_s \sum_\psi \sum_{\psi'} \langle\psi_s|\psi\rangle\langle\psi|A|\psi'\rangle\langle\psi'|\psi_s\rangle \\ &= \sum_\psi \sum_{\psi'} \left(\sum_s \langle\psi_s|\psi\rangle W_s \langle\psi'|\psi_s\rangle\right)\langle\psi|A|\psi'\rangle.\end{aligned} \tag{5.3}$$

Note that the expression in brackets in Equation (5.3) is just a matrix element of some operator in the $|\psi\rangle$ basis. In fact, this operator can be written as

$$\rho = \sum_s |\psi_s > W_s < \psi_s|. \tag{5.4}$$

This operator ρ is called the density operator. Carefully examining the expression in Equation (5.4), we can rewrite it as

$$\langle A\rangle = \sum_\psi \sum_{\psi'} \langle\psi'|\rho|\psi\rangle\langle\psi|A|\psi'\rangle = Tr(\rho A). \tag{5.5}$$

We consider the time evolution of the density operator. First, recall that in the Schrödinger picture, the states depend on time, and their evolution is given by the Schrödinger equation:

$$\frac{d}{dt}\rho = \frac{d}{dt}(|\psi(t)\rangle\langle\psi(t)|) = \left(\frac{d}{dt}|\psi(t)\rangle\right)\langle\psi(t)| + |\psi(t)\rangle\left(\frac{d}{dt}\langle\psi(t)|\right). \tag{5.6}$$

Since

$$\begin{aligned}\frac{d}{dt}|\psi(t)\rangle &= \frac{1}{i\hbar}H|\psi(t)\rangle \\ \frac{d}{dt}\langle\psi(t)| &= -\frac{1}{i\hbar}H\langle\psi(t)|,\end{aligned} \tag{5.7}$$

$$\frac{d}{dt}\rho = \frac{1}{i\hbar}(H|\psi(t)\rangle\langle\psi(t)| - |\psi(t)\rangle\langle\psi(t)|H) = \frac{1}{i\hbar}(H\rho - \rho H). \tag{5.8}$$

Therefore,

$$i\hbar \frac{d}{dt}\rho = [H,\rho]. \tag{5.9}$$

The equation is the quantum analogue of Liouville's theorem in statistical mechanics. Liouville's theorem describes the evolution in time of an ensemble of identical classical systems.

5.2 THE TIME-CONVOLUTIONLESS EQUATION

In this section, we derive TCL quantum kinetic equations for electron–hole pairs near the band edge [2]. These equations are the generalization of the semiconductor Bloch equations. The optical gain and the line shape function of the QW under an external optical field are derived from TCL quantum kinetic equations. In this section, we use a unit where $\hbar = 1$.

5.2.1 Time-Convolutionless Equation for a Reduced Density Operator of an Arbitrary Driven System

The Hamiltonian of the total system is assumed to be

$$\begin{aligned} H_T(t) &= H_0(t) + H_i(t) + H_{ext}(t) \\ &= H(t) + H_{ext}(t) \\ &= H_S(t) + H_i(t), \end{aligned} \tag{5.10}$$

where $H_0(t)$ is the Hamiltonian of the system, $H_{ext}(t)$ is the the interaction of the system with the external driving field, and $H_i(t)$ is the the Hamiltonian for the interaction of the system with its stochastic reservoir. The equation of motion for the density operator $\rho_T(t)$ of the total system is given by the stochastic Liouville equation

$$\frac{d\rho_T(t)}{dt} = -i[H_T(t), \rho_T(t)] \tag{5.11}$$

$$= -iL_T(t)\rho_T(t), \tag{5.12}$$

where $L_T(t) = L_0(t) + L_i(t) + L_{ext}(t) = L(t) + L_{ext}(t) = L_S(t) + L_i(t)$ is the Liouville super operator in one-to-one correspondence with the Hamiltonian. It is convenient to introduce the projection operators, which decompose the total system by eliminating the dynamical variables of the stochastic reservoir. We define time-independent projection operators $\underline{P}$ and $\underline{Q}$ as

$$\underline{P}X = p_0(R)\langle X\rangle_i \text{ and } \underline{Q} = 1 - \underline{P} \tag{5.13}$$

for any dynamical variable X. Here $p_0(R)$ is the initial distribution function of the random variable R, and $\langle\ldots\ldots\rangle_i$ is the average over the stochastic process $H_i(t)$. Projection operators $\underline{P}$ and $\underline{Q}$ satisfy the operator identity $\underline{P}^2 = \underline{P}$, $\underline{Q}^2 = \underline{Q}$, and $\underline{P}\underline{Q} = \underline{Q}\underline{P} = 0$. The information of the system is then contained in the reduced density operator $r(t)$, which is defined by

$$r(t) = \langle \underline{P}\rho_T(t)\rangle_i. \tag{5.14}$$

In order to derive the TCL equation, we first multiply Equation (5.10) by $\underline{P}$ and $\underline{Q}$ to obtain coupled equations for $\underline{P}\rho_T(t)$ and $\underline{Q}\rho_T(t)$:

$$\frac{d}{dt}\underline{P}\rho_T(t) = -i\underline{P}L_T(t)\underline{P}\rho_T(t) - i\underline{P}L_T(t)\underline{Q}\rho_T(t), \tag{5.15}$$

and

$$\frac{d}{dt}\underline{Q}\rho_T(t) = -i\underline{Q}L_T(t)\underline{Q}\rho_T(t) - i\underline{Q}L_T(t)\underline{P}\rho_T(t), \tag{5.16}$$

where we used the identity $\underline{P} + \underline{Q} = 1$.

We assume that the external driving field is turned on at $t = 0$, and the total system was in an arbitrary initial condition $\rho_T(t)$. It can be shown that the formal solution of Equation (5.16) is given by

$$\underline{Q}\rho_T(t) = -i\int_0^t d\tau \underline{H}(t,\tau)\underline{Q}L_T(\tau)\underline{P}\rho_T(\tau) + \underline{H}(t,0)\underline{Q}\rho_T(0), \tag{5.17}$$

where the projected propagator $\underline{H}(t,\tau)$ of the total system is defined as

$$\underline{H}(t,\tau) = \underline{T}\exp\left\{-i\int_\tau^t ds\underline{Q}L_T(s)\underline{Q}\right\}. \tag{5.18}$$

Here, $\underline{T}$ denotes the time-ordering operator. Next, we transform the memory kernel in Equation (5.17) into the TCL form by substituting the formal solution of Equation (5.12)

$$\rho_T(\tau) = \underline{G}(t,\tau)\rho_T(t) \tag{5.19}$$

into Equation (5.17).

Here, the evolution operator $\underline{G}(t,\tau)$ of the total system is given by

$$\underline{G}(t,\tau) = \underline{T}^c \exp\left\{i\int_\tau^t ds L_T(s)\right\}, \tag{5.20}$$

where $\underline{T}^c$ is the antitime-ordering operator. Evolution operators $\underline{G}(t,\tau)$ and $\underline{H}(t,\tau)$ satisfy the following relations:

$$\underline{G}(t,\tau)\underline{G}(s,t) = \underline{G}(s,\tau) \tag{5.21}$$

and

$$\underline{H}(t,\tau)\underline{H}(t,s) = \underline{H}(t,s). \tag{5.22}$$

From Equations (5.17) and (5.19), we obtain

$$\underline{Q}\rho_T(t) = \{\theta(t)-1\}\underline{P}\rho_T(t) + \theta(t)\underline{H}(t,0)\underline{Q}\rho_T(0), \tag{5.23}$$

where $\theta(t)^{-1} = g(t)$. Thus,

$$\underline{Q}\rho_T(t) = 1 + i\int_0^t d\tau \underline{H}(t,\tau)\underline{Q}L_T(\tau)\underline{P}\,\underline{G}(t,\tau). \tag{5.24}$$

The TCL equation of motion for $\underline{P}\rho_T(t)$ can be obtained from Equations (5.15) and (5.23) as

$$\begin{aligned}\frac{d}{dt}\underline{P}\rho_T(t) &= -i\underline{P}L_T(t)\underline{P}\rho_T(t) - i\underline{P}L_T(t)\{\theta(t)-1\}\underline{P}\rho_T(t)\\ &\quad - i\underline{P}L_T(t)\theta(t)\underline{H}(t,0)\underline{Q}\rho_T(0)\end{aligned} \tag{5.25}$$

It is now straightforward to obtain the TCL equation of motion for a reduced density operator $\rho(t)$. By taking an average of Equation (5.25) over the stochastic process $H_i(t)$, we get

$$\frac{d}{dt}\rho(t) = -i(L_S(t) + \langle L_i(t)\rangle_i)\rho(t) + C(t)\rho(t), \tag{5.26}$$

where the generalized collision operator $C(t)$ is defined by

$$C(t) = -i\langle L_i(t)\{\theta(t)-1\}\rangle_i = -i\left\langle L_i(t)\Sigma(t)\{1-\Sigma(t)\}^{-1}\right\rangle_i, \tag{5.27}$$

in which

$$\begin{aligned}\Sigma(t) &= 1-\theta(t)^{-1} \\ &= -i\int_0^t d\tau \underline{H}(t,\tau)\underline{Q}L_T(\tau)\underline{P}\underline{G}(t,\tau) \\ &= -i\int_0^t d\tau \underline{U}(t)\underline{S}(t,\tau)\underline{U}^{-1}(\tau)\underline{Q}L_T(\tau)\underline{P}\underline{U}(\tau)\underline{R}(t,\tau)\underline{U}^{-1}(t).\end{aligned} \tag{5.28}$$

Here, we define

$$\underline{U}(t) = \underline{T}\exp\left\{-i\int_\tau^t ds L_S(s)\right\}, \tag{5.29}$$

$$\underline{S}(t,\tau) = \underline{T}\exp\left\{-i\int_\tau^t ds \underline{Q}\underline{U}^{-1}L_i(s)\underline{U}(s)\underline{Q}\right\}, \tag{5.30}$$

and

$$\underline{R}(t,\tau) = \underline{T}^c \exp\left\{i\int_\tau^t ds \underline{U}^{-1}L_i(s)\underline{U}(s)\right\}, \tag{5.31}$$

where $\underline{U}(t)$ is the evolution operator of the system with driving field, and $\underline{R}(t,\tau)$ and $\underline{S}(t,\tau)$ are the evolution operator and the projected propagator of the total system in the interaction picture, respectively. In Equation (5.26), we assumed that the initial condition $\rho_T(0)$ is given by

$$\rho_T(0) = \rho(0)p_0(R), \tag{5.32}$$

which means that the system and the reservoir were uncoupled before the external driving field was turned on, and that the system was in an arbitrary state $\rho(0)$ at $t = 0$. Then, it is obvious that $\underline{Q}\rho_T(0) = 0$.

We now consider the case where the system is interacting weakly with the stochastic reservoir and expand Equation (5.26) up to the second order in powers of the stochastic Hamiltonian $H_i(t)$. We assume that the random force vanishes on the average over the stochastic process, that is,

$$\underline{P}L_i(t)\underline{P} = 0. \tag{5.33}$$

We further assume that the stochastic process is stationary.

The equation of motion for $\rho(t)$ up to the second order expansion in $H_i(t)$ becomes

$$\frac{d}{dt}\rho(t) = -iL_S(t)\rho(t) + C^{(2)}(t)\rho(t), \tag{5.34}$$

where

$$\begin{aligned} C^{(2)}(t) &= -i\langle L_i(t)\Sigma^{(1)}(t)\rangle_i \\ &= -\int_0^t d\tau \langle L_i(t)\underline{U}(t,\tau)L_i(\tau)\underline{U}^{-1}(t,\tau)\rangle_i \end{aligned} \tag{5.35}$$

with

$$\Sigma^{(1)}(t) = -i\int_0^t d\tau \underline{U}(t,\tau)L_i(\tau)\underline{U}^{-1}(t,\tau) \tag{5.36}$$

and

$$\underline{U}(t,\tau) = \underline{U}(t)\underline{U}^{-1}(\tau). \tag{5.37}$$

We transform $C^{(2)}(t)$ to more suitable form for the perturbation expansions with respect to $H_{ext}(t)$:

$$C^{(2)}(t) = -\int_0^t d\tau \langle L_i(t)\underline{U}_0(t)\underline{U}_{ext}(t,\tau)\underline{U}_0^{-1}(\tau)L_i(t)\underline{U}_0(\tau)\underline{U}_{ext}^{-1}(t,\tau)\underline{U}_0^{-1}(t)\rangle_i. \tag{5.38}$$

We can expand $C^{(2)}(t)$ in powers of the driving field as

$$C^{(2)}(t) = \sum_{m=0}^{\infty} C_m^{(2)}(t), \tag{5.39}$$

where $C_m^{(2)}(t)$ is the mth order term given by

$$\begin{aligned} C_m^{(2)}(t) = -\sum_{k=0}^{m}(-1)^k(i)^{m-k}\int_0^t d\tau\int_\tau^t d\tau_1\int_\tau^{\tau_1} d\tau_2\ldots\int_\tau^{\tau_{k-1}} d\tau_k\int_\tau^t d\tau_{k+1}\ldots\int_\tau^{\tau_{m-1}} d\tau_m \\ \times\langle L_i(t)\Phi_k(t,\tau_1,\ldots,\tau_k,\tau)L_i(\tau)\Phi_{m-k}(t,\tau_{k+1},\ldots,\tau_m,\tau)\rangle_i, \end{aligned} \tag{5.40}$$

with

$$\begin{aligned} \Phi_0(t,\tau) &= \underline{U}_0(t-\tau), \\ \Phi_k(t,\tau_1,\ldots,\tau_k,\tau) &= \underline{U}_0(t-\tau_1)L_{ext}(\tau_1)\underline{U}_0(\tau_1-\tau_2)L_{ext}(\tau_2) \\ &\quad\times\underline{U}_0(\tau_2-\tau_3)\ldots\underline{U}_0(\tau_{k-1}-\tau_k)L_{ext}(\tau_k)\underline{U}_0(\tau_k-\tau), \end{aligned} \tag{5.41}$$

$$\Psi_0(t,\tau) = \underline{U}_0(\tau - t), \tag{5.42}$$

and

$$\begin{aligned}\Psi_k(t,\tau_1,\ldots,\tau_k,\tau) = &\underline{U}_0(t-\tau_k)L_{ext}(\tau_k)\underline{U}_0(\tau_k-\tau_{k-1})L_{ext}(\tau_{k-1})\\ &\underline{U}_0(\tau_{k-2}-\tau_{k-3})\ldots\underline{U}_0(\tau_2-\tau_1)L_{ext}(\tau_1)\underline{U}_0(\tau_1-t).\end{aligned} \tag{5.43}$$

The TCL equation of motion for a reduced density operator given in Equation (5.34) with Equations (5.39)–(5.43) can be used in any time scale such as the femtosecond regime, and is valid up to the second order in powers in the interaction between the system and the stochastic reservoir. In the next subsection, a TCL equation for a reduced density operator is used to obtain quantum kinetic equations for the system of interacting electron–hole pairs near the band edge in semiconductors under an arbitrary optical field.

5.2.2 Time-Convolutionless Quantum Kinetic Equations for Interacting Electron–Hole Pairs in Semiconductors

Here, we apply the TCL equation (5.34) to the system of interacting electron–hole pairs in semiconductors with an external driving field. We assume that the system is weakly interacting with its stochastic reservoir. Many-body effects such as bandgap renormalization and phase space filling are included by taking into account the Coulomb interaction in the Hartree–Fock approximation. The stochastic Hamiltonian $H_i(t)$ may include electron–electron interaction and electron–LO phonon interaction for both conduction and valence electrons. We will not specify the explicit forms of H_i here. Instead, we obtain the intraband relaxation and the dephasing as correlation functions of $H_i(t)$.

We employ the two-band model for the semiconductor, and introduce two short-handed notations $|ck\rangle$ and $|vk\rangle$ such that

$$|ck\rangle = |c,k\rangle \text{ and } |vk\rangle = |v,k\rangle \tag{5.44}$$

where c and v denote conduction and valence bands, respectively, and k is the electron wave vector. In the following, we suppress the vector notation for simplicity.

In the time-dependent Hartree–Fock approximation, the unperturbed Hamiltonian $H_0(t)$ is given by

$$\langle\alpha k|H_0(t)|\beta k\rangle = E_\alpha^0(k)\delta_{\alpha\beta} - \sum_{k'} V(k-k')\langle\alpha k'|\rho(t)|\beta k'\rangle, \tag{5.45}$$

where $\alpha,\beta = c$ or v and $V(k-k)$ is the Coulomb interaction. The interaction of the system with an external driving field gives the interaction Hamiltonian H_{ext}, which is given by

$$H_{ext}(t) = -ME_p(t), \tag{5.46}$$

where M is the dipole operator, and E_p is the electric field strength of the optical radiation. The equation of motion for the reduced density operator becomes

$$\frac{d}{dt}\rho(t) = -i(L_0(t) + L_{ext}(t))\rho(t) + C_0^{(2)}(t)\rho(t) + D_1^{(2)}, \tag{5.47}$$

where

$$C_0^{(2)}(t)\rho(t) = -\int_0^t d\tau \left\langle L_i(t)\underline{U}_\tau L_i(t-\tau)\underline{U}_0^{-1}(\tau)\right\rangle_1 \rho(t), \tag{5.48}$$

and

$$D_1 = C_1^{(2)}(t)\underline{U}_0(t)\rho(0). \tag{5.49}$$

$C_1^{(2)}(t)$ is responsible for the intracollisional field effects, and can be derived from Equation (5.40):

$$\begin{aligned} C_1^{(2)}(t) = i\int_0^t d\tau \int_\tau^t ds\{&\langle L_i(t)\underline{U}_0(t-s)L_{ext}(s)\underline{U}_0(s-\tau)L_i(\tau)\underline{U}_0(\tau-t)\rangle_i \\ &-\langle L_i(t)\underline{U}_0(t-\tau)L_i(\tau)\underline{U}_0(\tau-s)L_{ext}(s)\underline{U}_0(s-t)\rangle_i\}. \end{aligned} \tag{5.50}$$

It can be shown that $D_1^{(2)}$ contains information on the effects of the interference of the external driving field with the stochastic reservoir of the system, and is the renormalization of the memory effects.

Nonequilibrium distributions $n_{ck}(t)$ and $n_{vk}(t)$ for electrons in the conduction and valence bands and the nondiagonal interband matrix element $p_k^*(t)$, which describes the interband polarization induced by the optical field, are the matrix elements of the reduced density operator, and are given by

$$\begin{aligned} n_{ck}(t) &= \rho_{cck}(t) \\ &= \langle ck|\rho(t)|ck\rangle, \end{aligned} \tag{5.51}$$

$$\begin{aligned} n_{vk}(t) &= \rho_{vvk}(t) \\ &= \langle vk|\rho(t)|vk\rangle, \end{aligned} \tag{5.52}$$

and

$$\begin{aligned} p_k^*(t) &= \rho_{vck}(t) \\ &= \langle vk|\rho(t)|ck\rangle. \end{aligned} \tag{5.53}$$

Next, we calculate the matrix elements of the collision term $C_0^{(2)}(t)\rho(t)$, $\langle ck|C_0^{(2)}(t)\rho(t)|ck\rangle$, $\langle vk|C_0^{(2)}(t)\rho(t)|vk\rangle$, and $\langle vk|C_0^{(2)}(t)\rho(t)|ck\rangle$ to obtain the non-Markovian intraband relaxation and dephasing. After some mathematical manipulations, we obtain:

$$\begin{aligned} \langle ck|C_0^{(2)}(t)\rho(t)|ck\rangle &= -\int_0^t d\tau \langle\langle ck|L_i(t)\underline{U}_0(\tau)L_i(t-\tau)\underline{U}_0^{-1}(\tau)\rho(t)|ck\rangle\rangle_i \\ &= -\int_0^t d\tau \langle\langle ck\|[H_i(t),[\underline{U}_0(\tau)H_i(t-\tau),\rho(t)]]\|ck\rangle\rangle_i \\ &\simeq -2\int_0^t d\tau Re\{\langle\langle ck\|[H_i(t)(\underline{U}_0(\tau)H_i(t-\tau))|ck\rangle\rangle_i\} \\ &\quad \{n_{ck}(t)-(\rho_0^{(0)})_{cck}(t)\}, \end{aligned} \tag{5.54}$$

$$\begin{aligned} \langle vk|C_0^{(2)}(t)\rho(t)|vk\rangle &= -\int_0^t d\tau \langle\langle vk|L_i(t)\underline{U}_0(\tau)L_i(t-\tau)\underline{U}_0^{-1}(\tau)\rho(t)|vk\rangle\rangle_i \\ &= -\int_0^t d\tau \langle\langle vk\|[H_i(t),[\underline{U}_0(\tau)H_i(t-\tau),\rho(t)]]\|vk\rangle\rangle_i \\ &\simeq -2\int_0^t d\tau Re\{\langle\langle vk\|[H_i(t)(\underline{U}_0(\tau)H_i(t-\tau))|vk\rangle\rangle_i\} \\ &\quad \{n_{vk}(t)-(\rho_0^{(0)})_{vvk}(t)\} \end{aligned} \tag{5.55}$$

and

$$\begin{aligned} \langle vk|C_0^{(2)}(t)\rho(t)|ck\rangle &= -\int_0^t d\tau \langle\langle vk|L_i(t)\underline{U}_0(\tau)L_i(t-\tau)\underline{U}_0^{-1}(\tau)\rho(t)|ck\rangle\rangle_i \\ &= -\int_0^t d\tau \langle\langle vk\|[H_i(t),[\underline{U}_0(\tau)H_i(t-\tau),\rho(t)]]\|ck\rangle\rangle_i \\ &= -\int_0^t d\tau \{\langle\langle vk|H_i(t)(\underline{U}_0(\tau)H_i(t-\tau)|vk\rangle\rangle_i \\ &\quad + \langle\langle ck|\underline{U}_0(\tau)H_i(t-\tau)H_i(t)|ck\rangle\rangle_i\} p_k^*(t) \end{aligned} \tag{5.56}$$

where $\rho_0^{(0)}(t) = \underline{U}_0(t)\rho(0)$. Equation (5.56) is the non-Markovian optical dephasing, which is the temporal decay of the interband polarization due to scattering processes.

Similarly,

$$\begin{aligned}&\langle ck|-i(L_0(t)+L_{ext}(t))\rho(t)|ck\rangle\\&=-2Im\left\{\left[\mu(k)E_p(t)+\sum_{k'}V(k-k')p_{k'}(t)\right]p_k^*(t)\right\},\end{aligned} \tag{5.57}$$

$$\begin{aligned}&\langle vk|-i(L_0(t)+L_{ext}(t))\rho(t)|vk\rangle\\&=2Im\left\{\left[\mu(k)E_p(t)+\sum_{k'}V(k-k')p_{k'}(t)\right]p_k^*(t)\right\},\end{aligned} \tag{5.58}$$

and

$$\begin{aligned}&\langle vk|-i(L_0(t)+L_{ext}(t))\rho(t)|ck\rangle=i[E_c(k)-E_v(k)]p_k^*\\&+i\left[\mu(k)E_p(t)+\sum_{k'}V(k-k')p_{k'}(t)\right][n_{ck}(t)-n_{vk}(t)],\end{aligned} \tag{5.59}$$

where $\mu(k)=\langle ck|M|ck\rangle$, $E_c(k)$, and $E_v(k)$ are renormalized single particle energies given by

$$E_c(k)=E_c^0(k)-\sum_{k'}V(k-k')n_{ck'}^0 \tag{5.60}$$

and

$$E_v(k)=E_v^0(k)-\sum_{k'}V(k-k')n_{vk'}^0. \tag{5.61}$$

In Equations (5.54)–(5.58), *Re* and *Im* denote the real and the imaginary part of complex variable, respectively. The last term $D_1^{(2)}$ of Equation (5.47) is the interference of the driving field with the surroundings, and is given by

$$\begin{aligned}D_1^{(2)}(t)&=C_1^{(2)}(t)\rho_0^{(0)}(t)\\&=i\int_0^t d\tau\int_0^\tau ds\{\langle L_i(t)\underline{U}_0(t-\tau)L_{ext}(\tau)\underline{U}_0(t-s)L_i(s)\underline{U}_0(s)\rho(0)\rangle_i\\&\quad-\langle L_i(t)\underline{U}_0(t-s)L_i(s)\underline{U}_0(s-\tau)L_{ext}(\tau)\underline{U}_0(\tau)\rho(0)\rangle_i\}\\&=i\int_0^t d\tau\int_0^\tau ds\langle[H_i(t),[\underline{U}_0(s)H_{ext}(t-s)),(\underline{U}_0(\tau)H_i(t-\tau))],\rho_0^{(0)}(t)]]\rangle,\end{aligned} \tag{5.62}$$

where we have made use of the transformation of integral variables and the commutation relation between the operators A, B, and C

$$[A,[B,C]]-[B,[A,C]]=[[A,B],C]. \tag{5.63}$$

It can be shown that the interference effects on the intraband collisions vanish to the first order if $\langle vk|\rho_0^{(0)}(t)|ck\rangle$ vanishes. In other word, we assume that

$$\langle ck|D_1^{(2)}(t)|ck\rangle = \langle vk|D_1^{(2)}(t)|vk\rangle = 0. \tag{5.64}$$

In the following, we include the effects of the interference between the stochastic reservoir and the external optical field only in the interband kinetics. It is involved but straightforward to evaluate the matrix element of $D_1^{(2)}$. The result is

$$\begin{aligned}
&\langle vk|D_1^{(2)}(t)|ck\rangle \\
&= i\int_0^t d\tau \int_0^\tau ds exp\{-i[E_v(k)-E_c(k)](t-\tau)\} \\
&\quad \{\langle\langle vk\|[H_i(t)(\underline{U}_0(t-s)H_i(s))|vk\rangle\rangle_i \\
&\quad + \langle\langle ck\|[\underline{U}_0(t-s)H_i(s))H_i(t)|ck\rangle\rangle_i\}\mu^*(k)E_p(t)\{(\rho_0^{(0)})_{cck}(t)-(\rho_0^{(0)})_{vvk}(t)\} \\
&\simeq i\int_0^t d\tau \int_0^\tau ds exp\{-i[E_v(k)-E_c(k)]s\} \\
&\quad \{\langle\langle vk\|[H_i(t)(\underline{U}_0(\tau)H_i(t-\tau))|vk\rangle\rangle_i \\
&\quad + \langle\langle ck\|[(\underline{U}_0(\tau)H_i(t-\tau))H_i(t)|ck\rangle\rangle_i\, \mu^*(k)E_p(t-s) \\
&\quad \{(\rho_0^{(0)})_{cck}(t)-(\rho_0^{(0)})_{vvk}(t)\},
\end{aligned} \tag{5.65}$$

where we have made use of the transformation of integral variables and dropped the Coulomb exchange term between electrons in the conduction and the valence bands. Using Equations (5.51)–(5.65), we finally obtain TCL quantum kinetic equations for $n_{ck}(t)$, $n_{vk}(t)$, and $p_k^*(t)$:

$$\begin{aligned}
\frac{\partial}{\partial t}n_{ck}(t) &= -2Im\left\{\left[\mu(k)E_p(t)+\sum_{k'}V(k-k')p_{k'}(t)\right]p_k^*(t)\right\} \\
&\quad -2\int_0^t d\tau Re\{\langle\langle ck\|[H_i(t)(\underline{U}_0(\tau)H_i(t-\tau))|ck\rangle\rangle_i\} \\
&\quad \{n_{ck}(t)-(\rho_0^{(0)})_{cck}(t)\},
\end{aligned} \tag{5.66}$$

$$\begin{aligned}
\frac{\partial}{\partial t}n_{vk}(t) &= 2Im\left\{\left[\mu(k)E_p(t)+\sum_{k'}V(k-k')p_{k'}(t)\right]p_k^*(t)\right\} \\
&\quad -2\int_0^t d\tau Re\{\langle\langle ck\|[H_i(t)(\underline{U}_0(\tau)H_i(t-\tau))|ck\rangle\rangle_i\} \\
&\quad \{n_{ck}(t)-(\rho_0^{(0)})_{cck}(t)\},
\end{aligned} \tag{5.67}$$

and

$$
\begin{aligned}
\frac{\partial}{\partial t}p_k^*(t) = {} & i[E_c(k)-E_v(k)]p_k^*(t) + i\left[\mu^*(k)E_p(t) + \sum_{k'} V(k-k')p_{k'}^*(t)\right] \\
& [n_{ck}(t)-n_{vk}(t)] \\
& -\int_0^t d\tau\{\langle\langle vk\|[H_i(t)(\underline{U}_0(\tau)H_i(t-\tau))|vk\rangle\rangle_i\} \\
& +\langle\langle ck\|[\underline{U}_0(\tau)H_i(t-\tau))H_i(t)|ck\rangle\rangle_i\}p_k^*(t) \\
& +i\int_0^t d\tau\int_0^\tau ds exp\{-i[E_c(k)-E_v(k)]s\} \\
& \{\langle\langle vk\|[H_i(t)(\underline{U}_0(\tau)H_i(t-\tau))|vk\rangle\rangle_i\} \\
& +\langle\langle ck\|[(\underline{U}_0(\tau)H_i(t-\tau))H_i(t)|ck\rangle\rangle_i\}.
\end{aligned}
\tag{5.68}
$$

The last term of Equation (5.68) modulates the interband polarization due to the interference of the driving optical field and the stochastic reservoir of the system, and gives the renormalized memory effects. TCL equations give the non-Markovian relaxation (both intraband relaxation and dephasing) and the renormalized memory effects through the modulation of the interband polarization. Moreover, TCL equations are in a form convenient for perturbation expansions in powers of the stochastic Hamiltonian H_i and the interaction with the driving field H_{ext}, and are valid to a very short time scale.

It can be shown that quantum kinetic equations are reduced to the conventional density matrix equations in the Markovian limit. In order to analyze the collision term $C_0^{(2)}(t)\rho(t)$ and the interference term $D_1^{(2)}(t)$ in the Markovian limit, we put

$$\langle\langle \alpha k\|[H_i(t)(\underline{U}_0(\tau)H_i(t-\tau))|\alpha k\rangle\rangle_i = \frac{1}{2\tau_\alpha}\delta(|\tau|). \tag{5.69}$$

Then, the intraband relaxation and the dephasing terms become

$$\langle ck|C_0^{(2)}(t)\rho(t)|ck\rangle = -\frac{1}{\tau_c(k)}\{n_{ck}(t)-(\rho_0^{(0)})_{cck}(t)\}, \tag{5.70}$$

$$\langle vk|C_0^{(2)}(t)\rho(t)|vk\rangle = -\frac{1}{\tau_v(k)}\{n_{vk}(t)-(\rho_0^{(0)})_{vvk}(t)\}, \tag{5.71}$$

and

$$\langle vk|C_0^{(2)}(t)\rho(t)|ck\rangle = -\frac{1}{\tau_{vc}(k)}p_k^*(t), \tag{5.72}$$

with

$$\frac{1}{\tau_{vc}(k)} = \frac{1}{2}\left\{\frac{1}{\tau_v(k)} + \frac{1}{\tau_c(k)}\right\}. \tag{5.73}$$

In the Markovian approximation, the interference term $\langle vk|D_1^{(2)}(t)|ck\rangle$ vanishes because, in the integration over ds, the upper limit τ is zero, and the resulting integral in Equation (5.65) becomes zero because of Equation (5.69). Intraband relaxation Equations (5.70) and (5.71), and dephasing Equation (5.72), characterized by the dephasing time Equation (5.73), agree well with the conventional density matrix theory with Markovian relaxation. As a result, the memory effects vanish in the Markovian limit because there is no correlation between the stochastic process $H_i(t-\tau)$ and $H_i(t)$.

Physically, $\langle\langle\alpha k\|[H_i(t)(\underline{U}_0(\tau)H_i(t-\tau))|\alpha k\rangle\rangle_i$ represents the averaged probability amplitude of finding a particle in the state $|\alpha k\rangle$ after being scattered at τ by $H_i(t)$ when it is initially in the state $|\alpha k\rangle$, gets scattered at $t-\tau$ by $H_i(t-\tau)$, and goes as a free particle for the time interval t. The Markovian approximation assumes that each interaction with the reservoir randomizes the previous information contained in the wave function, and treats scattering processes $H_i(t-\tau)$ and $H_i(t)$ as independent stochastic processes. In the non-Markovian theory, the memory effects extend over the time interval τ_c, the correlation time of the stochastic processes. For example, if we assume the simplest form of the non-Markovian correlation function $\langle\langle\alpha k\|[H_i(t)(\underline{U}_0(\tau)H_i(t-\tau))|\alpha k\rangle\rangle_i$ as

$$\langle\langle\alpha k\|[H_i(t)(\underline{U}_0(\tau)H_i(t-\tau))|\alpha k\rangle\rangle_i = \frac{1}{\tau_c\tau_\alpha(k)}\exp\left[-\frac{|\tau|}{\tau_c}\right], \tag{5.74}$$

the intraband relaxation and the dephasing terms are reduced to the results of the Markovian approximation when $t=\infty$, while the interference term retains infinitesimal phase shift in the interband polarization as a result of the renormalized memory effects.

In the derivation of ordinary semiconductor Bloch equations, all the kinetic effects of $D_1^{(2)}(t)$ are completely neglected, and the Markovian approximation in the narrowing limit $\tau_r >> \tau_c$ in the collision operator $C_0^{(2)}(t)\rho(t)$ is used. Here, τ_r denotes the macroscopic relaxation time of the system. In this way, we get the familiar semiconductor Bloch equations, which describe the particles as free particles in between collisions or interactions with photons. The more general Equations (5.66)–(5.68) include the effects of the non-Markovian relaxation on the motion of particles between collisions. Interband kinetic equations incorporate additional interference effects between the system–reservoir interaction and the external driving field. When the system is fairly dense, the particle never gets away from the other particles in the system, and we cannot really think of the particles as being "in between collisions." Quantum mechanically, the wave functions of the particles are smeared out so that there is always some overlap of wave functions, and, as a result, the particle retains some memory of the collisions it has experienced through its correlation with other particles in the system. These memory effects are the characteristics of the quantum kinetic equations. Equations (5.66) and (5.67) can be used to describe the quantum transport phenomena, and are the generalization of the Boltzmann equations. Equation (5.68) is the quantum kinetic equation for the interband process, and is valid and useful in any time scale. The TCL nature of these equations will reduce the computation time required for the self-consistent numerical solution of the equations.

5.3 THE THEORY OF NON-MARKOVIAN OPTICAL GAIN IN SEMICONDUCTOR LASERS

In this section, the non-Markovian optical gain of a semiconductor laser is derived from TCL quantum kinetic equations for electron–hole pairs, including the many-body effects. Plasma screening and excitonic effects are taken into account using an effective Hamiltonian in the time-dependent Hartree–Fock approximation. In order to calculate the optical gain, the equation of motion for the interband pair amplitude is integrated directly. Quantum kinetic equations containing unscreened Coulomb potential are shown to be inadequate for high-density systems near a quasi-equilibrium state in the high-density limit. In this case, it is important to include the effect of plasma screening of the Coulomb interaction between the charged particles. The screening can be described in a self-consistent way by replacing the unscreened potential $V(k)$ by a screened one, $V_s(k)$ which has a reduced interaction strength at the long-distance and single-particle energies $E_c(k)$, $E_v(k)$ by $E_c^{sc}(k)$, $E_v^{sc}(k)$. The replacement must be done in the original Hamiltonian for the system H_0. In the following, we include the screening in the kinetic equations by replacing $V(k)$ by $V_s(k)$, and $E_c(k)$, $E_v(k)$ by $E_c^{sc}(k)$, $E_v^{sc}(k)$. Here, $E_c^{sc}(k) = E_c^0(k) - \Sigma_{k'} V_s(k-k')n_{ck'}^0$, and $E_v^{sc}(k) = E_v^0(k) + \Sigma_q[V_s(q) - V(q)] - \Sigma_{k'} V_s(k-k')n_{vk'}^0$, where $\Sigma_q[V_s(q) - V(q)]$ is the Coulomb-hole part of the single-particle energy renormalization, which is the change of electrostatic energy (or self-energy) due to the presence of the electron–hole plasma.

Now, we derive the optical gain and the line shape function from the TCL quantum kinetic equations. We simplify the theory by considering the case of quasi-equilibrium and steady-state interband polarization only. Nonlinear effects caused by population modulation such as spectral hole burning are ignored here. We consider the system of interacting electron–hole pairs in semiconductors in the presence of coherent monochromatic radiation. The optical field $E_p(t)$ is given by

$$E_p(t) = E_p e^{-i\omega t} + E_p^* e^{i\omega t}. \tag{5.75}$$

In the rotating wave approximation, we rewrite the interband pair amplitude $p_k^*(t)$ as

$$p_k^*(t) = e^{i\omega t} p_k^*(t), \tag{5.76}$$

for electron–hole pairs in the band edge semiconductors driven by the coherent light field. Linearizing the equation of motion by assuming the electron–hole pairs are initially in quasi-equilibrium, we get:

$$\begin{aligned}
\frac{\partial}{\partial t} p_k^*(t) &= i\Delta_k p_k^*(t) \\
&\quad + i\left[\mu_k^* E_p^* + \sum_{k'} V_s(k-k') p_{k'}^*(t)\right][n_{ck}^0 - n_{vk}^0] \\
&\quad - \int_0^t d\tau \{\langle\langle vk \| [H_i(t)(\underline{U}_0(\tau) H_i(t-\tau)) | vk \rangle\rangle_i \\
&\quad + \langle\langle ck \| [(\underline{U}_0(\tau) H_i(t-\tau)) H_i(t) | ck \rangle\rangle_i \} p_k^*(t) \\
&\quad + i\int_0^t d\tau \int_0^\tau ds \exp\{-i[E_v^{sc}(k) - E_c^{sc}(k)]s\} \\
&\quad \{\langle\langle vk \| [H_i(t)(\underline{U}_0(\tau) H_i(t-\tau)) | vk \rangle\rangle_i \\
&\quad + \langle\langle ck \| [(\underline{U}_0(\tau) H_i(t-\tau)) H_i(t) | ck \rangle\rangle_i \} \mu^*(k) E_p(t-s)[n_{ck}^0 - n_{vk}^0],
\end{aligned} \tag{5.77}$$

where $\Delta_k = E_c^{sc}(k) - E_v^{sc}(k) - \omega$, and n_{ck}^0 and n_{vk}^0 are the quasi-equilibrium distribution of electrons in the conduction band and the valence band, respectively. To derive Equation (5.77), we assumed that the interband pair amplitude follows the temporal variation of the polarization and the field amplitude adiabatically, and suppressed the dynamics of the particle distributions (or assumed the particle distribution is determined mostly by injection) by replacing $n_{\alpha k}(t) \rightarrow n_{\alpha k}^{0}(t)$. We also ignore the nonlinear effects due to the population beating caused by a resonant four-wave mixing.

We introduce functions $g_1(t)$ and $g_2(t, \Delta k)$ as

$$\begin{aligned} g_1(t) = \int_0^t d\tau \langle\langle vk \| [H_i(t)(\underline{U}_0(\tau)H_i(t-\tau))| vk\rangle\rangle_i \\ + \langle\langle ck \| [(\underline{U}_0(\tau)H_i(t-\tau))H_i(t)| ck\rangle\rangle_i, \end{aligned} \tag{5.78}$$

and

$$\begin{aligned} g_2(t, \Delta_k) = \int_0^t d\tau \int_0^\tau ds \exp\{i\Delta_k s\}\{\langle\langle vk \| [H_i(t)(\underline{U}_0(\tau)H_i(t-\tau))| vk\rangle\rangle_i \\ + \langle\langle ck \| [(\underline{U}_0(\tau)H_i(t-\tau))H_i(t)| ck\rangle\rangle_i. \end{aligned} \tag{5.79}$$

Integrating Equation (5.77) directly using the integrating factor, we obtain

$$\begin{aligned} p_k^*(t) = i\int_0^\tau d\tau \exp\left[-\int_\tau^t dt'(-i\Delta_k + g_1(t'))\right] \\ \times [\mu_k^* E_p^* \{1 + g_2(\tau, \Delta_k)\} + \sum_{k'} V_s(k-k') p_{k'}^*(\tau)][n_{ck}^0 - n_{vk}^0]. \end{aligned} \tag{5.80}$$

Equation (5.80) is the generalized form of the optical dipole with phase damping. Mathematical manipulations can be simplified considerably by taking the Laplace transformation of Equation (5.80). In Equations (5.78)–(5.80), $g_1(t)$ is the optical dephasing, which is the temporal decay of the interband polarization. Usually, dephasing is studied using the semiconductor Bloch equations by replacing dynamical self-energies with phenomenological relaxation times neglecting memory effects and the kinetic equations for the nonequilibrium Green's functions. The latter approach shows that it is time dependent, and non-Markovian dephasing processes are expected to be a source for nonexponential decay. Experimental results indicated that the decay dynamics show strongly nonexponential behavior. It is suggested that the dynamics of both polarization and population are mainly determined by frequency fluctuations, and not by scattering processes or by recombination. The physical nature of the dephasing is due to the interactions of the interband polarization, or the excited system with its surroundings. These interactions cause the fluctuation of the transition frequency within a certain spectral distribution and the homogeneous dephasing, spectral migration of the excited state population or phase, and the decay of the excited state population are suggested to contribute to the optical dephasing.

We define the following Laplace transformations:

$$\Psi_k(s) = L\{\underline{p}_k^*(t)\}, \tag{5.81}$$

$$\Xi(s,\Delta_k) = L\left\{\int_0^t dt'(i\Delta_k - g_1(t'))\right\}, \tag{5.82}$$

and

$$G_2(s,\Delta_k) = L\{g_2(t,\Delta_k)\}, \tag{5.83}$$

where $L\{f(t)\}$ denotes the Laplace transformation of $f(t)$. We obtain

$$\Psi_k(s) = \Psi_k^{(0)}(s)Q_k(s), \tag{5.84}$$

where

$$\Psi_k^{(0)}(s) = i\Xi(s,\Delta_k)\mu_k^*(k)E_p^*\left[\frac{1}{s} + G_2(s,\Delta_k)\right][n_{ck}^0 - n_{vk}^0], \tag{5.85}$$

and

$$Q_k(s) = 1 + \frac{s}{\mu_k^*(k)E_p^*[1 + sG_2(s,\Delta_k)]}\sum_{k'} V_s(k-k')\Psi_{k'}^{(0)}(s)Q_{k'}(s). \tag{5.86}$$

In the simplest Padé approximation, the vertex function $Q_k(s)$ can be approximated by

$$Q_k(s) = \frac{1}{1 - q_{1k}(s)}, \tag{5.87}$$

where

$$q_{1k}(s) = \frac{s}{\mu^*(k)E_p^*(1 + sG_2(s,\Delta_k))}\sum_{k'} V_s(k-k')\Psi_{k'}^{(0)}(s). \tag{5.88}$$

The vertex function can be obtained exactly by a simple manipulation. First, we make a summation of both sides of Equation (5.86) over k

$$\begin{aligned}\sum_k Q_k(s) &= \sum_{k'} Q_{k'}(s) \\ &= \sum_k 1 + \sum_{kk'} \frac{s}{\mu_k^*(k)E_p^*[1+sG_2(s,\Delta_k)]} V_s(k-k')\Psi_{k'}^{(0)}(s)Q_{k'}(s) \\ &= \sum_k 1 + \sum_{kk'} \frac{s}{\mu_k^*(k')E_p^*[1+sG_2(s,\Delta_{k'})]} V_s(k'-k)\Psi_k^{(0)}(s)Q_k(s).\end{aligned} \tag{5.89}$$

or

$$\sum_k \left[1 - \sum_{k'} \frac{s}{\mu^*(k')E_p^*(1+sG_2(s,\Delta_{k'})} V_s(k'-k)\Psi_k^{(0)}(s)\right] Q_k(s) = \sum_k 1. \tag{5.90}$$

In Equation (5.90), the range of k is restricted to the first Brillouin zone. Then, the definition of $Q_k(s)$ is obvious and given by

$$\begin{aligned}Q_k(s) &= 1 + \frac{s}{\mu^*(k)E_p^*(1+sG_2(s,\Delta_k))} \sum_{k'} V_s(k-k')\Psi_{k'}^{(0)}(s)Q_{k'}(s) \\ &= \frac{1}{1-q_k(s)},\end{aligned} \tag{5.91}$$

where

$$\begin{aligned}q_k(s) &= \sum_{k'} \frac{s}{\mu^*(k')E_p^*(1+sG_2(s,\Delta_{k'}))} V_s(k'-k)\Psi_k^{(0)}(s) \\ &= s\Psi_k^{(0)}(s) \sum_{k'} \frac{V_s(k'-k)}{\mu^*(k')E_p^*(1+sG_2(s,\Delta_{k'}))}.\end{aligned} \tag{5.92}$$

Note that the expression for $q_k(s)$ is exact, and is different from $q_{1k}(s)$ given in Equation (5.88), which is the first-order term in the Padé approximation. The vertex function based on the TCL reduced-density operator formalism is equivalent to the one derived from the solution of the Bethe–Salpeter equation obtained from the many-body Green's function approach, because Green's functions can be interpreted as the matrix elements of the reduced-density operator.

The factor $Q_k(s)$ describes the excitonic enhancement of interband polarization. The steady state interband pair amplitude is determined by

$$\begin{aligned} \underline{p}_k^*(\infty) &= lim_{s\to 0} s\Psi_k(s) \\ &= lim_{s\to 0} s \frac{s\Psi_k(s)}{1-q_k(s)} \\ &= i\frac{\Xi(0,\Delta_k)}{1-q_0(s)}\mu^*(k)E_p^*[1+g_2(\infty,\Delta_k)][n_{ck}^0 - n_{vk}^0], \end{aligned} \tag{5.93}$$

with

$$\begin{aligned} q_k(0) &= \lim_{s\to 0} s\Psi_k^{(0)}(s)\sum_{k'}\frac{V_s(k'-k)}{\mu^*(k')E_p^*(1+sG_2(s,\Delta_{k'}))} \\ &= i\Xi(0,\Delta_k)\mu^*(k)E_p^*(1+g_2(\infty,\Delta_k))[n_{ck}^0 - n_{vk}^0] \\ &\times \sum_{k'}\frac{V_s(k'-k)}{\mu^*(k')E_p^*(1+sG_2(s,\Delta_{k'}))}. \end{aligned} \tag{5.94}$$

The interband polarization P and the susceptibility χ can be expressed through the dipole operator and the interband pair amplitude as

$$P = \frac{1}{V}Tr\{\mu(k)\underline{p}_k^*(\infty)\}, \tag{5.95}$$

or

$$\varepsilon_0\chi(\omega) = \frac{1}{V}Tr\left\{i\frac{\Xi(0,\Delta_k)}{1-q_{1k}(s)}|\mu(k)|^2[1+g_2(\infty,\Delta_k)][n_{ck}^0 - n_{vk}^0]\right\}. \tag{5.96}$$

The optical gain $g(\omega)$ is

$$\begin{aligned} g(\omega) &= \frac{\omega\mu c}{n_r} Im\varepsilon_0\chi(\omega) \\ &= \frac{\omega\mu c}{n_r}\frac{1}{V}TrRe\left\{\frac{\Xi(0,\Delta_k)}{1-q_k(0)}|\mu(k)|^2[1+g_2(\infty,\Delta_k)][n_{ck}^0 - n_{vk}^0]\right\} \\ &= \frac{\omega\mu c}{n_r}\frac{1}{V}TrRe\left\{\frac{\mu(k)|^2\,[1+g_2(\infty,\Delta_k)][n_{ck}^0 - n_{vk}^0]}{(1-Req_k(0))^2+(Imq_k(0)^2)}\right. \\ &\left.\times(Re\Xi(0,\Delta_k)+iIm\Xi(0,\Delta_k))(1-Req_k(0)+iImq_k(0))\right\} \\ &\simeq \frac{\omega\mu c}{n_r}\frac{1}{V}Tr\left\{\frac{\mu(k)|^2\,[1+Reg_2(\infty,\Delta_k)][n_{ck}^0 - n_{vk}^0]}{(1-Req_k(0))^2+(Imq_k(0)^2)}\right. \\ &\left.\times(Re\Xi(0,\Delta_k)(1-Req_k(0))-Im\Xi(0,\Delta_k)Imq_k(0))\right\}, \end{aligned} \tag{5.97}$$

where

$$Req_k(0) = -Im\Xi(0,\Delta_k)\mu^*(k)[n_{ck}^0 - n_{vk}^0]\sum_{k} \frac{V_s(k'-k)}{\mu^*(k')}, \tag{5.98}$$

and

$$Imq_k(0) = Re\Xi(0,\Delta_k)\mu^*(k)[n_{ck}^0 - n_{vk}^0]\sum_{k'} \frac{V_s(k'-k)}{\mu^*(k')}. \tag{5.99}$$

Here, μ is the permeability, n_r is the refractive index, c is the speed of light in free space, V is the volume, Tr denotes the trace, and ε_0 is the permittivity of free space. $\chi(\omega)$ is the Fourier component of $\chi(t)$ with $e^{i\omega t}$ dependence. In Equation (5.97), $Re\ \Xi(0,\Delta_k)$ is the line shape function that describes the spectral shape of the optical gain in a driven semiconductor. It will be shown that the line shape function becomes Gaussian for the simplest non-Markovian relaxation, and Lorentzian for Markovian relaxation, later in this section. The denominator $[1 - Req_{1k}(0)]$ describes the gain enhancement due to the excitonic effects caused by the attractive Coulomb interaction. The excitonic enhancement can be neglected at room temperature for carrier densities above the Mott density, but one needs to keep the density dependent bandgap renormalization. On the other hand, excitonic enhancement has a noticeable contribution if the carrier density approaches the Mott density.

The factor $(1 + Reg_2(\infty,\Delta_k))$ in Equation (5.97) describes the gain (or line shape) enhancement due to the interaction between the optical field and the stochastic reservoir of the system. This enhancement of gain is due to the absence of a strict energy conservation in the non-Markovian quantum kinetics. It can be shown that $Reg_2(\infty,\Delta_k)$ vanishes in the Markovian limit. We now turn our attention to the study of line shape function and its enhancement for the cases of Markovian and non-Markovian relaxations.

5.3.1 The Markovian Limit

In order to analyze the collision term and the interference term in the Markovian limit, we put

$$\langle\langle \alpha k \| [H_i(t)(\underline{U}_0(\tau)H_i(t-\tau))| \alpha k \rangle\rangle_i = \frac{1}{2\tau_\alpha(k)}\delta(|\tau|). \tag{5.100}$$

In the Markovian limit, the interference term $g_2(t,\Delta_k)$ vanishes. The calculation of the line shape function is straightforward, and is given by

$$Re\Xi(0,\Delta_{k'}) = \frac{\gamma_{cv}(k)}{\Delta_k^2 + \gamma_{cv}^2(k)}, \tag{5.101}$$

which is Lorentzian used in most calculations. Here,

$$\gamma_{vc}(k) = \gamma_{cv}(k) = \frac{1}{2}\left\{\frac{1}{\tau_v(k)} + \frac{1}{\tau_c(k)}\right\}. \tag{5.102}$$

5.3.2 Simple Non-Markovian Relaxation

We assume the simplest form of the non-Markovian correlation function $\langle\langle \alpha k \| [H_i(t)(\underline{U}_0(\tau)H_i(t-\tau))| \alpha k\rangle\rangle_i$ as

$$\langle\langle \alpha k \| [H_i(t)(\underline{U}_0(\tau)H_i(t-\tau))| \alpha k\rangle\rangle_i = \frac{1}{2\tau_c\tau_\alpha(k)} exp\left[-\frac{|\tau|}{\tau_c}\right], \tag{5.103}$$

where τ_c is the correlation time for the intraband relaxation.

We obtain

$$\begin{aligned}\Xi(0,\Delta_k) &= \int_0^\infty dtexp[i\Delta_k t - \gamma_{cv}(k)(t + \tau_c exp[-t/\tau_c] - \tau_c)] \\ &= \tau_c I_0(-i\Delta_k\tau_c, \gamma_{cv}(k)\tau_c),\end{aligned} \tag{5.104}$$

where

$$I_0(A,B) = \int_0^\infty dt \exp[-At - B(t + exp(-t) - 1)]. \tag{5.105}$$

In general, $I_0(A,B)$ is evaluated by the continued fraction representation. If we expand the argument of the exponential function in Equation (5.105) up to the second order in t (weak modulation limit), we get the Gaussian line shape function:

$$Re\Xi(0,\Delta_k) = \sqrt{\frac{\tau_c\pi}{2\gamma_{cv}(k)}} \exp\left(-\frac{\tau_c\Delta_k^2}{2\gamma_{cv}(k)}\right). \tag{5.106}$$

When $\tau_c = 0$ (strong modulation limit), Equation (5.104) yields the Markovian line shape given by Equations (5.101) and (5.102).

The line shape enhancement (or gain enhancement) due to the interference between the driving field and the stochastic reservoir is given by

$$1+Reg_2(\infty,\Delta_k)=1+\frac{\gamma_{cv}(k)\tau_c}{1+\Delta_k^2\tau_c^2}. \qquad (5.107)$$

The gain (or line shape) enhancement factor at resonance ($\Delta_k=0$) is

$$1+Reg_2(\infty,\Delta_k)=1+\gamma_{cv}(k)\tau_c. \qquad (5.108)$$

Gain enhancement is due to the renormalized memory effects, and may be caused by the absence of a strict energy conservation on the time scale as compared with the correlation time of the system governed by non-Markovian quantum kinetics. Assume that an electron at $k=0$ in the conduction band was scattered at $t-\tau$ by the reservoir. For time t in the time interval τ_c, an electron still may retain its previous wave function corresponding to the state $k=0$ as a partial wave that decays exponentially with time t. When there is an incident photon at $t \ll \tau_c$ to this system, there will be a nonzero probability of transition for this electron to the state $k=0$ in the valence band. As a result, the stimulated emission probability for the state $k=0$ is enhanced by the presence of the memory effect. The probability that the enhanced transition occurs is proportional to the number of nonrandomizing scattering events per second, $\gamma_{cv}(k)$, times the time interval τ_c for which the memory effects extend. Non-Markovian enhancement of optical gain becomes significant as the correlation time increases. For example, when τ_c is 50 fs, the peak value of optical gain at $\Delta_k=0$ is predicted to be enhanced by as much as 50% for the typical intraband relaxation time of 100 fs.

The correlation time τ_c and the intraband relaxation $\gamma_{cv}(k)$ are related to each other by the frequency fluctuations, and their relation is described as follows.

Substituting $\tau=0$ in Equation (5.103), we get

$$\langle\langle \alpha k|H_i^2(t)|\alpha k\rangle\rangle_i=\frac{1}{2\tau_c\tau_\alpha(k)}. \qquad (5.109)$$

Then, from Equations (5.102) and (5.109) and integrating Equation (5.103) over the time t, we obtain

$$\frac{\gamma_{cv}(k)}{\tau_c} = \langle\langle ck|H_i^2(t)|ck\rangle\rangle_i + \langle\langle vk|H_i^2(t)|vk\rangle\rangle_i, \tag{5.110}$$

where the right half side of Equation (5.110) is the total intraband frequency fluctuation. We assume that the intraband relaxation $\gamma_{cv}(k)$ used in the line shape functions Equations (5.101) and (5.106) is mostly due to elastic or randomizing collisions, that is, Markovian processes. The correlation time is then inversely proportional to the energy transfer in the intraband frequency fluctuations.

5.4 OPTICAL GAIN OF A QUANTUM WELL LASER WITH NON-MARKOVIAN RELAXATION AND MANY-BODY EFFECTS

In this section, we present a theoretical model of the optical gain of a QW laser, taking into account the non-Markovian relaxation and many-body effects [3–5]. Many-body effects such as bandgap renormalization and excitonic enhancement are included by taking into account the Coulomb interaction in the Hartree–Fock approximation. Unnormalized single-particle energies are obtained using the multiband effective approximation, that is, a parabolic band structure for electrons, and a Luttinger–Kohn model for holes. The latter includes the spin–orbit split-off (SO) band coupling effects on the valence band structure. The account of the SO coupling effects would be particularly important for semiconductors with narrow SO splitting such as GaN, InGaP, and SiGe, the heterostructures with relatively large valence band edge discontinuity, such as InGaAs/InP system, and the strained-layer QWs, as the strain causes additional coupling between the Γ_7 and the Γ_8 bands.

5.4.1 Conventional Model for Gain

We will first discuss the conventional theory assuming the Markovian relaxation, which gives the Lorentzian line shape and introduces a phenomenological damping term in the density matrix formalism. Conventionally, the optical gain of a QW laser is calculated using the density matrix formalism. The linear gain is the result of $k_{\|}$-space integration of the product of the absolute square of the dipole matrix element, the Fermi function difference between electrons and hole, and the line shape function. The linear gain $g(\omega)$ is given by

$$g(\omega)=\frac{\omega\mu c}{n_r}\frac{1}{V}\sum_{\sigma\eta lm}\sum_{k_\parallel}\left|\hat{\varepsilon}\cdot M_{lm}^{\eta\sigma}(k_\parallel)\right|^2 \frac{(f_c^l-f_{h\sigma}^m)(\hbar/\tau_{in})}{(E_l^c(k_\parallel)-E_m^{h\sigma}(k_\parallel)-E_G-\hbar\omega)^2+(\hbar/\tau_{in})^2} \tag{5.111}$$

where f_c^l and $f_{h\sigma}^m$ are the Fermi functions for the lth subband in the conduction band and the mth subband in the valence band of the 3×3 block-diagonalized Hamiltonian H^σ, respectively; $\sigma=U$ (or L) refers to the upper (or lower) blocks, respectively; $E_l^c(k_\parallel)$ and $E_m^{h\sigma}(k_\parallel)$ are the lth subband energies in the conduction band and mth subband energy in the valence band of H^σ at $k_\parallel$, respectively; $M_{lm}^{\eta\sigma}(k_\parallel)|^2$ is the dipole matrix element between the lth subband in the conduction band with a spin state η, and the mth subband in the valence band; E_G is the bandgap energy; ω is the angular frequency of photons; V is the volume; and τ_{in} is the intraband relaxation time.

We use the envelope function approximation to calculate the valence-band structure of a QW within the 6×6 Luttinger–Kohn model. The simplest approach, as commonly used by most researchers, is to assume a constant intraband relaxation time. Various scattering mechanisms, such as carrier–carrier scattering and carrier–phonon scattering, contribute to the relaxation time. When we apply the Lorentzian line shape function to quantum well lasers, the gain coefficients are underestimated, and an anomalous absorption region appears at photon energy below the bandgap in the gain spectra. Thus, the tail ends of the Lorentzian line shape may not be reliable. This is so because the initial decay or the temporal behavior of the polarization is often characterized by the presence of memory effects.

5.4.2 Non-Markovian Model for Gain with Many-Body Effects

We now extend the above results given in Section 5.3 to the case of two-dimensional electron–hole pairs in a QW. The linear gain is then given by

$$\begin{aligned} g(\omega)=\frac{\omega\mu c}{n_r}\frac{1}{V}\sum_{\sigma\eta lm}\sum_{k_\parallel}\left|\hat{\varepsilon}\cdot M_{lm}^{\eta\sigma}(k_\parallel)\right|^2 &\frac{(f_c^l-f_{h\sigma}^m)[1+Reg_2(\infty,\Delta_{lm}^{\eta\sigma}(k_\parallel))]}{(1-Req_{lm}^{\eta\sigma}(k_\parallel))^2+(Imq_{lm}^{\eta\sigma}(k_\parallel))^2}\\ &\times\{Re\Xi_{lm}^{\eta\sigma}(0,\Delta_{lm}^{\eta\sigma}(k_\parallel))(1-Req_{lm}^{\eta\sigma}(k_\parallel))\\ &-Im\Xi_{lm}^{\eta\sigma}(0,\Delta_{lm}^{\eta\sigma}(k_\parallel))Imq_{lm}^{\eta\sigma}(k_\parallel)\}, \end{aligned} \tag{5.112}$$

where

$$Req_{lm}^{\eta\sigma}(k_{\|}') = -\sum_{k_{\|}} \frac{V_{lmml}^{eh\sigma}(|k_{\|}' - k_{\|}|)}{|\hat{\varepsilon} \cdot M_{lm}^{\eta\sigma}(k_{\|}')|} |\hat{\varepsilon} \cdot M_{lm}^{\eta\sigma}(k_{\|})| (f_c^l(k_{\|}) - f_{h\sigma}^m(k_{\|})) \\ Im\Xi_{lm}^{\eta\sigma}(0, \Delta_{lm}^{\eta\sigma}(k_{\|})), \tag{5.113}$$

and

$$Imq_{lm}^{\eta\sigma}(k_{\|}) = \sum_{k_{\|}'} \frac{V_{lmml}^{eh\sigma}(|k_{\|}' - k_{\|}|)}{|\hat{\varepsilon} \cdot M_{lm}^{\eta\sigma}(k_{\|}')|} |\hat{\varepsilon} \cdot M_{lm}^{\eta\sigma}(k_{\|})| (f_c^l(k_{\|}) - f_{h\sigma}^m(k_{\|})) \\ Re\Xi_{lm}^{\eta\sigma}(0, \Delta_{lm}^{\eta\sigma}(k_{\|})). \tag{5.114}$$

We also have

$$\Delta_{lm}^{\eta\sigma}(k_{\|}) = E_l^c(k_{\|}) - E_m^{h\sigma}(k_{\|}) + E_G + \Delta E_{CH} + \Delta E_{SX} - \hbar\omega, \tag{5.115}$$

and

$$Re\Xi_{lm}^{\eta\sigma}(0, \Delta_{lm}^{\eta\sigma}(k_{\|})) = \sqrt{\frac{\tau_c \pi}{2\gamma_{cv}(k)}} \exp\left(-\frac{\tau_c \Delta_{lm}^{\eta\sigma}(k_{\|})^2}{2\gamma_{cv}(k_{\|})} \right). \tag{5.116}$$

$Im\Xi_{lm}^{\eta\sigma}$ is calculated numerically from the Kramers–Kronig relations. That is,

$$Im\Xi_{lm}^{\eta\sigma}(0, \Delta_{lm}^{\eta\sigma}(k_{\|})) = \frac{\tau_c}{\hbar} \int_0^\infty \exp\left(-\frac{\gamma_{cv}(k_{\|})\tau_c}{2\hbar} t^2 \right) \\ \times \sin\left(\frac{\tau_c \Delta_{lm}^{\eta\sigma}(k_{\|})}{\hbar} t \right) dt. \tag{5.117}$$

In the following, we assume that the correlation time τ_c and the intraband relaxation $\gamma_{cv}(k_{\|})$ are constants. The bandgap renormalization is given as a summation of the Coulomb-hole self energy and the screened-exchange shift. The Coulomb-hole contribution to the bandgap renormalization is written as

$$\Delta E_{CH} = -2E_R a_o \lambda_s \ln\left(1 + \sqrt{\frac{32\pi N L_w}{C\lambda_s^3 a_o}} \right), \tag{5.118}$$

where N is the carrier density and C is a constant usually taken between 1 and 4. The Rydberg constant E_R and the exciton Bohr radius a_o are given by

$$\Delta E_R(eV) = 13.6\frac{m_o/m_r}{(\varepsilon/\varepsilon_o)^2}, \tag{5.119}$$

and

$$a_o(\text{Å}) = 0.53\frac{\varepsilon/\varepsilon_o}{m_o/m_r}, \tag{5.120}$$

where m_r is the reduced electron–hole mass defined by $1/m_r = 1/m_e + 1/m_h$. The exchange contribution to the bandgap renormalization is given by

$$\Delta E_{SX} = -\frac{2E_R a_o}{\lambda_s}\int_0^\infty dk_\| k_\| \frac{1+\dfrac{C\lambda_s a_o k_\|^2}{32\pi N L_w}}{1+\dfrac{k_\|}{\lambda_s}+\dfrac{C a_o k_\|^3}{32\pi N L_w}}\left[f_n^c(k_\|)+1-f_m^v(k_\|)\right]. \tag{5.121}$$

Also, a simpler expression for $q(k_\|)$ than Equations (5.113) and (5.114) is often used in the calculation. That is, the factor $q(k_\|,\hbar\omega)$ is given by [4,5]

$$q(k_\|,\hbar\omega) = i\frac{a_o E_R}{\pi\lambda_s|M_{nm}(k_\|)|}\int_0^\infty dk_\|' k_\|' |M_{nm}(k_\|')|(f_n^c(k_\|') - f_m^v(k_\|')) \\ \Xi(E_{lm}(k_\|',\hbar\omega))\Theta(k_\|,k_\|'), \tag{5.122}$$

where

$$\Theta(k_\|,k_\|') = \int_0^\infty d\theta\left(1+\frac{C\lambda_s a_o q_\|^2}{32\pi N L_w}\right)\left(1+\frac{q_\|}{\lambda_s}+\frac{C a_o q_\|^3}{32\pi N L_w}\right)^{-1}. \tag{5.123}$$

and

$$q_\| = |k_\| - k_\||. \tag{5.124}$$

5.5 NUMERICAL METHODS FOR VALENCE BAND STRUCTURE IN NANOSTRUCTURES

5.5.1 Finite Difference Method

In the case of QW, the effective mass equations can be usually solved using the finite difference method (FDM), the propagation matrix method, and the basis expansion method [6]. Each method has its advantages and disadvantages. Here, we consider numerical solutions for the Schrödinger equation under the framework of the FDM. The first-order total derivative $d\Psi/dx$ at a point $x = x_i$ can be approximated as

$$\left(\frac{d\Psi}{dx}\right)_i = \frac{\Psi_{i+1} - \Psi_{i-1}}{2h}, \tag{5.125}$$

using a three-point central difference representation. For the second-order derivatives, the central difference is

$$\left(\frac{d^2\Psi}{dx^2}\right)_i = \frac{\Psi_{i+1} - 2\Psi_i + \Psi_{i-1}}{h^2}. \tag{5.126}$$

As an example, consider a Schrödinger equation for the conduction band of a GaAs/Al_xGa_{1-x}As QW structure. Here, x is the mole fraction of aluminum.

$$-\frac{\hbar^2}{2m_0 m_c^*}\frac{d^2\Psi}{dx^2} + V(x)\Psi = E\Psi, \tag{5.127}$$

where m_c^* is the effective mass of the electron in the conduction band. For a finite barrier QW, we have

$$V(z) = \begin{cases} \Delta E_c & |x| \geq L_w/2 \\ 0 & |x| < L_w/2 \end{cases}, \tag{5.128}$$

where L_w is the width of a QW. The band edge discontinuity is $\Delta E_c = 0.33\Delta E_g$, where ΔE_g is the bandgap difference between GaAs and AlGaAs.

Dividing Equation (5.129) by $\hbar^2/2m_0$, we get

$$\frac{1}{m_C^*}\frac{d^2\Psi}{dx^2} - \frac{2m_0}{\hbar^2}V(x)\Psi = -\frac{2m_0}{\hbar^2}E\Psi. \tag{5.129}$$

In solving the Schrödinger equation by an FDM, we first discretize the solution domain with a grid system. Let us suppose that the domain is discretized by an even grid system with N nodes and a constant grid size $h' = \Delta x$ with an angstrom (Å) unit. By replacing the derivatives in the Schrödinger equation with the corresponding finite difference representation, the equation is reduced to a so-called difference equation. The difference equation is then applied at the interior grid points (sometimes called the mesh points or the nodes) within the solution domain. That is, using central difference, we have

$$\frac{1}{m_i}\frac{\Psi_{i+1} - 2\Psi_i + \Psi_{i-1}}{h^2} - V_i'\Psi_i = -E'\Psi_i, \tag{5.130}$$

where $V_i' = (2m_0\text{Å}^2/\hbar^2)V(x_i)$, $E' = (2m_0\text{Å}^2/\hbar^2)E$, and $h = h'/\text{Å}$.

Substituting node numbers from $i = 1$ to $i = N$ into Equation (5.130), we get the eigenvalue matrix equation

$$([K] - E')[\Psi] = 0, \tag{5.131}$$

where

$$[K] = \begin{pmatrix} a_{11} & a_{12} & 0 & 0 & & & & \cdots & 0 \\ a_{21} & a_{22} & a_{23} & 0 & & & & \cdots & 0 \\ 0 & a_{32} & a_{33} & a_{33} & \cdots & & & & 0 \\ 0 & \cdots & & & & \cdots & \cdots & \cdots & 0 \\ 0 & \cdots & & & & a_{N-2N-3} & a_{N-2N-2} & a_{N-2N-1} & 0 \\ 0 & \cdots & & & & \cdots & a_{N-1N-2} & a_{N-1N-1} & a_{N-1N} \\ 0 & \cdots & \cdots & & & \cdots & \cdots & a_{NN-1} & a_{NN} \end{pmatrix}, \tag{5.132}$$

and

$$[\Psi] = \begin{pmatrix} \Psi_1 \\ \Psi_2 \\ \cdots \\ \cdots \\ \Psi_{N-1} \\ \Psi_N \end{pmatrix}. \tag{5.133}$$

We apply the Dirichlet condition that the wave functions at the boundaries are set to zero:

$$\Psi_i = 0. \tag{5.134}$$

This requires a further process. Since Equation (5.130) has to hold for the ith components Ψ_i of eigenvector $[\Psi]$, some matrix elements other than the diagonal terms for matrixes $[K]$ have to be set as zero:

$$[K] = \begin{pmatrix} & & & 0 & & & \\ & & & 0 & & & \\ 0 & \cdots & 0 & K_{ii} & 0 & \cdots & 0 \\ & & & 0 & & & \\ & & & 0 & & & \end{pmatrix}. \tag{5.135}$$

Then, $([K]-E')[\Psi]=0$ satisfies the Dirichlet condition.

Next, we consider the upper 3×3 Hamiltonian for a (001)-oriented zinc blende (ZB) crystal of Equation (3.294) in Chapter 3, which is given by

$$H^U=\begin{pmatrix} P+Q & S_t & -iR_t \\ S_t & P+Q-\frac{2}{3}\Delta_{so} & \Delta_{so}-iR_t \\ iR_t & \Delta_{so}+iR_t & P-2Q-\frac{1}{3}\Delta_{so} \end{pmatrix}, \tag{5.136}$$

where

$$\begin{aligned} P&=-\frac{\hbar^2}{2m_o}\gamma_1(k_x^2+k_y^2+k_z^2)+a_v(\varepsilon_{xx}+\varepsilon_{yy}+\varepsilon_{zz}),\\ Q&=\frac{\hbar^2}{2m_o}\gamma_2(k_x^2+k_y^2-2k_z^2)+\frac{b}{2}(\varepsilon_{xx}+\varepsilon_{yy}-2\varepsilon_{zz}),\\ S_t&=\frac{\hbar^2}{2m_o}\left[-\frac{3}{2}(\gamma_2+\gamma_3)\right]k_t^2,\\ R_t&=\frac{\hbar^2}{2m_o}\left[-3\sqrt{2}\gamma_3\right]k_tk_z, \end{aligned} \tag{5.137}$$

with $k_t=\sqrt{k_x^2+k_y^2}$ being the magnitude of the wavevector in the k_x-k_y plane, ε_{ij} the symmetric strain tensor, γ_2, γ_2, and γ_3 the Luttinger parameters, a_v and b the deformation potentials, and Δ_{so} the SO energy.

The Hamiltonian has the following form

$$[K][\Psi]=\begin{pmatrix} \begin{pmatrix} \cdots & \cdots & \cdots \\ \cdots & H^{11} & \cdots \\ \cdots & \cdots & \cdots \end{pmatrix} & \begin{pmatrix} \cdots & \cdots & \cdots \\ \cdots & H^{12} & \cdots \\ \cdots & \cdots & \cdots \end{pmatrix} & \begin{pmatrix} \cdots & \cdots & \cdots \\ \cdots & H^{13} & \cdots \\ \cdots & \cdots & \cdots \end{pmatrix} \\ \begin{pmatrix} \cdots & \cdots & \cdots \\ \cdots & H^{21} & \cdots \\ \cdots & \cdots & \cdots \end{pmatrix} & \begin{pmatrix} \cdots & \cdots & \cdots \\ \cdots & H^{22} & \cdots \\ \cdots & \cdots & \cdots \end{pmatrix} & \begin{pmatrix} \cdots & \cdots & \cdots \\ \cdots & H^{23} & \cdots \\ \cdots & \cdots & \cdots \end{pmatrix} \\ \begin{pmatrix} \cdots & \cdots & \cdots \\ \cdots & H^{31} & \cdots \\ \cdots & \cdots & \cdots \end{pmatrix} & \begin{pmatrix} \cdots & \cdots & \cdots \\ \cdots & H^{32} & \cdots \\ \cdots & \cdots & \cdots \end{pmatrix} & \begin{pmatrix} \cdots & \cdots & \cdots \\ \cdots & H^{33} & \cdots \\ \cdots & \cdots & \cdots \end{pmatrix} \end{pmatrix}\begin{pmatrix} \begin{pmatrix} \cdots \\ \Psi^{(1)} \\ \cdots \end{pmatrix} \\ \begin{pmatrix} \cdots \\ \Psi^{(2)} \\ \cdots \end{pmatrix} \\ \begin{pmatrix} \cdots \\ \Psi^{(3)} \\ \cdots \end{pmatrix} \end{pmatrix}. \tag{5.138}$$

Here, H^{ij} takes the form similar to Equation (5.139) below.

$$H^{ij} = \begin{pmatrix} a_{11}^{ij} & a_{12}^{ij} & 0 & 0 & & & & \dots & 0 \\ a_{21}^{ij} & a_{22}^{ij} & a_{23}^{ij} & 0 & & & & \dots & 0 \\ 0 & a_{32}^{ij} & a_{33}^{ij} & a_{33}^{ij} & \dots & & & & 0 \\ 0 & \dots & & & & \dots & \dots & \dots & 0 \\ 0 & \dots & & & & a_{N-2N-3}^{ij} & a_{N-2N-2}^{ij} & a_{N-2N-1}^{ij} & 0 \\ 0 & \dots & & & & \dots & a_{N-1N-2}^{ij} & a_{N-1N-1}^{ij} & a_{N-1N}^{ij} \\ 0 & \dots & \dots & & & \dots & \dots & a_{NN-1}^{ij} & a_{NN}^{ij} \end{pmatrix}, \tag{5.139}$$

and

$$\Psi^{(i)} = \begin{pmatrix} \Psi_1^{(i)} \\ \dots \\ \Psi_N^{(i)} \end{pmatrix}. \tag{5.140}$$

We have to apply boundary conditions to each element H^{ij} before constructing the total Hamiltonian equation (5.138).

If there is no band-mixing effect (off-diagonal terms = 0), bands are decoupled and only one component $\Psi^{(i)}$ of the wavefunction is nonzero. Here, $i = 1,2,3$ means heavy-hole (HH), light-hole (LH), and SO bands. However, the band-mixing effect usually exists, and one band is coupled to the rest of the bands. The naming of the subband follows the dominant component of the wavefunction in terms of HH, LH, and SO bands. If components corresponding to HH and LH bands are nonzero and the HH component among them is dominant, the band is called the HL subband.

5.5.2 Finite Element Method

Two-dimensional and three-dimensional quantum-confined structures, called QWRs and quantum dot (QD), are innovative materials applicable in optical devices, such as laser diodes. The application of lower dimensional semiconductor structures as active regions of laser diodes promises improved device performances compared with conventional QW laser diodes. For low-dimensional structures, the density of states at the band edge is extremely high. This should lead to higher optical gain, and thus very low threshold currents are predicted.

However, the calculations of the energy band diagram and the wave function of QWRs and QDs are generally very complicated. In general, they cannot be done analytically, except for under special circumstances. By now, several numerical techniques have been developed for the analysis of these structures. For example, in the case of QWRs, they include effective bond orbital method (EBOM), tight binding method (TBM), finite FDM, and finite element method (FEM). Among them, FEM and FDM methods are relatively simpler, and require only 6 or 8 basis function. The advantage of FEM as a numerical technique over FDM is that it can utilize a nonuniform mesh; hence, the energy eigenstate and wave functions of arbitrarily shaped geometries with a wide range of lateral dimensions can be analyzed accurately. In this section, we introduce a FEM method related to a band structure calculation of the QWR and QD nanostructures.

5.5.2.1 Quantum Wire We assume that the system is quantized in the x and z directions. Then, the wave function Ψ_h of holes in the valence band for a QWR is expressed with the envelope function F_v and the basis functions $|v\rangle$ as

$$\Psi_h(k_y,x,y,z)=\sum_{\nu=1}^{6}F_\nu(k_y,x,z)\exp(k_y y)|\nu\rangle. \tag{5.141}$$

The energies and the envelope functions of valence subbands can be obtained by solving the effective mass equation

$$\sum_{v}[H_{\mu v}+V_h(x,z)\delta_{\mu v}]F_v(k_y,x,z)=E(k_y)F_\mu(k_y,x,z),$$
$$\nu,\mu\in\{|1\rangle,|2\rangle,|3\rangle,|4\rangle,|5\rangle,|6\rangle\}, \tag{5.142}$$

where H is the Luttinger–Kohn Hamiltonian equation (3.263) in Chapter 3, and $V_h(x,z)$ is the QWR potential for holes. The envelope function F_v follows the normalization rules

$$\sum_{i=1}^{6}\int dy|F_{(\nu)}^{m}(y)|^2=1, \tag{5.143}$$

where $m = \{m1, m2\}$ denotes the subband states in the valence band. In the QWR, the wave numbers k_x and k_z are substituted by operators as $k_x \to -i\partial/\partial x$ and $k_z \to -i\partial/\partial z$, respectively We have to be careful that these substitutions are done to preserve the hermiticity of the Hamiltonian.

The FEM is used to solve Equation (5.142) for the energy levels, as well as the eigenfunctions F_ν. The region of interest is divided into a number of linear triangles. We applied the Galerkin method to each element, and the rank of the derivatives of the partial differential equations in Equation (3.263) in Chapter 3 is decreased from 2 to 1 as a result of partial integration, the derivatives of which are called the weak forms. The envelope function F_e in an element e is expanded as

$$F_e = \sum_{i=1}^{M_e} N_i F_i = (N_{e1} \quad N_{e2} \quad N_{e3}) \begin{pmatrix} F_{e1} \\ F_{e2} \\ F_{e3} \end{pmatrix} = \{N_e\}^T \{F_e\}, \tag{5.144}$$

where M_e is the number of nodes in element e, which is 3 for a linear triangular element, $\{N_e\}$ is the shape function as will be defined in Equation (5.168), and the expansion coefficient F_i corresponds to the value of the envelope function at each node.

Substituting Equation (5.144) into multiband effective mass Equation (5.142), we get

$$\sum_{\mu=1}^{6} H_{\nu\mu} \{N_e\}^T \{F_{\nu e}\} = E\{N_e\}^T \{F_{\nu e}\}, \tag{5.145}$$

where $\mu, \nu = 1,2,3,4,5,6$. Multiplying the left-hand side of this equation by the shape function $\{N_e\}$, and integrating it for element e, we get

$$\sum_{\mu}^{6} \iint_e \{N_e\} H_{\nu\mu} \{N_e\}^T \{F_{\nu e}\} dxdz = E \iint_e \{N_e\}\{N_e\}^T \{F_{\nu e}\}. \tag{5.146}$$

We define the matrices $\{H_{\nu\mu}\}$ and $\{M\}$

$$\begin{aligned} \{H_{\nu\mu}\} &= \iint_e \{N_e\} H_{\nu\mu} \{N_e\}^T dxdz, \\ \{M\} &= \iint_e \{N_e\}\{N_e\}^T dxdz. \end{aligned} \tag{5.147}$$

Then, we get

$$\sum_{\mu}^{6}\{H_{\nu\mu}\}\{F_{\nu e}\}=E\{M\}\{F_{\nu e}\}. \tag{5.148}$$

The shape function for the linear triangular element $\{N_e\}$ is given by

$$\{N_e\}=\frac{1}{A_e}\begin{pmatrix} a_1 & b_1 & c_1 \\ a_2 & b_2 & c_2 \\ a_3 & b_3 & c_3 \end{pmatrix}\begin{pmatrix} 1 \\ x \\ z \end{pmatrix}, \tag{5.149}$$

where

$$\begin{aligned} a_1 &= x_2z_3 - x_3z_2, b_1 = z_2 - z_3, c_1 = x_3 - x_2, \\ a_2 &= x_3z_1 - x_1z_3, b_2 = z_3 - z_1, c_2 = x_1 - x_3, \\ a_3 &= x_1z_2 - x_2z_1, b_3 = z_1 - z_2, c_3 = x_2 - x_1, \end{aligned} \tag{5.150}$$

and

$$A_e=\frac{1}{2}\begin{pmatrix} 1 & 1 & 1 \\ x_1 & x_2 & x_3 \\ z_1 & z_2 & z_3 \end{pmatrix}, \tag{5.151}$$

where A_e is equal to the area of the linear triangular element.

The matrix elements for differentiations such as Ck_x^2, $Ck_y^2, Ck_xk_z, Ck_x, Ck_z$ and C, where C is the material parameter, are obtained by substituting the shape function into integration of weighted residual of the differentiation as follows:

$$\begin{aligned} \{Ck_x^2\} &= -\left[\frac{\partial}{\partial x}C\frac{\partial}{\partial x}\right]\{F_e\}=\int_{\Omega^e}\begin{pmatrix} \dfrac{\partial N_{e1}}{\partial x} \\ \dfrac{\partial N_{e2}}{\partial x} \\ \dfrac{\partial N_{e3}}{\partial x} \end{pmatrix} C\begin{pmatrix} \dfrac{\partial N_{e1}}{\partial x} & \dfrac{\partial N_{e2}}{\partial x} & \dfrac{\partial N_{e3}}{\partial x} \end{pmatrix}\begin{pmatrix} F_{e1} \\ F_{e2} \\ F_{e3} \end{pmatrix} d\Omega \\ &= C\frac{1}{4A_e}\begin{pmatrix} b_1b_1 & b_1b_2 & b_1b_3 \\ b_2b_1 & b_2b_2 & b_2b_3 \\ b_3b_1 & b_3b_2 & b_3b_3 \end{pmatrix}\begin{pmatrix} F_{e1} \\ F_{e2} \\ F_{e3} \end{pmatrix}, \end{aligned} \tag{5.152}$$

$$\{Ck_xk_z\} = -\frac{1}{2}\left[\frac{\partial}{\partial x}C\frac{\partial}{\partial z} + \frac{\partial}{\partial z}C\frac{\partial}{\partial x}\right]\{F_e\}$$

$$= \frac{1}{2}\int_{\Omega^e}\begin{pmatrix}\frac{\partial N_{e1}}{\partial x}\\ \frac{\partial N_{e2}}{\partial x}\\ \frac{\partial N_{e3}}{\partial x}\end{pmatrix}C\begin{pmatrix}\frac{\partial N_{e1}}{\partial z} & \frac{\partial N_{e2}}{\partial z} & \frac{\partial N_{e3}}{\partial z}\end{pmatrix}\begin{pmatrix}F_{e1}\\ F_{e2}\\ F_{e3}\end{pmatrix}d\Omega$$

$$+ \frac{1}{2}\int_{\Omega^e}\begin{pmatrix}\frac{\partial N_{e1}}{\partial z}\\ \frac{\partial N_{e2}}{\partial z}\\ \frac{\partial N_{e3}}{\partial z}\end{pmatrix}C\begin{pmatrix}\frac{\partial N_{e1}}{\partial x} & \frac{\partial N_{e2}}{\partial x} & \frac{\partial N_{e3}}{\partial x}\end{pmatrix}\begin{pmatrix}F_{e1}\\ F_{e2}\\ F_{e3}\end{pmatrix}d\Omega$$

$$= \frac{C}{8A_e}\begin{pmatrix}b_1c_1 & b_1c_2 & b_1c_3\\ b_2c_1 & b_2c_2 & b_2c_3\\ b_3c_1 & b_3c_2 & b_3c_3\end{pmatrix}\begin{pmatrix}F_{e1}\\ F_{e2}\\ F_{e3}\end{pmatrix} + \frac{C}{8A_e}\begin{pmatrix}c_1b_1 & c_1b_2 & c_1b_3\\ c_2b_1 & c_2b_2 & c_2b_3\\ c_3b_1 & c_3b_2 & c_3b_3\end{pmatrix}\begin{pmatrix}F_{e1}\\ F_{e2}\\ F_{e3}\end{pmatrix}, \tag{5.153}$$

$$\{Ck_x\} = -\frac{i}{2}\left[C\frac{\partial}{\partial x} + \frac{\partial}{\partial x}C\right]\{F_e\}$$

$$= -\frac{i}{2}\int_{\Omega^e}\begin{pmatrix}N_{e1}\\ N_{e2}\\ N_{e3}\end{pmatrix}C\begin{pmatrix}\frac{\partial N_{e1}}{\partial x} & \frac{\partial N_{e2}}{\partial x} & \frac{\partial N_{e3}}{\partial x}\end{pmatrix}\begin{pmatrix}F_{e1}\\ F_{e2}\\ F_{e3}\end{pmatrix}d\Omega$$

$$+ \frac{i}{2}\int_{\Omega^e}\begin{pmatrix}\frac{\partial N_{e1}}{\partial x}\\ \frac{\partial N_{e2}}{\partial x}\\ \frac{\partial N_{e3}}{\partial x}\end{pmatrix}C\begin{pmatrix}N_{e1} & N_{e2} & N_{e3}\end{pmatrix}\begin{pmatrix}F_{e1}\\ F_{e2}\\ F_{e3}\end{pmatrix}d\Omega \tag{5.154}$$

$$= -\frac{iC}{12}\begin{pmatrix}b_1 & b_2 & b_3\\ b_1 & b_2 & b_3\\ b_1 & b_2 & b_3\end{pmatrix}\begin{pmatrix}F_{e1}\\ F_{e2}\\ F_{e3}\end{pmatrix} + \frac{iC}{12}\begin{pmatrix}b_1 & b_1 & b_1\\ b_2 & b_2 & b_2\\ b_3 & b_3 & b_3\end{pmatrix}\begin{pmatrix}F_{e1}\\ F_{e2}\\ F_{e3}\end{pmatrix},$$

$$
\begin{aligned}
\{C\} &= [C]\{F_e\} \\
&= \int_{\Omega^e} \begin{pmatrix} N_{e1} \\ N_{e2} \\ N_{e3} \end{pmatrix} C(N_{e1} \quad N_{e2} \quad N_{e3}) \begin{pmatrix} F_{e1} \\ F_{e2} \\ F_{e3} \end{pmatrix} d\Omega \\
&= C\frac{A_e}{12}\begin{pmatrix} 2 & 1 & 1 \\ 1 & 2 & 1 \\ 1 & 1 & 2 \end{pmatrix}\begin{pmatrix} F_{e1} \\ F_{e2} \\ F_{e3} \end{pmatrix} = C[M_e]\{F_e\},
\end{aligned} \tag{5.155}
$$

where matrix elements for $\{Ck_z^2\}$ and $\{Ck_z\}$ are obtained by replacing b with c in Equations (5.152) and (5.154). Thus, the matrix elements in Equation (3.263) in Chapter 3 are given by

$$
\begin{aligned}
H_{11} = \{P+Q\}_{ij} &= \frac{\hbar^2}{2m_o}\left[\frac{b_i b_j}{4A_e}(\gamma_1+\gamma_2) + \frac{c_i c_j}{4A_e}(\gamma_1 - 2\gamma_2) \right. \\
&\left. + \frac{(1+\delta_{ij})A_e}{12}(\gamma_1+\gamma_2)k_y^2\right] \\
&+ \frac{(1+\delta_{ij})A_e}{12}(P_\varepsilon + Q_\varepsilon + V_h(x,z)),
\end{aligned} \tag{5.156}
$$

$$
\begin{aligned}
H_{22} = \{P-Q\}_{ij} &= \frac{\hbar^2}{2m_o}\left[\frac{b_i b_j}{4A_e}(\gamma_1-\gamma_2) + \frac{c_i c_j}{4A_e}(\gamma_1 + 2\gamma_2) \right. \\
&\left. + \frac{(1+\delta_{ij})A_e}{12}(\gamma_1-\gamma_2)k_y^2\right] \\
&+ \frac{(1+\delta_{ij})A_e}{12}(P_\varepsilon - Q_\varepsilon + V_h(x,z)),
\end{aligned} \tag{5.157}
$$

$$
H_{12} = \{-S\}_{ij} = -\frac{\sqrt{3}\hbar^2}{m_o}\gamma_3\left[\frac{b_i c_j + c_i b_j}{8A_e} + \frac{c_i - c_j}{12}k_y\right] - \frac{(1+\delta_{ij})A_e}{12}S_\varepsilon, \tag{5.158}
$$

$$
\begin{aligned}
H_{13} = \{R\}_{ij} &= \frac{\sqrt{3}\hbar^2}{2m_o}\left[\frac{b_i b_j}{4A_e}\gamma_2 + \frac{b_i - b_j}{6}\gamma_3 k_y - \frac{(1+\delta_{ij})A_e}{12}k_y^2\gamma_2\right] \\
&+ \frac{(1+\delta_{ij})A_e}{12}Q_\varepsilon,
\end{aligned} \tag{5.159}
$$

$$\{P+\Delta\}_{ij} = \frac{\hbar^2}{2m_o}\left[\frac{b_i b_j}{4A_e}\gamma_1 + \frac{c_i c_j}{4A_e}\gamma_1 + \frac{(1+\delta_{ij})A_e}{12}\gamma_1 k_y^2\right] + \frac{(1+\delta_{ij})A_e}{12}(P_\varepsilon + \Delta + V_h(x,z)), \tag{5.160}$$

where $\{R^\dagger\}_{ij} = \{R\}^*_{ji}$ and $\{-S^\dagger\}_{ij} = \{-S\}^*_{ji}$.

Finally, we need to sum matrix elements Equations (5.156)–(5.160) for each element to make a $n \times n$ subtotal matrix with n nodes in the system under consideration. The boundary condition should be applied to n_{bc} nodes on the edges of the analysis region for a subtotal matrix. The Dirichlet condition requires that the wave functions at the boundaries be set to zero:

$$\Psi_i = 0. \tag{5.161}$$

A simple way to implement the Dirichlet condition is to omit the nodes at the boundaries, because, under the Dirichlet condition, they do not influence the other nodes. Then, the net effect is to reduce the dimensions of the matrix by the number of nodes at the boundaries, to $n - n_{bc}$.

The subtotal matrices with the dimension $n_s = n - n_{bc}$ are then summed up to make a total generalized eigenvalue equation:

$$[H][F] = E[M][F]. \tag{5.162}$$

$$[H] = \begin{pmatrix} \{H^{ij}_{11}\} & \{H^{ij}_{12}\} & \{H^{ij}_{13}\} & \{H^{ij}_{14}\} & \{H^{ij}_{15}\} & \{H^{ij}_{16}\} \\ \{H^{ij}_{21}\} & \{H^{ij}_{22}\} & \{H^{ij}_{23}\} & \{H^{ij}_{24}\} & \{H^{ij}_{25}\} & \{H^{ij}_{26}\} \\ \cdots & \cdots & \cdots & \cdots & \cdots & \cdots \\ \{H^{ij}_{61}\} & \{H^{ij}_{62}\} & \{H^{ij}_{63}\} & \{H^{ij}_{64}\} & \{H^{ij}_{65}\} & \{H^{ij}_{66}\} \end{pmatrix} \tag{5.163}$$

$$[M] = \begin{pmatrix} \{M^{ij}_{11}\} & \{M^{ij}_{12}\} & \{M^{ij}_{13}\} & \{M^{ij}_{14}\} & \{M^{ij}_{15}\} & \{M^{ij}_{16}\} \\ \{M^{ij}_{21}\} & \{M^{ij}_{22}\} & \{M^{ij}_{23}\} & \{M^{ij}_{24}\} & \{M^{ij}_{25}\} & \{M^{ij}_{26}\} \\ \cdots & \cdots & \cdots & \cdots & \cdots & \cdots \\ \{M^{ij}_{61}\} & \{M^{ij}_{62}\} & \{M^{ij}_{63}\} & \{M^{ij}_{64}\} & \{M^{ij}_{65}\} & \{M^{ij}_{66}\} \end{pmatrix} \tag{5.164}$$

$$[F]=[F_1^1F_1^2\ldots F_1^{n_s}F_2^1F_2^2\ldots F_2^{n_s}\ldots\ldots\ldots F_6^1F_6^2\ldots F_6^{n_s}]^T, \tag{5.165}$$

where the subscript denotes the Bloch wave function.

For an FDM, a rectangular grid with square unit cells and $N = N_xN_y$ nodes can be used. Here, we have to be careful to preserve the Hermiticity of the Hamiltonian. To accomplish this, we introduce two different matrices $D^{(+)}$ and $D^{(-)}$ for both ∂_x and ∂_z with $[D^{(+)}]^T = -D^{(-)}$. Here, $D^{(+)}$ and $D^{(-)}$ are used in the upper right half and the lower left half of each matrix element in Equation (5.142), respectively. The discretizations of the symmetrized expressions for $C\partial_x$, $C\partial_z$, $C\partial_{xx}$, $C\partial_{xz}$, $C\partial_{zz}$ are then obtained from the standard first-order forward (backward) difference quotients $\partial_x^{(+)}, \partial_z^{(+)}$ ($\partial_x^{(-)}, \partial_z^{(-)}$) by

$$\begin{aligned}
D_x^{(+)}(C)&=[C\partial_x^{(+)}+\partial_x^{(+)}C]/2,\\
D_x^{(-)}(C)&=[C\partial_x^{(-)}+\partial_x^{(-)}C]/2,\\
D_y^{(+)}(C)&=[C\partial_y^{(+)}+\partial_y^{(+)}C]/2,\\
D_y^{(-)}(C)&=[C\partial_y^{(-)}+\partial_y^{(-)}C]/2,\\
D_{xx}(C)&=[\partial_x^{(+)}C\partial_x^{(-)}+\partial_x^{(-)}C\partial_x^{(+)}]/2,\\
D_{yy}(C)&=[\partial_y^{(+)}C\partial_y^{(-)}+\partial_y^{(-)}C\partial_y^{(+)}]/2,\\
D_{xy}(C)&=[\partial_x^{(+)}C\partial_y^{(+)}+\partial_y^{(+)}C\partial_x^{(+)}+\partial_x^{(-)}C\partial_y^{(-)}+\partial_y^{(-)}C\partial_x^{(-)}]/4.
\end{aligned} \tag{5.166}$$

In the case of $\partial_x^{(+)}$, $\partial_z^{(+)}$ in Q and S, the central difference method is used to prevent abnormal behavior in the valence subband energy.

5.5.2.2 *Quantum Box* The envelope function F_e in an element e is expanded as

$$F_e=\sum_{i=1}^{M_e}N_iF_i=(N_{e1}\quad N_{e2}\quad N_{e3}\quad N_{e4})\begin{pmatrix}F_{e1}\\F_{e2}\\F_{e3}\\F_{e4}\end{pmatrix}=\{N_e\}^T\{F_e\}, \tag{5.167}$$

where M_e is the number of nodes in element e, which is 4 for a linear triangular element, and the expansion coefficient F_i corresponds to the value of the envelope function at each node. The shape function for the tetrahedral element $\{N_e\}$ is given by

$$\{N_e\} = \begin{pmatrix} a_{11} & a_{21} & a_{31} & a_{41} \\ a_{12} & a_{22} & a_{32} & a_{42} \\ a_{13} & a_{23} & a_{33} & a_{43} \\ a_{14} & a_{24} & a_{34} & a_{44} \end{pmatrix} \begin{pmatrix} 1 \\ x_1 \\ x_2 \\ x_3 \end{pmatrix}, \qquad (5.168)$$

and

$$\begin{pmatrix} a_{11} & a_{12} & a_{13} & a_{14} \\ a_{21} & a_{22} & a_{23} & a_{24} \\ a_{31} & a_{32} & a_{33} & a_{34} \\ a_{41} & a_{42} & a_{43} & a_{44} \end{pmatrix} = \begin{pmatrix} 1 & (x_1)_1 & (x_2)_1 & C(x_3)_1 \\ 1 & (x_1)_2 & (x_2)_2 & (x_3)_2 \\ 1 & (x_1)_3 & (x_2)_3 & (x_3)_3 \\ 1 & (x_1)_4 & C(x_2)_4 & (x_3)_4 \end{pmatrix}^{-1}, \qquad (5.169)$$

where $(x_1)_i$, $(x_2)_i$, and $(x_3)_i$ are coordinates for four nodes in the tetrahedral element, as shown in Figure 5.1.

Substituting Equation (5.144) into multivalley effective mass Equation (3.263) in Chapter 3, we get

$$\sum_{\mu=1}^{6} H_{\nu\mu}\{N_e\}^T\{F_{\nu e}\} = E\{N_e\}^T\{F_{\nu e}\}, \qquad (5.170)$$

where μ and $\nu = 1, 2, 3, 4, 5, 6$.

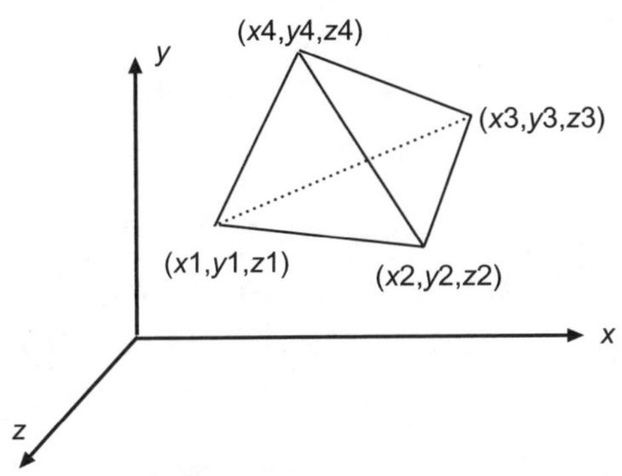

Figure 5.1. Tetrahedral element for finite element calculation.

Multiplying the left-hand side of this equation by the shape function $\{N_e\}$ and integrating it for element e, we get

$$\sum_{\mu}^{6}\iiint_e \{N_e\}H_{\nu\mu}\{N_e\}^T\{F_{\nu e}\}dxdydz = E\iiint_e \{N_e\}\{N_e\}^T\{F_{\nu e}\}. \quad (5.171)$$

We define the matrices $\{H_{\nu\mu}\}$ and $\{M\}$

$$\begin{aligned}\{H_{\nu\mu}\} &= \iiint_e \{N_e\}H_{\nu\mu}\{N_e\}^T\,dxdz,\\ \{M\} &= \iiint_e \{N_e\}\{N_e l\}^T\,dxdydz.\end{aligned} \quad (5.172)$$

Then, we get

$$\sum_{\mu}^{6}\{H_{\nu\mu}\}\{F_{\nu e}\} = E\{M\}\{F_{\nu e}\}. \quad (5.173)$$

The matrix elements for differentiations such as $Ck_{x_i}^2, Ck_{x_i}k_{x_j}, Ck_{x_i}, Cx_ik_{x_j}$, $F(x_1,x_2,x_3)$, and C, where C is the material parameter, are obtained by substituting the shape function into integration of weighted residual of the differentiation. For example,

$$\begin{aligned}\{Ck_{x_1}^2\} &= -\left[\frac{\partial}{\partial x_1}C\frac{\partial}{\partial x_1}\right]\{F_e\}\\ &= \int_{\Omega^e}\begin{pmatrix}\dfrac{\partial N_{e1}}{\partial x_1}\\ \dfrac{\partial N_{e2}}{\partial x_1}\\ \dfrac{\partial N_{e3}}{\partial x_1}\\ \dfrac{\partial N_{e4}}{\partial x_1}\end{pmatrix} C\begin{pmatrix}\dfrac{\partial N_{e1}}{\partial x_1} & \dfrac{\partial N_{e2}}{\partial x_1} & \dfrac{\partial N_{e3}}{\partial x_1} & \dfrac{\partial N_{e4}}{\partial x_1}\end{pmatrix}\begin{pmatrix}F_{e1}\\ F_{e2}\\ F_{e3}\\ F_{e4}\end{pmatrix}d\Omega \quad (5.174)\\ &= CV_e[a_{2i}a_{2j}]\{F_{ej}\},(i,j=1,2,3),\end{aligned}$$

where

$$[a_{2i}a_{2j}]=\begin{pmatrix} a_{21}a_{21} & a_{21}a_{22} & a_{21}a_{23} & a_{21}a_{24} \\ a_{22}a_{21} & a_{22}a_{22} & a_{22}a_{23} & a_{22}a_{24} \\ a_{23}a_{21} & a_{23}a_{22} & a_{23}a_{23} & a_{23}a_{24} \\ a_{24}a_{21} & a_{24}a_{22} & a_{24}a_{23} & a_{24}a_{24} \end{pmatrix}. \tag{5.175}$$

and V_e is the volume of the element. Then, $\{Ck_{x2}^2\}$ and $\{Ck_{x3}^2\}$ are obtained by replacing $a_{2i}a_{2j}$ in Equation (5.174) with $a_{3i}a_{3j}$ and $a_{4i}a_{4j}$, respectively.

$$\begin{aligned}\{2Ck_{x_1}k_{x_3}\} &= -\frac{1}{2}\left[\frac{\partial}{\partial x_1}C\frac{\partial}{\partial Gx_3}+\frac{\partial}{\partial x_3}C\frac{\partial}{\partial x_1}\right]\{F_e\} \\ &= \frac{1}{2}\int_{\Omega^e}\begin{pmatrix}\frac{\partial N_{e1}}{\partial x_1}\\ \frac{\partial N_{e2}}{\partial x_1}\\ \frac{\partial N_{e3}}{\partial x_1}\\ \frac{\partial N_{e4}}{\partial x_1}\end{pmatrix}C\begin{pmatrix}\frac{\partial N_{e1}}{\partial x_3} & \frac{\partial N_{e2}}{\partial x_3} & \frac{\partial N_{e3}}{\partial x_3} & \frac{\partial N_{e4}}{\partial x_3}\end{pmatrix}\begin{pmatrix}F_{e1}\\ F_{e2}\\ F_{e3}\\ F_{e4}\end{pmatrix}d\Omega \\ &\quad+\frac{1}{2}\int_{\Omega^e}\begin{pmatrix}\frac{\partial N_{e1}}{\partial x_3}\\ \frac{\partial N_{e2}}{\partial x_3}\\ \frac{\partial N_{e3}}{\partial x_3}\\ \frac{\partial N_{e4}}{\partial x_3}\end{pmatrix}C\begin{pmatrix}\frac{\partial N_{e1}}{\partial x_1} & \frac{\partial N_{e2}}{\partial x_1} & \frac{\partial N_{e3}}{\partial x_1}G & \frac{\partial N_{e4}}{\partial x_1}\end{pmatrix}\begin{pmatrix}F_{e1}\\ F_{e2}\\ F_{e3}\\ F_{e4}\end{pmatrix}d\Omega \\ &= \frac{C}{2}V_e[a_{2i}a_{4j}]\{F_{ej}\}+\frac{C}{2}V_e[a_{4i}a_{2j}]\{F_{ej}\},\end{aligned} \tag{5.176}$$

where $\{Ck_{x_1}k_{x_2}\}$ and $\{Ck_{x_2}k_{x_3}\}$ are obtained by replacing $a_{2i}a_{4j}$ $(a_{4i}a_{2j})$ in Equation (5.176) by $a_{2i}a_{3j}$ $(a_{3i}a_{2j})$ and $a_{3i}a_{4j}$ $(a_{4i}a_{3j})$, respectively.

$$\{Ck_{x_1}\} = -\frac{i}{2}\left[C\frac{\partial}{\partial x_1} + \frac{\partial}{\partial x_1}C\right]\{F_e\}$$

$$= -\frac{i}{2}\int_{\Omega^e}\begin{pmatrix} N_{e1} \\ N_{e2} \\ N_{e3} \\ N_{e4} \end{pmatrix} C\begin{pmatrix} \frac{\partial N_{e1}}{\partial x_1} & \frac{\partial N_{e2}}{\partial x_1} & \frac{\partial N_{e3}}{\partial x_1} & \frac{\partial N_{e3}}{\partial x_1} \end{pmatrix}\begin{pmatrix} F_{e1} \\ F_{e2} \\ F_{e3} \\ F_{e4} \end{pmatrix} d\Omega$$

$$+\frac{i}{2}\int_{\Omega^e}\begin{pmatrix} \frac{\partial N_{e1}}{\partial x_1} \\ \frac{\partial N_{e2}}{\partial x_1} \\ \frac{\partial N_{e3}}{\partial x_1} \\ \frac{\partial N_{e4}}{\partial x_1} \end{pmatrix} C(N_{e1} \quad N_{e2} \quad N_{e3} \quad N_{e4})\begin{pmatrix} F_{e1} \\ F_{e2} \\ F_{e3} \\ F_{e4} \end{pmatrix} d\Omega$$

$$= -\frac{iC}{8}V_e\begin{pmatrix} a_{21} & a_{22} & a_{23} & a_{24} \\ a_{21} & a_{22} & a_{23} & a_{24} \\ a_{21} & a_{22} & a_{23} & a_{24} \\ a_{21} & a_{22} & a_{23} & a_{24} \end{pmatrix}\begin{pmatrix} F_{e1} \\ F_{e2} \\ F_{e3} \\ F_{e4} \end{pmatrix}$$

$$+\frac{iC}{8}V_e\begin{pmatrix} a_{21} & a_{21} & a_{21} & a_{21} \\ a_{22} & a_{22} & a_{22} & a_{22} \\ a_{23} & a_{23} & a_{23} & a_{23} \\ a_{24} & a_{24} & a_{24} & a_{24} \end{pmatrix}\begin{pmatrix} F_{e1} \\ F_{e2} \\ F_{e3} \\ F_{e4} \end{pmatrix}, \tag{5.177}$$

In the above, we used the following relation

$$\int_{\Omega^e}(N_{e1})^k(N_{e2})^l(N_{e3})^m(N_{e4})^n dV = \frac{k!l!m!n!}{(k+l+m+n+3)!}6V_e. \tag{5.178}$$

$$\{C\} = [C]\{F_e\}$$

$$= \int_{\Omega^e}\begin{pmatrix} N_{e1} \\ N_{e2} \\ N_{e3} \\ F_{e4} \end{pmatrix} C(N_{e1} \quad N_{e2} \quad N_{e3} \quad N_{e4})\begin{pmatrix} F_{e1} \\ F_{e2} \\ F_{e3} \\ F_{e4} \end{pmatrix} d\Omega \tag{5.179}$$

$$= C\frac{V_e}{20}\begin{pmatrix} 2 & 1 & 1 & 1 \\ 1 & 2 & 1 & D1 \\ 1 & 1 & 2 & 1 \\ 1 & 1 & 1 & 2 \end{pmatrix}\begin{pmatrix} F_{e1} \\ F_{e2} \\ F_{e3} \\ F_{e4} \end{pmatrix} = C[M_e]\{F_e\}.$$

For the cases $Cx_i k_{x_j}$ and $F(x_1, x_2, x_3)$, we should numerically integrate to obtain matrix elements. For example,

$$
\begin{aligned}
\{Cx_2 k_{x1}\} &= -i\frac{1}{2}\left[x_2 C\frac{\partial}{\partial x_1} + \frac{\partial}{\partial x_1}(Cx_2)\right]\{F_e\} \\
&= -i\frac{C}{2}\int_{\Omega^e}\begin{pmatrix} N_{e1} \\ N_{e2} \\ N_{e3} \\ N_{e4} \end{pmatrix} x_2 \begin{pmatrix} \frac{\partial N_{e1}}{\partial x_1} & \frac{\partial N_{e2}}{\partial x_1} & \frac{\partial N_{e3}}{\partial x_1} & \frac{\partial N_{e4}}{\partial x_1} \end{pmatrix}\begin{pmatrix} F_{e1} \\ F_{e2} \\ F_{e3} \\ F_{e4} \end{pmatrix} d\Omega \\
&\quad + i\frac{C}{2}\int_{\Omega^e}\begin{pmatrix} \frac{\partial N_{e1}}{\partial x_1} \\ \frac{\partial N_{e2}}{\partial x_1} \\ \frac{\partial N_{e3}}{\partial x_1} \\ \frac{\partial N_{e4}}{\partial x_1} \end{pmatrix} x_2 \begin{pmatrix} N_{e1} & N_{e2} & N_{e3} & N_{e4} \end{pmatrix}\begin{pmatrix} F_{e1} \\ F_{e2} \\ F_{e3} \\ F_{e4} \end{pmatrix} d\Omega.
\end{aligned}
\tag{5.180}
$$

Here, we need to numerically integrate the above equation by using the relation Equation (5.178) to obtain the matrix element.

Finally, we sum matrix elements for each element to make an $n \times n$ subtotal matrix, with n nodes in the system under consideration. The boundary condition should be applied to n_{bc} nodes on the edges of the analysis region for a subtotal matrix. The Dirichlet condition requires that the wave functions at the boundaries be set to zero:

$$\Psi_i = 0. \tag{5.181}$$

A simple way to implement the Dirichlet condition is to omit the nodes at the boundaries, because under the Dirichlet condition, they do not influence the other nodes, as we did in Equation (5.135). Then, the net effect is to reduce the dimension of the matrix by the number of nodes at the boundaries, to $n - n_{bc}$. The subtotal matrices with the dimension $n_s = n - n_{bc}$ are then summed up to make a total generalized eigenvalue equation:

$$[H][F] = E[M][F], \tag{5.182}$$

where

$$[H]=\begin{pmatrix}\{H_{11}^{ij}\} & \{H_{12}^{ij}\}\\ \{H_{21}^{ij}\} & \{H_{22}^{ij}\}\end{pmatrix},[M]=\begin{pmatrix}\{M_{11}^{ij}\} & \{M_{12}^{ij}\}\\ \{M_{21}^{ij}\} & \{M_{22}^{ij}\}\end{pmatrix}, \tag{5.183}$$

and

$$[F]=[F_1^1F_1^2\ldots F_1^{n_s}F_2^1F_2^2\ldots F_2^{n_s}]^T, \tag{5.184}$$

where the subscript denotes the Bloch wave function.

5.6 ZINC BLENDE BULK AND QUANTUM WELL STRUCTURES

In this section, we discuss electronic and optical properties of strained zinc blende semiconductors [7]. First, we review developments of conventional III-V-based heterostructure lasers. In the 1970s, a pair of semiconductors with different bandgaps but with nearly the same lattice constants were investigated because a good epilayer of one could be grown on the other. Thus, the GaAs/AlGaAs system has been a major player in the field of heterostructure devices because AlAs and GaAs are closely lattice matched but have different bandgaps. The first semiconductor laser diodes (LDs) were GaAs homojunction devices fabricated in 1962. But room temperature operation of these lasers was not possible. It was only in 1969 that room temperature operation of a heterostructure semiconductor laser was demonstrated. However, methods using the GaAs/AlGaAs system did not produce greatly improved devices until late 1970s, mainly because of the limitations of the growth technology. The improvement of novel growth techniques, such as molecular beam epitaxy (MBE) and metalorganic chemical vapor deposition (MOCVD) growth methods have made it possible to produce high-performance AlGaAs lasers and other heterostructure lasers. GaAs QWs cladded with AlAs or AlGaAs were easily formed. The QW structures reduce the threshold current, and enhance performance of lasers. However, the GaAs/AlGaAs system has some problems, because Al has high affinity for oxygen.

Considerable research has also been done on the ternary InGaAs and the quaternary InGaAsP alloys grown on InP substrate. These devices are less sensitive to the environments because Al is absent. The ternary alloy $Ga_{0.47}In_{0.53}As$ is lattice matched to InP, and the quaternary InGaAsP alloys can be lattice matched to InP for several pairs of values of Ga and As. The quaternary InGaAsP lasers emitting 1.3 to 1.5 μm are suitable for optical communication. However, the technology of fabricating the quaternary epilayers on InP substrate is involved because of the vapor pressure of P. In addition, the use of lattice-matched heterostructures severely limited the choice of conventional III-V semiconductors. Thus, the LEDs and LDs could emit light only in a narrow region of the spectrum.

After that, the improvement of novel growth techniques has made it possible to produce extremely high-quality expitaxial interfaces, not only between lattice-matched semiconductors, but even between materials that differ in lattice constant by several percent. In particular, the growth of strained-layer structures has had several advantages. Stained layers allow for new materials combinations on established substrates. Also, with the independent variation of bandgap and lattice constant, it is possible to access new band regimes. Here, we investigate valence band structures and optical gain of strained zinc blende semiconductors.

5.6.1 Bulk Zinc Blende Structure

Strain deforms the crystal lattice. However, if a crystal is under uniform deformation, it will be still periodic such that Bloch theorem is still applicable. In the following, we consider the common case of biaxial strain, namely,

$$\begin{aligned} \varepsilon_{xx} &= \varepsilon_{yy} \neq \varepsilon_{zz} \\ \varepsilon_{xy} &= \varepsilon_{yz} = \varepsilon_{zx}. \end{aligned} \tag{5.185}$$

For mismatched superlattices or QWs grown in the z direction, we obtain

$$\begin{aligned} \varepsilon_{xx} &= \varepsilon_{yy} = \frac{a_s - a_e}{a_e} \\ \varepsilon_{zz} &= -\frac{2C_{12}}{C_{11}}\varepsilon_{xx}, \end{aligned} \tag{5.186}$$

where a_s and a_e are the lattice constants of the substrate and the epi-layer, respectively. For compressive strain, $a_s < a_e$, $\varepsilon_{xx} = \varepsilon_{yy} < 0$, and $\varepsilon_{zz} > 0$. In Figure 5.2, we plotted a schematic representation of an epi-layer a_e grown on a substrate a_s: (a) tensile stained, (b) unstrained, and (c) compressively strained semiconductors.

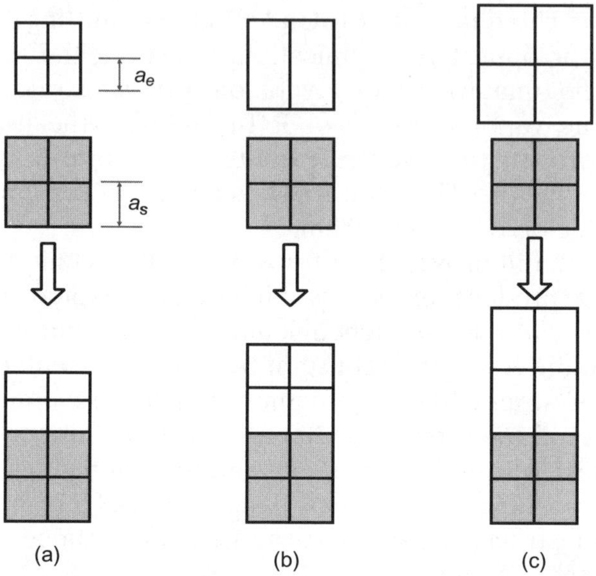

Figure 5.2. A schematic representation of an epilayer a_e grown on a substrate a_s: (a) tensile stained; (b) unstrained; (c) compressively strained semiconductors.

At a zone center ($\mathbf{k}=0$), the 6×6 Hamiltonian given in Equation (3.263) in Chapter 3 is simplified to

$$H_{\text{zinc blende}}^{(001)}(\mathbf{k}=0)$$

$$=-\begin{pmatrix} P_\varepsilon+Q_\varepsilon & 0 & 0 & 0 & 0 & \\ 0 & P_\varepsilon-Q_\varepsilon & 0 & 0 & -\sqrt{2}Q_\varepsilon & 0 \\ 0 & 0 & P_\varepsilon-Q_\varepsilon & 0 & 0 & \sqrt{2}Q_\varepsilon \\ 0 & 0 & 0 & P_\varepsilon+Q_\varepsilon & 0 & 0 \\ 0 & -\sqrt{2}Q_\varepsilon & 0 & 0 & P_\varepsilon+\Delta & 0 \\ 0 & 0 & \sqrt{2}Q_\varepsilon & 0 & 0 & P_\varepsilon+\Delta \end{pmatrix} \begin{matrix} \left|\frac{3}{2},\frac{3}{2}\right\rangle \\ \left|\frac{3}{2},\frac{1}{2}\right\rangle \\ \left|\frac{3}{2},-\frac{1}{2}\right\rangle \\ \left|\frac{3}{2},-\frac{3}{2}\right\rangle \\ \left|\frac{1}{2},\frac{1}{2}\right\rangle \\ \left|\frac{1}{2},-\frac{1}{2}\right\rangle \end{matrix}. \tag{5.187}$$

The nonvanishing terms P_ε and Q_ε in the Hamiltonian equation (5.187) are

$$\begin{aligned} P_\varepsilon &= -a_v(\varepsilon_{xx} + \varepsilon_{yy} + \varepsilon_{zz}) \\ Q_\varepsilon &= -\frac{b}{2}(\varepsilon_{xx} + \varepsilon_{yy} - 2\varepsilon_{zz}). \end{aligned} \tag{5.188}$$

As a result of this modification, we obtain the band edge energies of the HH, LH, and split-off bands as follows:

$$\begin{aligned} E_{HH}(0) &= -P_\varepsilon - Q_\varepsilon \\ E_{LH}(0) &= -P_\varepsilon + \frac{1}{2}\left(Q_\varepsilon - \Delta + \sqrt{\Delta^2 + 2\Delta Q_\varepsilon + 9Q_\varepsilon^2}\right) \\ E_{SO}(0) &= -P_\varepsilon + \frac{1}{2}\left(Q_\varepsilon - \Delta - \sqrt{\Delta^2 + 2\Delta Q_\varepsilon + 9Q_\varepsilon^2}\right). \end{aligned} \tag{5.189}$$

The conduction band edge is given by

$$E_c(0) = E_g + a_c(\varepsilon_{xx} + \varepsilon_{yy} + \varepsilon_{zz}). \tag{5.190}$$

The factors a_c and a_v are so-called hydrostatic deformation potentials; b is the shear deformation potential.

As an example, shift and deformation of energy bands for a bulk zinc blende InGaAsP semiconductor under (a) biaxial compression, (b) lattice-matched condition, and (c) biaxial tension (C, conduction band; HH, heavy-hole band; LH, light-hole band) are shown in Figure 5.3. The vertical axes give the energy in electron volts, the lateral axes give the wave number k_x and k_y, respectively. In the case of unstrained InGaAsP, the LH and HH bands degenerate at the Brillouin zone center Γ, and the SO band lies Δ lower in energy. Under biaxial compression, the axial component splits the degeneracy of the valence band maximum, and introduces an anisotropic valence band structure, with the highest band being heavy along the strain (k_x or k_y) axis and comparatively light perpendicular to that axis (along k_z). Under biaxial tension, the valence splitting is reversed so that the highest band is now light along the strain axis (k_x or k_y), and comparatively heavy along k_z. Strained layers retain all the advantages of the lattice-matched heterostructures, and have, in addition, further features. Strain has relatively little effect on the conduction band of direct gap semiconductors, but splits the degeneracy of the valence band maximum, significantly affecting its properties. The lowest conduction band (CB) is separated by the bandgap energy E_g from the valence bands. It is approximately

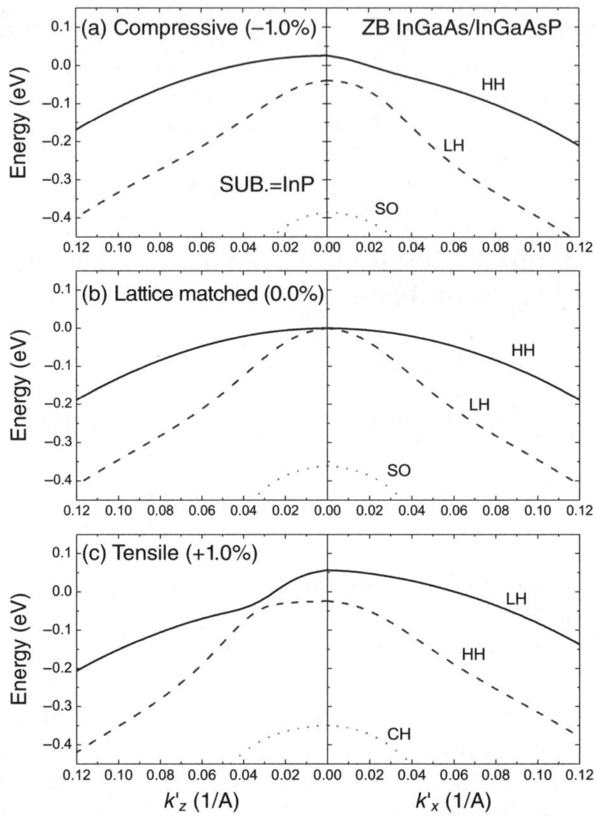

Figure 5.3. Shift and deformation of energy bands for a bulk zinc blende (ZB) InGaAsP semiconductor under (a) biaxial compression, (b) lattice-matched condition, and (c) biaxial tension (C, conduction band; HH, heavy-hole band; LH, light-hole band). InGaAsP lattice matched to InP is used as a substrate. The vertical axes give the energy in electron volts, the lateral axes give the wave number k_x and k_y respectively.

parabolic near the zone center, and the electron dispersion at small k can be described by

$$E_c(k) = E_g + \frac{\hbar^2 k^2}{2m_c^*}. \tag{5.191}$$

Next, we shall discuss strain effects for QW structures.

5.6.2 Zinc Blende Quantum Well Structures

In a QW heterostructure, the carriers are quantized along the growth direction, but remain free to move in the well plane with wavevector

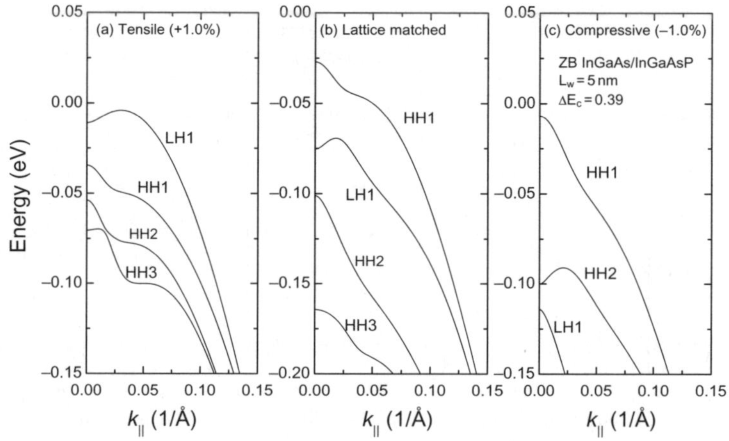

Figure 5.4. Valence band structure with (a) compressive (left), (b) no (middle), and (c) tensile strain (right) for 5-nm zinc blende InGaAsP/InGaAsP quantum well. The InGaAsP barrier is lattice matched to the InP substrate.

$k_{||}$. The ability to confine carriers along one dimension has opened the possibility of band structure engineering in lattice-matched heterostructures. In lattice-matched QWs, the conduction band properties can be readily modified by applying hydrostatic pressure or varying QW layer thickness. The valence band behavior is in general dominated by the HH dispersion, which is much less sensitive to variations. The growth of strained-layer structures introduces a new degree of freedom to the valence bands, thereby accessing hole-based effects. The confinement energies are determined by the dispersion along the growth direction. Figure 5.4 shows valence band structure with (a) compressive (left), (b) no (middle), and (c) tensile strain (right) for 5-nm zinc blende InGaAsP/InGaAsP QW. For a layer under biaxial compression, the HH band edge shifts upwards with respect to the LH band edge, which can increase the energy splitting ($E_{HH1} - E_{LH1}$) between the highest confined HH and LH states. For a layer under biaxial tension, the splitting is reversed. In a strained QW, it is then possible to vary independently the splitting between the two highest HH states predominantly through the well width, and the splitting between the highest HH and LH states predominantly via the axial strain.

The highest valence subband in a strained-layer structure can have a relatively low in-plane effective mass. This low effective mass could have significant advantages for device applications, including long-wavelength lasers with substantially reduced threshold currents.

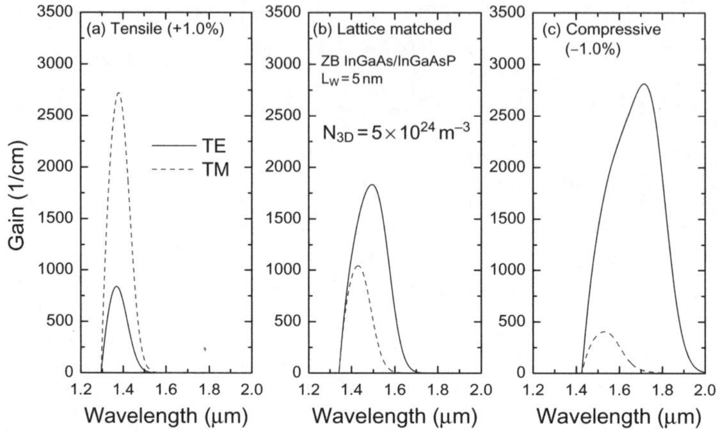

Figure 5.5. Gain spectra for both TE and TM polarizations of (a) a compressive strain ($x = 0.320$), (b) an unstrained (0.468), and (c) a tensile strain ($x = 0.624$) $Ga_xIn_{1-x}As$/InGaAsP (barrier bandgap 0.987 eV and lattice matched to InP) quantum well laser. Here, $\Delta E_c = 0.39\Delta E_g$.

Strained QWs are widely utilized in optoelectronics, as they allow for improvements of material properties, such as the optical gain. Commonly, strain is generated by varying the composition of the materials so that the lattice constant slightly deviates from that of the substrate.

As an example, Figure 5.5 shows the gain spectra for light with a TE ($\hat{x}$ or $\hat{y}$) or TM ($\hat{z}$) polarization for (a) a compressive strain ($x = 0.320$), (b) an unstrained (0.468), and (c) a tensile strain ($x = 0.624$) GaxIn$_{1-x}$As/GaInAsP QW. The barrier is lattice matched to InP, and the bandgap is 0.987 eV. For the compressive strain and the lattice-matched cases, the gain for the TE polarization is always larger than that of the TM polarization, because the top valence subband for these two cases is always heavy hole in nature, and the optical matrix element for C1-HH1 transition prefers TE polarization. The injected holes populate mostly the ground (HH1) subband. On the other hand, in the case of the tensile strain, the TM polarization is usually dominant.

5.7 WURTZITE BULK AND QUANTUM WELL STRUCTURES

Wurtzite GaN-based semiconductors have wide bandgaps, and have the other properties that are necessary for making high-performance semiconductor devices. Wide bandgap semiconductors, including GaN, AlN, InN, and their alloys, have great potential for applications in high-power optoelectronic devices in the blue and ultraviolet regions. Their bandgap values are 0.76 eV for InN, 3.4 eV for GaN, and 6.2 eV for AlN. III-Nitrides are stable because the bonds between nitrogen and group III elements are very strong. They have ideal properties for high-temperature and high-power electronics.

We provide here a short history of the research on GaN. Juza and Hahn synthesized small needles and platelets of GaN by passing ammonia over hot gallium in 1938. Grimmeiss et al. used the same method to produce small crystals of GaN, and measured their PL spectra in 1959. The first large area GaN layer was deposited on sapphire by Maruska and Tietjen in 1969. This success created widespread interest in GaN research. The first blue LED using Zn-doped GaN was fabricated by Pankove et al. in 1972. It was a metal/insulating GaN:Zn/n-GaN structure. However, although a number of other important discoveries were made, further research on GaN was hampered because it could not be doped p-type. The quality of the GaN was also not very good.

Breakthrough results in growth technology were obtained in the 1980s. Akasaki, Amano, and coworkers in late 1980s showed that high-quality GaN layers can be grown by MOCVD on a sapphire substrate if the layer is grown in two steps. In the first step, a thin buffer layer of AlN is grown on the sapphire substrate at a low temperature of ~500°C. The main GaN layer is then grown on the buffer layer at a much higher temperature.

The second breakthrough was made by Amano et al. when they obtained p-type conductivity in GaN for the first time. They found that p-type conductivity can be obtained by low-energy electron beam irradiation of Mg-doped GaN. In 1992, Nakamura et al. discovered that annealing the Mg-doped GaN layers at ≥750°C in nitrogen or vacuum also activated the Mg acceptors.

The first InGaN/GaN-layered structures were grown by Nakamura et al. in 1993. Akasaki and coworkers have done extensive work on GaN-based LEDs. They made the first p-n junction GaN LED in 1989. The success of bringing the blue and green LEDs to a level at which they could be commercialized is due to Nakamura and his colleagues at Nichia Laboratories. The work done by Nakamura et al. is very extensive. Onset of stimulated emission by current injection was first observed by Akasaki and collaborators in 1995. A nitride-based injection laser was fabricated by Nakamura and coworkers in 1996. Akasaki et al. fabricated an LD emitting at 376 nm, also in 1996. A room temperature CW laser emitting at ~400 nm with a lifetime greater than 6000 hours has been demonstrated by Nakamura.

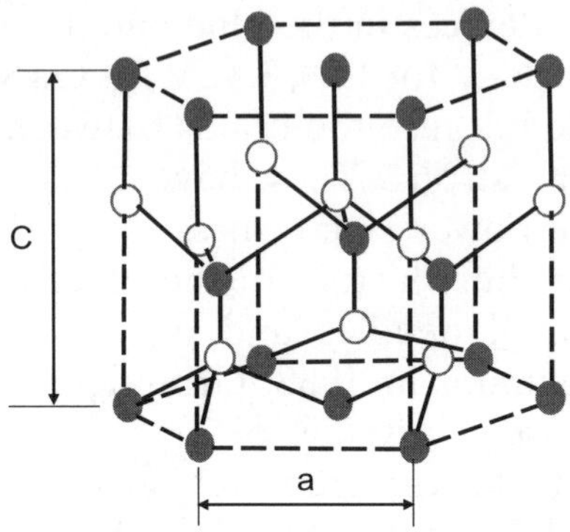

Figure 5.6. A wurtzite crystal consists of two interpenetrating hexagonal closely packed sublattices, displaced by 5/8 of the $c(0001)$-axis.

5.7.1 Bulk Wurtzite Structures

The wurtzite crystal structure with lattice constants c_0 and a_0 is illustrated in Figure 5.6. The structure is formed by two intertwined hexagonal sublattices of Ga and N atoms. For wurtzite structures, all three valence bands are strongly coupled, and valence band mixing must be taken into account. The three valence bands are referred to as HH, LH, and crystal-field split-hole (CH) band. Spin–orbit interaction leads to a slight separation of all three band edges.

For the biaxial strain, the strain tensor has only the following nonvanishing diagonal elements:

$$\begin{aligned}
\varepsilon_{xx} &= \varepsilon_{yy} = \frac{a_s - a_e}{a_e}, \\
\varepsilon_{xy} &= \varepsilon_{yz} = \varepsilon_{zx}, \\
\varepsilon_{zz} &= -\frac{2C_{13}}{C_{33}}\varepsilon_{xx}.
\end{aligned} \tag{5.192}$$

With strain effects, the valence band edge energies are

$$\begin{aligned}
E_1^0 &= E_v^0 + \Delta_1 + \Delta_2 + \theta_\varepsilon + \lambda_\varepsilon, \\
E_2^0 &= E_v^0 + \frac{\Delta_1 - \Delta_2 + \theta_\varepsilon}{2} + \lambda_\varepsilon + \sqrt{\left(\frac{\Delta_1 - \Delta_2 + \theta_\varepsilon}{2}\right)^2 + 2\Delta_3^2}, \\
E_3^0 &= E_v^0 + \frac{\Delta_1 - \Delta_2 + \theta_\varepsilon}{2} + \lambda_\varepsilon - \sqrt{\left(\frac{\Delta_1 - \Delta_2 + \theta_\varepsilon}{2}\right)^2 + 2\Delta_3^2},
\end{aligned} \tag{5.193}$$

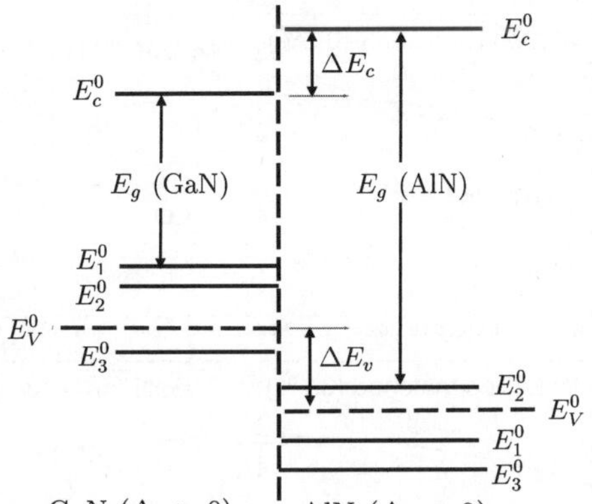

Figure 5.7. The band lineup at the GaN/AlN interface.

with

$$\begin{aligned} \theta_\varepsilon &= D_3\varepsilon_{zz} + D_4(\varepsilon_{xx} + \varepsilon_{yy}), \\ \lambda_\varepsilon &= D_1\varepsilon_{zz} + D_2(\varepsilon_{xx} + \varepsilon_{yy}). \end{aligned} \tag{5.194}$$

Here, it is important to note that energies are measured from a reference energy E_v^0, which is taken as zero, of a given semiconductor, as shown in Figure 5.7. The conduction band edge is above the top valence band with a bandgap energy E_g. For GaN, the top valence band is E_1^0, since Δ_1 is positive. For AlN, on the other hand, with a negative Δ_1, the top valence band is E_2^0. Therefore, the conduction band edge considering a hydrostatic energy shift $P_{c\varepsilon}$ for GaN is taken as

$$E_c^0 = E_V^0 + \Delta_1 + \Delta_2 + E_g + P_{c\varepsilon} \tag{5.195}$$

with

$$P_{c\varepsilon} = a_{cz}\varepsilon_{zz} + a_{ct}(\varepsilon_{xx} + \varepsilon_{yy}). \tag{5.196}$$

For AlN, the conduction band edge is

$$E_c^0 = E_2^0 + E_g. \tag{5.197}$$

As an example, energy bands for a bulk wurtzite GaN semiconductor under (a) biaxial compression, (b) lattice-matched conditions, and (c) biaxial tension are shown in Figure 5.8. The vertical axes give the energy in electron volts, and the lateral axes give the wave number k_z and k_y, respectively. The difference in symmetry between the wurtzite and zinc blende structures causes some fundamental differences in

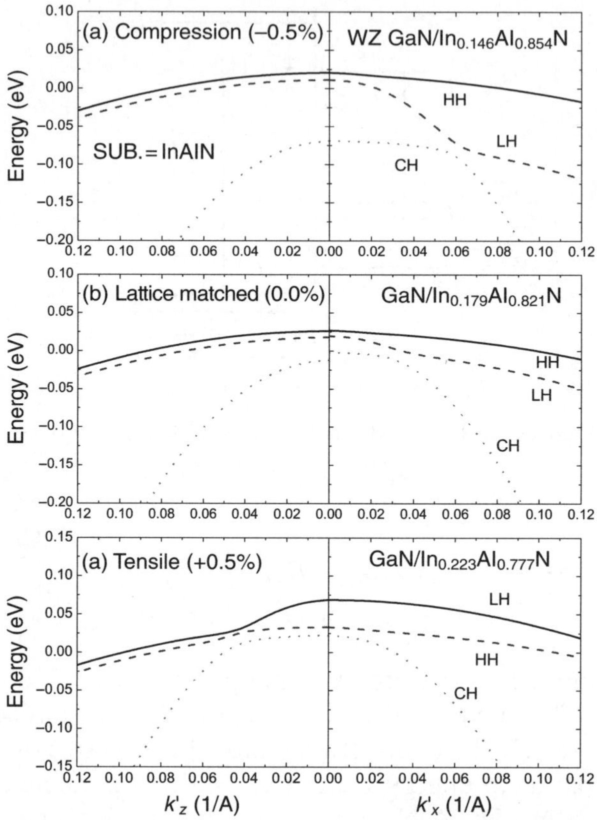

Figure 5.8. Energy bands for a bulk wurtzite GaN semiconductor under (a) biaxial compression, (b) lattice-matched condition, and (c) biaxial tension. The vertical axes give the energy in electron volts, the lateral axes give the wave number k_x and k_y, respectively. InAlN is used as a substrate.

their band structures. The degeneracy between the HH and LH bands at the zone center (Γ point) for the zinc blende structure is broken for the wurtzite structure. For a zinc blende layer grown along the (001) direction with a compressive strain, the deformation potentials lift the HH band up and reduce its in-plane effective mass. However, for a wurtzite layer grown along the *c*-axis, the compressive strain shifts both the HH and LH bands by almost the same amount, and the in-plane effective masses remain almost the same as those in the unstrained case. The polarization selection rules for the conduction band to the LH band are also changed. For the zinc blende structure, the dominant C-LH transitions are TM polarization (along the growth axis). However, for the wurtzite structure, TE polarization is favored for the C-LH transitions. In the case of tensile strain, the LH band edge shifts upwards with respect to the HH band edge. The energy splitting ($E_{HH1} - E_{LH1}$) between the highest confined HH and LH states is improved compared with that for the compressive strain. The hole effective masses at the Γ-point can be approximated as

$$
\begin{aligned}
m_{hh}^{z} &= -(A_1 + A_3)^{-1},\\
m_{hh}^{t} &= -(A_2 + A_4)^{-1},\\
m_{lh}^{z} &= -\left[A_1 + \left(\frac{E_2^0 - \lambda_\varepsilon}{E_2^0 - E_3^0}\right)A_3\right],\\
m_{lh}^{z} &= -\left[A_2 + \left(\frac{E_2^0 - \lambda_\varepsilon}{E_2^0 - E_3^0}\right)A_4\right],\\
m_{lh}^{z} &= -\left[A_1 + \left(\frac{E_3^0 - \lambda_\varepsilon}{E_3^0 - E_2^0}\right)A_3\right],\\
m_{lh}^{z} &= -\left[A_2 + \left(\frac{E_3^0 - \lambda_\varepsilon}{E_3^0 - E_2^0}\right)A_4\right].
\end{aligned}
\tag{5.198}
$$

5.7.2 Wurtzite Quantum Well Structures

Figure 5.9 shows valence band structure with (a) biaxial compression (left), (b) lattice-matched conditions (middle), and (c) biaxial tension (right) for 5-nm wurtzite GaN/InAlN QW. Here, we used the flat-band model for simplicity, ignoring the internal field effect due to the

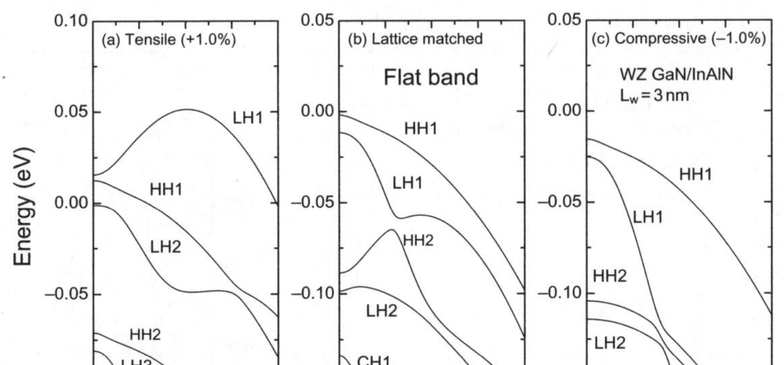

Figure 5.9. Valence band structures for a wurtzite GaN/InAlN QW under (a) biaxial compression, (b) lattice-matched condition, and (c) biaxial tension. Here, the internal field effects due to the piezoelectric and spontaneous polarizations are neglected and InAlN is used as a substrate.

piezoelectric and spontaneous polarizations. In general, the compressive strain separates the HH and the LH subbands by pulling down the HH subbands, and pulling away the LH subbands from the valence band edge. However, as shown in the bulk case, the introduction of biaxial strain into the (0001) plane of a wurtzite GaN-based crystal does not effectively reduce the effective masses in the transverse direction, unlike in other III-V semiconductors. Similarly, we know that in the case of a QW structure, both the HH and LH bands are shifted by almost the same amount for a layer under biaxial compression. For a layer under biaxial tension, the LH band edge shifts upwards with respect to the HH band edge. The combination of strain and mixing gives rise to an electron-like effective mass at the zone center.

In Figure 5.10, we plotted the gain spectra for light with a TE ($\hat{x}$ or $\hat{y}$) or TM ($\hat{z}$) polarization for (a) a compressive strain, (b) an unstrained, and (c) a tensile strain GaN/AlInN QW. Here, we used a flat-band model. Both the C1-HHI and Cl-LHI transitions have strong TE components near the zone center, while the Cl-CHI transition has a dominant TM component, which can be larger than the other transitions for the TE polarizations. However, since the compressive strain in the QW makes the top few valence subbands HH- or LH-like, the dominant transitions for the compressive QW will be TE polarization.

This result is different from that of a zinc blende structure, for which the conduction to LH subband transition has a dominant TM component, and the LH subband can be shifted downward, away from the top HH subband, with a compressive strain. For a wurtzite structure, the

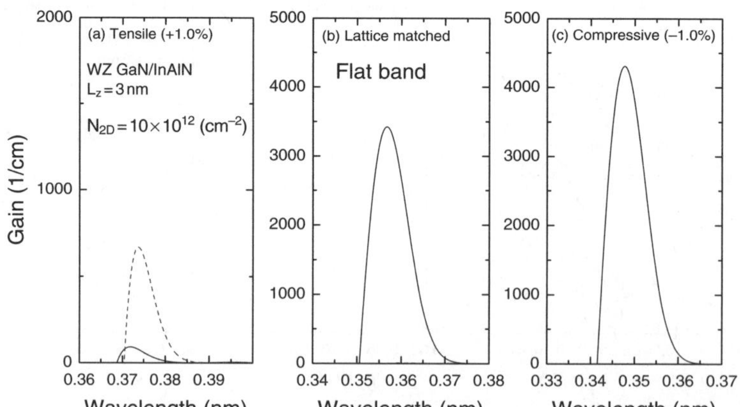

Figure 5.10. Gain spectra for light with a TE ($\hat{x}$ or $\hat{y}$) or TM ($\hat{z}$) polarization for (a) a compressive strain, (b) an unstrained, and (c) a tensile strain GaN/AlInN quantum well. The dashed line corresponds to the TM polarization.

HH and LH subbands tend to shift in the same direction by almost the same amount of energy, resulting in competition for the hole population in each subband. For the compressive strain and the lattice-matched cases, it is clearly seen that the TM-polarized gain is very small, as expected from the valence band structures, since the HH and LH subbands contribute mostly to TE polarization. On the other hand, in the case of the tensile strain, the TM polarization is usually dominant.

5.8 QUANTUM WIRES AND QUANTUM DOTS

The basic theory for calculating the stress and strain fields for a QWR and a QD [8–11] can be obtained from the work of Eshelby. First, we will consider strain for a QWR.

5.8.1 Strain in Quantum Wires

Here, the QWR is assumed to be buried in an infinite medium so that the effect of free surfaces is neglected, and materials are assumed to be continuous, linear, isotropic, and to obey Hooke's laws.

The stress components at position (x, y) are determined by evaluating

$$\sigma_{ij}(x,y) = \oint \mathbf{A}_{ij}(x_0 - x, y_0 - y) \cdot d\mathbf{r} \tag{5.199}$$

where the contour integral is performed around the boundary of the QWR in an anticlockwise sense, and (x_0, y_0) are points on the boundary. Equation (5.199) provides a simple method of numerically calculating stress fields inside and outside a QWR of arbitrary shape by discretizing the QWR boundary and converting the contour integral to a summation. The choice of vectors $\mathbf{A}$ is not unique, because other expressions for $\mathbf{A}$ could also yield the given Green's function stress components when the curl is taken.

Three vectors may be defined as follows:

$$\begin{aligned}
\mathbf{A}_{xx}(x,y) &= -\Lambda \frac{y}{x^2+y^2}\hat{i}, \\
\mathbf{A}_{yy}(x,y) &= \Lambda \frac{x}{x^2+y^2}\hat{j}, \\
\mathbf{A}_{xy}(x,y) &= \frac{\Lambda}{2}\left(\frac{x}{x^2+y^2}\hat{i} - \frac{y}{x^2+y^2}\hat{j}\right),
\end{aligned} \tag{5.200}$$

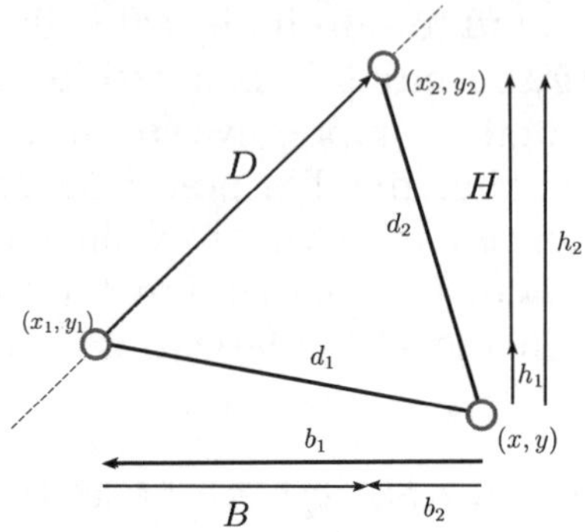

Figure 5.11. Line element forming part of the boundary of a QWR extending from coordinate (x_1, y_1) to (x_2, y_2).

where Λ is equal to $\varepsilon_0 E/[2\pi(1-\nu)]$, with ν the Poisson's ratio, and E the Young's modulus. ε_0 is the misfit strain of the QWR so that $\varepsilon_0 = (a_s - a_l)/a_l$, where a_l and a_s are the lattice parameters of the QWR material and surrounding matrix, respectively. The QWR is initially strained to ε_0 in each of the x, y, and z directions. For simple geometries, Equation (5.199) may be evaluated analytically. The calculation is completed by employing the principle of superposition to calculate the stress at a point due to several strained wires. Here, as an example, we consider analytic expressions for a line element.

Figure 5.11 illustrates a line element forming part of the boundary of a QWR extending from coordinate (x_1, y_1) to (x_2, y_2). We obtain the following stress components using Equation (5.199).

$$\begin{aligned}\sigma_{xx} &= \Lambda\left[\left(\frac{B}{D}\right)^2 (\tan^{-1} E + \tan^{-1} F) - \left(\frac{BH}{D^2}\right)\ln\left(\frac{d_2}{d_1}\right)\right],\\ \sigma_{yy} &= \Lambda\left[\left(\frac{H}{D}\right)^2 (\tan^{-1} E + \tan^{-1} F) + \left(\frac{BH}{D^2}\right)\ln\left(\frac{d_2}{d_1}\right)\right],\end{aligned} \tag{5.201}$$

$$\sigma_{xy} = \Lambda\left[\left(\frac{BH}{D^2}\right)^2 (\tan^{-1} E + \tan^{-1} F) + \left(\frac{B^2 - H^2}{2D^2}\right)\ln\left(\frac{d_2}{d_1}\right)\right], \tag{5.202}$$

where

$$\begin{aligned}&E = \frac{Bb_1 + Hh_1}{Bh_1 - Hb_1}, F = \frac{Hh_2 + Bb_2}{Bb_1 - Bh_1},\\ &H = y_2 - y_1, h_1 = y_1 - y, h_2 = y_2 - y,\\ &B = x_2 - x_1, b_1 = x_1 - x, b_2 = x_2 - x,\\ &d_1^2 = h_1^2 + b_1^2, d_2^2 = h_2^2 + b_2^2, \quad D^2 = H^2 + B^2.\end{aligned} \tag{5.203}$$

A stress component at an arbitrary point (x, y) inside or outside any QWR with a boundary comprised of line elements can be evaluated by adding the contributions due to each line element as given by Equations (5.201)–(5.203).

However, the denominator of E or F becomes zero when $Bh_1 = Hb_1$, corresponding to the locus of points along the gradient as indicated by the dotted line in Figure 5.11. Along this line, the stress components reduce to simpler forms, retaining only the log term of Equations (5.201)–(5.203),

$$\sigma_{xx} = -\Lambda\left(\frac{BH}{D^2}\right)\ln\left(\frac{d_2}{d_1}\right), \tag{5.204}$$

$$\sigma_{yy} = \Lambda\left(\frac{BH}{D^2}\right)\ln\left(\frac{d_2}{d_1}\right), \tag{5.205}$$

and

$$\sigma_{yy} = \Lambda\frac{B^2 - H^2}{2D^2}\ln\left(\frac{d_2}{d_1}\right). \tag{5.206}$$

Equations (5.201)–(5.203) also reduce to simpler forms for horizontal or vertical line elements. In particular, for a horizontal line element $(H = 0)$, it is found that $\sigma_{yy} = 0$, which indicates that horizontal line elements do not contribute to the y component of the stress. Similarly, for a vertical line element $(B = 0)$, the result $\sigma_{xx} = 0$ is obtained. These results are a consequence of the choice of the vectors **A**.

5.8.2 Strain in Quantum Dots

The stress components at position (x, y, z) in a QD are determined by evaluating

$$\sigma_{ij}(x, y, z) = \oint_S \mathbf{A}(x - x_0, y - y_0, z - z_0)\cdot d\mathbf{S}(x_0, y_0, z_0), \tag{5.207}$$

where the integral is over the surface of the QD, so that (x_0, y_0, z_0) is now a point on the surface. Equation (5.207) can be integrated numerically for an arbitrarily shaped QD, and in certain cases where the geometry is simple, an analytic result can be obtained. Here, we consider a cuboidal QD as an example. The stress distribution for a cuboidal QD is calculated analytically using Equation (5.207). The dot considered occupies a volume of $2a \times 2b \times 2c$ centered on the origin, and is aligned so that its faces are perpendicular to the direction of the axes. The length a refers to the x direction and so forth. As a concrete example, the stresses σ_{xx} and σ_{xz} are presented here, and the other stresses are easily obtained. The expressions have been compacted for display. The stress σ_{xx} is the sum of 8 arctan terms, one for each combination of $(a \pm x), (b \pm y)$, and $(c \pm z)$, and, similarly, σ_{xz} is a combination of eight log terms, but in this case, the sign of the log is positive for combinations of $(a \pm x), (b \pm y)$, and $(c \pm z)$ with either 1 or 3 plus signs, and negative for 0 or 2 plus signs.

$$\sigma_{xx}(x, y, z) = -\Lambda \tan^{-1}\left(\frac{(b \pm y)(c \pm z)}{(a \pm x)\sqrt{(a \pm x)^2 + (b \pm y)^2 + (c \pm z)^2}} \right). \tag{5.208}$$

$$\sigma_{xz}(x, y, z) = \Lambda \log\left\{ \sqrt{(a \pm x)^2 + (b \pm y)^2 + (c \pm z)^2} - (y \pm b) \right\}. \tag{5.209}$$

5.8.3 Optical Gain of a Quantum Wire

For a QWR, we solve one Schrödinger equation for electrons and 6×6 Hamiltonian for holes for simplicity. The electron and hole concentrations are calculated according to the following relations:

$$\begin{aligned} N &= 2\sum_p \int_0^\infty \rho(k_z) f_c[E_p(k_z)] dk_z \\ P &= 2\sum_q \int_0^\infty \rho(k_z) f_v[E_q(k_z)] dk_z, \end{aligned} \tag{5.210}$$

where

$$\begin{aligned} f_c[E_p(k_z)] &= \left[1 + exp\left[\frac{(E_p(k_z) - E_{fc})}{kT} \right] \right]^{-1} \\ f_v[E_q(k_z)] &= \left[1 + \exp\left[\frac{(E_q(k_z) - E_{fv})}{kT} \right] \right]^{-1} \end{aligned} \tag{5.211}$$

are Fermi–Dirac distributions of electrons for the conduction and the valence band, respectively, while E_{fc} and E_{fv} are the relevant quasi-Fermi levels. $\rho(k_z)$ is the density-of-state (DOS) in k space. For a QWR, $\rho(k_z)$ is given by

$$\rho(k_z) = \frac{1}{\pi A}. \tag{5.212}$$

The absorption coefficient becomes

$$\alpha(\hbar\omega) = C_0 \frac{1}{V} \sum_{k_a} \sum_{k_b} |\hat{e} \cdot \mathbf{p}_{ba}|^2 \, \delta(E_b - E_a - \hbar\omega)(f_a - f_b), \tag{5.213}$$

where

$$C_0 = \frac{\pi e^2}{n_r c \varepsilon_0 m_0^2 \omega}. \tag{5.214}$$

The summations over the quantum number $\mathbf{k}_a$ and $\mathbf{k}_b$ become summations over $(k_z', m_1 m_2)$ and $(k_z, n_1 n_2)$. The indices $n = \{n1n2\}$ and $m = \{m1m2\}$ denote the electron states in the conduction band and the subband states in the valence band, respectively. The matrix element contains a k-selection rule when we take the inner product. That is, $k_z = k_z'$. When the scattering relaxation is included, the delta function may be replaced by a Lorentzian function with a linewidth $\Gamma = 2\hbar/\tau_{in}$:

$$\delta(E_b - E_a - \hbar\omega) \rightarrow \frac{\hbar/(\tau_{in}\pi)}{(E_b - E_a - \hbar\omega)^2 + (\hbar/\tau_{in})^2} \tag{5.215}$$

where a factor π has been included such that the area under the function is properly normalized:

$$\int \delta(E_b - E_a - \hbar\omega) d\hbar\omega = 1 \tag{5.216}$$

Hence, we can write

$$\alpha(\hbar\omega) = C_0 \sum_{n1n2, m1m2} \frac{2}{V} \sum_{k_z} |\hat{e} \cdot \mathbf{p}_{m1m2}^{n1n2}|^2 \frac{\hbar/(\tau_{in}\pi)}{(E_{hm1m2}^{en1n2} - \hbar\omega)^2 + (\hbar/\tau_{in})^2} (f_v^{m1m2} - f_c^{n1n2}), \tag{5.217}$$

where E_{hm1m2}^{en1n2} is the energy difference between the electron subband and the hole subband. For example, within a two-band model, the quantizations of the electron and hole energies E_a and E_b are

$$E_a = E_{hm_1m_2} - \frac{\hbar^2 k_z^2}{2m_h^*} \tag{5.218}$$

and

$$E_b = E_g + E_{hn_1n_2} + \frac{\hbar^2 k_z^2}{2m_e^*}. \tag{5.219}$$

Note that $E_{hm_1m_2} < 0$ and

$$E_b - E_a = E_{hm_1m_2}^{en_1n_2} + \frac{\hbar^2 k_z^2}{2m_r} = E_{hm_1m_2}^{en_1n_2}(k_z), \tag{5.220}$$

where

$$\frac{1}{m_r} = \frac{1}{m_e^*} + \frac{1}{m_h^*}, \tag{5.221}$$

and

$$E_{hm_1m_2}^{en_1n_2} = E_g + E_{hn_1n_2} - E_{hm_1m_2} \tag{5.222}$$

is the band edge transition energy ($k_z = 0$). When we change the summation to the integral over k_z

$$\frac{2}{V}\sum_{k_z} = \frac{2}{V}\frac{L_z}{2\pi}\int dk_z = \frac{2}{A}\frac{1}{2\pi}\int dk_z, \tag{5.223}$$

we obtain the optical gain for the free carrier model

$$g(\hbar\omega) = \frac{e^2}{2A\pi m_0^2\omega}\sqrt{\frac{\mu_0}{\varepsilon}}\sum_{n1n2,m1m2}\int dk_z |\hat{e}\cdot\mathbf{p}_{m1m2}^{n1n2}|^2 \frac{\hbar/\tau_{in}}{(E_{hm1m2}^{en1n2} - \hbar\omega)^2 + (\hbar/\tau_{in})^2}(f_v^{m1m2} - f_c^{n1n2}), \tag{5.224}$$

where we used the relation $\frac{1}{\varepsilon_0 nc} = \sqrt{\frac{\mu_0}{\varepsilon}}$.

Assuming that the system is quantized in the x and y directions, the energies and the envelope functions of valence subbands can be obtained by solving the effective mass equation

$$\sum_{\nu}[H_{\mu\nu} + V_h(x,y)\delta_{\mu\nu}]F_{h,\nu}(k_z,x,y) = E(k_z)F_{h,\mu}(k_z,x,y), \quad \nu,\mu \in \{|h1\rangle,|l1\rangle,|l2\rangle,|h2\rangle,|s1\rangle,|s2\rangle\}, \tag{5.225}$$

where H is the Luttinger–Kohn Hamiltonian in Equation (3.263) in Chapter 3, and $V_h(x,y)$ is the QWR potential for holes. Then, the optical momentum matrix elements M for a QWR are given by

$$|\hat{\mathbf{e}}\cdot\mathbf{M}(k_z)|^2 = |\langle\Psi_c^{nn}|\hat{\mathbf{e}}\cdot\mathbf{p}|\Psi_h^m\rangle|^2, \tag{5.226}$$

where Ψ_c and Ψ_h are wave functions for the conduction and the valence bands, respectively, and $\eta = \uparrow$ and $\downarrow$ for both electron spins. The polarization-dependent interband optical matrix elements $\mathbf{M}$ can be written as

TE-polarization $\hat{\mathbf{e}} = \hat{\mathbf{x}}$:

$$\begin{aligned}|\hat{\mathbf{x}}\cdot\mathbf{M}^{\uparrow}(k_z)|^2 &= \left|\sum_{\nu=1}^{6}\langle F_c^{n\uparrow}|F_{h,\nu}^{m}\langle iS\uparrow|\hat{\mathbf{x}}\cdot\hat{\mathbf{p}}|\nu\rangle\right|^2 \\ &= \left|-\frac{1}{\sqrt{2}}\langle F_c^{n\uparrow}|F_{h,1}^{m}\rangle+\frac{1}{\sqrt{6}}\langle F_c^{n\uparrow}|F_{h,3}^{m}\rangle+\frac{1}{\sqrt{3}}\langle F_c^{n\uparrow}|F_{h,6}^{m}\rangle\right|^2 p_x^2,\end{aligned} \tag{5.227}$$

$$\begin{aligned}|\hat{\mathbf{x}}\cdot\mathbf{M}^{\downarrow}(k_z)|^2 &= \left|\sum_{\nu=1}^{6}\langle F_c^{n\downarrow}|F_{h,\nu}^{m}\langle iS\downarrow|\hat{\mathbf{x}}\cdot\mathbf{p}|\nu\rangle\right|^2 \\ &= \left|-\frac{1}{\sqrt{6}}\langle F_c^{n\downarrow}|F_{h,2}^{m}\rangle+\frac{1}{\sqrt{2}}\langle F_c^{n\downarrow}|F_{h,4}^{m}\rangle+\frac{1}{\sqrt{3}}\langle F_c^{n\downarrow}|F_{h,5}^{m}\rangle\right|^2 p_x^2,\end{aligned} \tag{5.228}$$

TE-polarization $\hat{\mathbf{e}} = \hat{\mathbf{y}}$:

$$\begin{aligned}|\hat{\mathbf{y}}\cdot\mathbf{M}^{\uparrow}(k_z)|^2 &= \left|\sum_{\nu=1}^{6}\langle F_c^{n\uparrow}|F_{h,\nu}^{m}\langle iS\uparrow|\hat{\mathbf{y}}\cdot\mathbf{p}|\nu\rangle\right|^2 \\ &= \left|-i\frac{1}{\sqrt{2}}\langle F_c^{n\uparrow}|F_{h,1}^{m}\rangle-i\frac{1}{\sqrt{6}}\langle F_c^{n\uparrow}|F_{h,3}^{m}\rangle-i\frac{1}{\sqrt{3}}\langle F_c^{n\uparrow}|F_{h,6}^{m}\rangle\right|^2 p_y^2,\end{aligned} \tag{5.229}$$

$$\begin{aligned}|\hat{\mathbf{y}}\cdot\mathbf{M}^{\downarrow}(k_z)|^2 &= \left|\sum_{\nu=1}^{6}\langle F_c^{n\downarrow}|F_{h,\nu}^{m}\langle iS\downarrow|\hat{\mathbf{y}}\cdot\mathbf{p}|\nu\rangle\right|^2 \\ &= \left|-i\frac{1}{\sqrt{6}}\langle F_c^{n\downarrow}|F_{h,2}^{m}\rangle-i\frac{1}{\sqrt{2}}\langle F_c^{n\downarrow}|F_{h,4}^{m}\rangle+i\frac{1}{\sqrt{3}}<\langle F_c^{n\downarrow}|F_{h,5}^{m}\rangle\right|^2 p_y^2,\end{aligned} \tag{5.230}$$

TM-polarization $\hat{\mathbf{e}} = \hat{\mathbf{z}}$:

$$\begin{aligned}|\hat{\mathbf{z}}\cdot\mathbf{M}^{\uparrow}(k_z)|^2 &= \left|\sum_{\nu=1}^{6}\langle F_c^{n\uparrow}|F_{h,\nu}^{m}\langle iS\uparrow|\hat{\mathbf{z}}\cdot\mathbf{p}|\nu\rangle\right|^2 \\ &= \left|\sqrt{\frac{2}{3}}\langle F_c^{n\uparrow}|F_{h,2}^{m}\rangle+\frac{1}{\sqrt{3}}\langle F_c^{n\uparrow}|F_{h,5}^{m}\rangle\right|^2 p_z^2,\end{aligned} \tag{5.231}$$

$$
\begin{aligned}
\left|\hat{\mathbf{z}}\cdot\mathbf{M}^{\downarrow}(k_z)\right|^2 &= \left|\sum_{\nu=1}^{6}\langle F_c^{n\downarrow}|F_{h,\nu}^{m}\langle iS\uparrow|\hat{\mathbf{z}}\cdot\mathbf{p}|\nu\rangle\right|^2 \\
&= \left|\sqrt{\frac{2}{3}}\langle F_c^{n\downarrow}|F_{h,3}^{m}\rangle - \frac{1}{\sqrt{3}}\langle F_c^{n\downarrow}|F_{h,5}^{m}\rangle\right|^2 p_z^2,
\end{aligned} \tag{5.232}
$$

where momentum matrix parameters P_x, P_y, and P_z are defined as

$$
\begin{aligned}
&P_x = \langle iS|p_x|X\rangle, P_y = \langle iS|p_y|Y\rangle, P_z = \langle iS|p_z|Z\rangle \\
&P_x = P_y = P_z.
\end{aligned} \tag{5.233}
$$

5.8.4 Optical Gain for Quantum Dots

In the case of a QD, the eight-band $\mathbf{k}\cdot\mathbf{p}$ Hamiltonian is often used for eliminating spurious solutions. The wave function Ψ of electrons and holes is expressed with the envelope function F_ν, and the basis functions $|\nu\rangle$ as

$$
\Psi(x,y,z) = \sum_{\nu=1}^{8} F_\nu(x,y,z)|\nu\rangle, \tag{5.234}
$$

where F_ν follow the normalization rules

$$
\sum_{i=1}^{8}\int dz|F_\nu^{\alpha}|^2 = 1, \tag{5.235}
$$

where $\alpha(=n \text{ or } m)$ denotes the subband states in the conduction or the valence band.

The energies and the envelope functions of conduction and valence subbands can be obtained by solving the effective mass equation

$$
\begin{aligned}
&\sum_{\nu}[H_{\mu\nu} + V(x,y,z)\delta_{\mu\nu}]F_\nu(x,y,z) = EF_\mu(x,y,z), \\
&\qquad \nu,\mu \in \{|c1\rangle,|c2\rangle,|h1\rangle,|l1\rangle,|l2\rangle,|h2\rangle,|s1\rangle,|s2\rangle\},
\end{aligned} \tag{5.236}
$$

where H is the Luttinger–Kohn 8×8 Hamiltonian in Equation (3.263) in Chapter 3, and $V(x,y,z)$ is the QD potential. The optical momentum matrix elements $\mathbf{M}$ for a QD are given by

$$
|\hat{\mathbf{e}}\cdot\mathbf{M}(k_z)|^2 = |\langle\Psi_c^n|\hat{\mathbf{e}}\cdot\mathbf{p}|\Psi_h^m\rangle|^2, \tag{5.237}
$$

where Ψ_c and Ψ_h are wave functions for the conduction and the valence bands, respectively. The indices $n = \{n1n2n3\}$ and $m = \{m1m2m3\}$ denote the electron states in the conduction band and the subband states in the valence band, respectively. That is,

$$\Psi_c^n = \sum_{\nu=1}^{8} F_\nu^n(x,y,z)|\nu\rangle$$
$$\Psi_h^m = \sum_{\nu=1}^{8} F_\nu^m(x,y,z)|\nu\rangle. \tag{5.238}$$

The electron and hole concentrations are calculated according to the following relations:

$$N = \sum_l \int_0^\infty \rho(E) f_c(E) dE$$
$$P = \sum_m \int_0^\infty \rho(E) f_v(E) dE, \tag{5.239}$$

where

$$f_c(E) = \left[1 + exp\left[\frac{(E - E_{fc})}{kT}\right]\right]^{-1}$$
$$f_v(E) = \left[1 + exp\left[\frac{(E - E_{fv})}{kT}\right]\right]^{-1}, \tag{5.240}$$

are Fermi–Dirac distributions of electrons for the conduction and the valence band, respectively, while E_{fc} and E_{fv} are the relevant quasi-Fermi levels. $\rho(E)$, the DOS for a QD, is given by

$$\rho(E) = \frac{1}{V}\delta(E - E_{cv}). \tag{5.241}$$

Similarly, the material gain for the QD is

$$g(\hbar\omega) = \frac{e^2}{m_0^2\omega}\sqrt{\frac{\mu_0}{\varepsilon}}\sum_{n,m}\frac{1}{V}|\hat{\mathbf{e}}\cdot\mathbf{M}_m^n|^2 \frac{\hbar/\tau_{in}}{(E_{hm}^{en} - \hbar\omega)^2 + (\hbar/\tau_{in})^2}(f_c^n - f_v^m). \tag{5.242}$$

Figure 5.12 shows a numerical example for the gain spectra calculated for $100\,\text{Å} \times 100\,\text{Å} \times 100\,\text{Å}$ cubic GaAs/$Al_{0.3}Ga_{0.7}As$ QD, $100\,\text{Å} \times 100\,\text{Å}$ square QWR, and $100\,\text{Å}$ QW. The dominant electric-field polarizations are chosen for each structures. For simplicity, a free

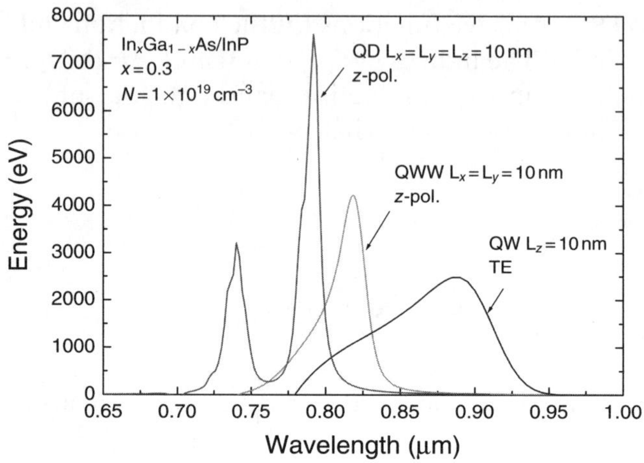

Figure 5.12. Gain spectra calculated for 100 Å × 100 Å × 100 Å cubic GaAs/$Al_{0.3}Ga_{0.7}As$ quantum dot, 100 Å × 100 Å square quantum wire and 100 Å quantum well.

carrier model for the optical gain was used for all cases. As can be seen, the shape of the spectrum becomes sharper with increasing quantization dimensions. This is due to the variation of the DOSs. The width of gain spectrum is determined only by the relaxation broadening for QD, since the DOS is given by the delta function, while for the other QWs, it is determined by all of the shapes of the DOS, the relaxation broadening, and the thermal distribution of carriers.

APPENDIX: FORTRAN 77 CODE FOR THE BAND STRUCTURE

```
C%%%%%%%%%%%%%%%%%%%%%%%%%%%%%%%%%%%%%%
C%%%   GaAs/AlGaAs/GaAs               %%%
C%%%   VALENCE BAND STRUCTURE         %%%
C%%%   BARRIER WIDTH=DIVISI*NW        %%%
C%%%   WELL WIDTH=DIVISI*(LW+1)       %%%
C%%%%%%%%%%%%%%%%%%%%%%%%%%%%%%%%%%%%%%
      IMPLICIT REAL*8 (A-H,O-Z)
C
      INTEGER   NB,NW,ND,NN,NP,NK
      PARAMETER (NB=30,NW=39,ND=NB+NW+NB,NN=ND+ND+ND,NP=ND+2,NK=50)
C
      REAL*8 CG2(NP),CF(NP),DIVISI
      REAL*8 AP1(NP),AP2(NP),AP3(NP),AP4(NP),AP5(NP),AP6(NP),GDELM(NP)
      REAL*8 FH(NP),FL(NP),FSO(NP)
      INTEGER NCB,NVB
C======= CB : EIGENVALUE, EIGENVECTOR =====
      OPEN (UNIT=10, FILE='CVADAS.DAT',STATUS='UNKNOWN')
      OPEN (UNIT=20, FILE='CVEDAS.DAT',STATUS='UNKNOWN')
C======= VB : EIGENVALUE =====
      OPEN (UNIT=50, FILE='HVADUAS.DAT',STATUS='UNKNOWN')
```

```
C====== VB STRUCTURE FOR PLOT =====
      OPEN (UNIT=80, FILE='GRPHUAS.DAT',STATUS='UNKNOWN')
C
      NCB=2
      NVB=4
C%%%%%% MATERIAL PARAMETERS %%%%%%%%%%%
      CALL MATER(CG2,CF,DIVISI,
     $ AP1,AP2,AP3,AP4,AP5,AP6,GDELM,FH,FL,FSO)
C%%%%%% OBTAINING CB EIGENVALUE %%%%%%
   CALL CBEIGEN(CG2,CF,DIVISI,NCB)
C
      CALL VBEIGEN(NVB,DIVISI,
     $ AP1,AP2,AP3,AP4,AP5,AP6,GDELM,FH,FL,FSO)
C
      STOP
      END
C%%%%%%%%%%%%%%%%%%%%%%%%%%%%%%%%%%%%%
C%% MATERIAL PARAMETERS              %%
C%%%%%%%%%%%%%%%%%%%%%%%%%%%%%%%%%%%%%
      SUBROUTINE MATER(CG2,CF,DIVISI,
     $ AP1,AP2,AP3,AP4,AP5,AP6,GDELM,FH,FL,FSO)
      INTEGER   NB,NW,ND,NN,NP,NK
      PARAMETER (NB=30,NW=39,ND=NB+NW+NB,NN=ND+ND+ND,NP=ND+2,NK=50)
C
      REAL*8 HBAR,EV,MO,AUNG,PI,T,KT
      REAL*8 DIVISI,TLL,TLB,TLW,RATIO,AC1
      INTEGER I1,I2
      REAL*8 BARX,GAIW,GA2W,GA3W,GAIB,GA2B,GA3B
      REAL*8 DELSOW,DELSOB
      REAL*8 WEG,BEG,WEMC,BEMC
      REAL*8 TLCWELL,TLCBARR,TLCSUB,C11,C12,C11B,C12B
      REAL*8 ACW,AVW,BW,ACB,AVB,BB,AO,EXX,EZZ,AOB,EXXB,EZZB
      REAL*8 CNITW,VNITW,ZETAW,CNITB,VNITB,ZETAB
      REAL*8 AP1(NP),AP2(NP),AP3(NP),AP4(NP),AP5(NP),AP6(NP)
      REAL*8 FH(NP),FL(NP),FSO(NP),GDELM(NP),CG2(NP),CF(NP)
C
      HBAR=1.0546D-34
      EV=1.60219D-19
      MO=9.1095D-31
      AUNG=1.0D-10
      PI=4.0*ATAN(1.0)
      T=300.0
      KT=T/300.0*0.026*EV
C ======== QW STRUCTURE =============
      DIVISI=2.0*AUNG
      TLB=DIVISI*(NB)      ! BARRIER LENGTH
      TLW=DIVISI*(NW+1)    ! WELL LENGTH
      I1=NB               ! BOUNDARY BETWEEN BARRIER AND WELL
      I2=NB+NW+1
      TLL=DIVISI*(NB+NW+NB+1) ! TOTAL LENGTH OF QW STRUCTURE
      RATIO=MO/HBAR*TLL/HBAR*TLL
      AC1=(ND+1)**2
C ========= GaAs/AlxGal-xAs ==========
      BARX=0.3 !Al COMPOSITION
C ========== PARAMETERS =============
      GA1W=6.8
      GA2W=1.9
      GA3W=2.73
      GA1B=6.8*(1.0-BARX)+3.45*BARX
      GA2B=1.9*(1.0-BARX)+0.68*BARX
      GA3B=2.73*(1.0-BARX)+1.29*BARX
```

```
C
      DELSOW=0.34*EV
      DELSOB=(0.34*(1.0-BARX)+0.28*BARX)*EV
C ====================================
      WEG=1.424*EV
      BEG=(1.424+1.247*BARX)*EV
      WEMC=0.067*M0
      BEMC=(0.067+0.083*BARX)*M0
C ===== STRAIN RELATED ================
      TLCWELL=5.6533*AUNG
      TLCBARR=(5.6533*(1.0-BARX)+5.6600*BARX)*AUNG
      TLCSUB=5.6533*AUNG
C
      C11=11.879
      C12=5.376
      C11B=(11.879*(1.0-BARX)+12.5*BARX)
      C12B=(5.376*(1.0-BARX)+5.34*BARX)
C
      ACW=-7.17*EV
      AVW=1.16*EV
      BW=-1.7*EV
C
      ACB=(-7.17*(1.0-BARX)-5.64*BARX)*EV
      AVB=(1.16*(1.0-BARX)+2.47*BARX)*EV
      BB=(-1.7*(1.0-BARX)-1.5*BARX)*EV
C
      A0=(TLCSUB-TLCWELL)/TLCWELL
      EXX=AO
      EZZ=(-2.0)*C12/C11*EXX
      A0B=(TLCSUB-TLCBARR)/TLCBARR
      EXXB=AOB
      EZZB=(-2.0)*C12B/C11B*EXXB
C
      ZETAW= -BW/2.0*(EXX+EXX-2.0*EZZ)
      CNITW=+(ACW)*(EXX+EXX+EZZ)
      VNITW=-(AVW)*(EXX+EXX+EZZ)
      ZETAB= -BB/2.0*(EXXB+EXXB-2.0*EZZB)
      CNITB=+(ACB)*(EXXB+EXXB+EZZB)
      VNITB=-(AVB)*(EXXB+EXXB+EZZB)
C
      QQ=0.65 ! BANDOFFSET IN CONDUCTION BAND
C ==== LUTTINGER PARAMETERS RELATED
      DO 45 I=1,NP,1
        IF ((I .GT. (I1+1)) .AND. (I .LT. (I2+1))) THEN
          AP1(I)=GA1W
          AP2(I)=GA2W
          AP3(I)=GA1W
          AP4(I)=GA2W
          AP5(I)=(GA2W+GA3W)/2.0
          AP6(I)=GA3W
          FH(I)=RATIO*(0.0+VNITW+ZETAW)/AC1
          FL(I)=RATIO*(0.0+VNITW-ZETAW)/AC1
          FSO(I)=RATIO*(0.0+VNITW+DELSOW)/AC1
          GDELM(I)=RATIO*(SQRT(2.0)*ZETAW)/AC1
          CG2(I)=M0/WEMC
          CF(I)=RATIO* (0.0+CNITW)/AC1
        ELSE
          AP1(I)=GA1B
          AP2(I)=GA2B
          AP3(I)=GA1B
```

```
         AP4(I)=GA2B
         AP5(I)=(GA2B+GA3B)/2.0
         AP6(I)=GA3B
         FH(I)=RATIO*((1.0-QQ)*(BEG-WEG)+VNITB+ZETAB)/AC1
         FL(I)=RATIO*((1.0-QQ)*(BEG-WEG)+VNITB-ZETAB)/AC1
         FSO(I)=RATIO*((1.0-QQ)*(BEG-WEG)+VNITW+DELSOB)/AC1
         GDELM(I)=RATIO*(SQRT(2.0)*ZETAB)/AC1
         CG2(I)=M0/BEMC
         CF(I)=RATIO*(QQ*(BEG-WEG)+CNITB)/AC1
       ENDIF
45    CONTINUE
C
      RETURN
      END
C%%%%%%%%%%%%%%%%%%%%%%%%%%%%%%%%%%%%%
C%% SUBROUTINES FOR CONDUCTION BAND %%
C%%%%%%%%%%%%%%%%%%%%%%%%%%%%%%%%%%%%%
      SUBROUTINE CBMAT(CG2,CF,AC1,CB)
      PARAMETER (NB=30,NW=39,ND=NB+NW+NB,NN=ND+ND+ND,NP=ND+2,NK=50)
      INTEGER   NB,NW,ND,NN,NP,NK
      INTEGER I1,I2
      REAL*8 CG2(NP),CF(NP),AC1
      REAL*8 CUNIM(ND,ND),CAVH2(NP),CAVDF2(ND,ND)
      REAL*8  CB(ND,ND)
C
      I1=NB   ! BOUNDARY BETWEEN BARRIER AND WELL
      I2=NB+NW+1
C ====== DEFINE MATRIX ===========================
      DO 52 I=1,ND,1
        DO 54 J=1,ND,1
         CUNIM(I,J)=0.0
54      CONTINUE
52    CONTINUE
      DO 56 I=1,ND,1
        CUNIM(I,I)=CUNIM(I,I)+1.0
56    CONTINUE
C
      DO 60 I=1,NP,1
       CAVH2(I)=0.5*CG2(I)
60    CONTINUE
      DO 62 I=1,NP,1
       IF ((I .EQ. (I1+1)) .OR. (I .EQ. (I2+1))) THEN
          CAVH2(I)=(CAVH2(I-1)+CAVH2(I+1))/2.0
       ENDIF
62    CONTINUE
C
      DO 64 I=1,ND,1
       DO 66 J=1,ND,1
         CAVDF2(I,J)=0.0
66      CONTINUE
64    CONTINUE
C
      DO 68 I=1,ND,1
        CAVDF2(I,I)=CAVDF2(I,I)-CAVH2(I+1)-((CAVH2(I)+CAVH2(I+2)))/2.0
68    CONTINUE
C
      DO 70 I=1,ND-1,1
        CAVDF2(I,I+1)=CAVDF2(I,I+1)+(CAVH2(I+1)+CAVH2(I+2))/2.0
        CAVDF2(I+1,I)=CAVDF2(I+1,I)+(CAVH2(I+2)+CAVH2(I+1))/2.0
70    CONTINUE
```

```
C
      DO 72 I=1,ND,1
        DO 74 J=1,ND,1
          CAVDF2(I , J)=CAVDF2(I ,J)*AC1
74      CONTINUE
72    CONTINUE
C
      DO 82 I=1,ND,1
        DO 84 J=1,ND,1
          CB(I,J)=0.0
84      CONTINUE
82    CONTINUE
C
      DO 90 I=1,ND,1
        DO 95 J=1,ND,1
          CB(I,J)=CB(I,J)-CAVDF2(I,J)+CF(I+1)*CUNIM(I,J)*AC1
95      CONTINUE
90    CONTINUE
C
      DO 96 I=1,ND,1
        DO 98 J=1,ND,1
          CB(I,J)=CB(I,J)/AC1
98      CONTINUE
96    CONTINUE
C
        RETURN
        END
C%%%%%%%%%%%%%%%%%%%%%%%%%%%%%%%%%%%%%%%%
C%% DEFINE UPPER & LOWER HAMILTONIAN %%
C%%%%%%%%%%%%%%%%%%%%%%%%%%%%%%%%%%%%%%%%
      SUBROUTINE VBMAT(KP,DIVISI,AP1,AP2,AP3,AP4,AP5,AP6,
     $ GDELM,FH,FL,FSO,HUPMAT)
      PARAMETER (NB=30,NW=39,ND=NB+NW+NB,NN=ND+ND+ND,NP=ND+2,NK=50)
      INTEGER   NB,NW,ND,NN,NP,NK
      REAL*8 KP,HUPMAT(NN,NN),DIVISI
      REAL*8 AP1(NP),AP2(NP),AP3(NP),AP4(NP),AP5(NP),AP6(NP),GDELM(NP)
      REAL*8 FH(NP),FL(NP),FSO(NP),AC1
      REAL*8 UNIM(ND,ND),AVH1(NP),AVH2(NP),BVH1(NP),BVH2(NP),GVH1(NP)
      REAL*8 GVH2(NP),CVH(NP),DVH(NP),FVH1(NP),FVH2(NP),AVHDF2(ND,ND)
      REAL*8 BVHDF2(ND,ND),FVHDF2(ND,ND),GVHDF2(ND,ND)
      REAL*8 DVHDF1(ND,ND)
      REAL*8 B(NN,NN),LL
      INTEGER I,J,I1,I2
C
      I1=NB
      I2=NB+NW+1
      LL=DIVISI*(NB+NW+NB+I)
      AC1=(ND+1)**2
C
      DO 305 I=1,ND,1
        DO 307 J=1,ND,1
          UNIM(I,J)=0.0
307     CONTINUE
305   CONTINUE
      DO 315 I=1,ND,1
          UNIM(I,I)=UNIM(I,I)+1.0
315   CONTINUE
C ======= DEFINE MATRIX =================
      DO 320 I=1,NP,1
        AVH1(I)=0.5*(AP3(I)+AP4(I))*((LL*KP)**2)
        AVH2(I)=0.5*(AP1(I)-2.0*AP2(I))
```

```
        BVH1(I)=0.5*(AP3(I)-AP4(I))*((LL*KP)**2)
        BVH2(I)=0.5*(AP1(I)+2.0*AP2(I))
        GVHI(I)=0.5*AP3(I)*((LL*KP)**2)
        GVH2(I)=0.5*AP1(I)
        CVH(I)=0.5*SQRT(3.0)*AP5(I)*((LL*KP)**2)
        DVH(I)=SQRT(3.0)*AP6(I)*LL*KP
        FVHI(I)=AP4(I)/SQRT(2.0)*((LL*KP)**2)
        FVH2(I)=SQRT(2.0)*AP2(I)
320   CONTINUE
      DO 325 I=1,NP,1
         IF ((I .EQ. (I1+1)) .OR. (I .EQ. (I2+1))) THEN
            AVH2(I)=(AVH2(I-1)+AVH2(I+1))/2.0
            BVH2(I)=(BVH2(I-1)+BVH2(I+1))/2.0
            GVH2(I)=(GVH2(I-1)+GVH2(I+1))/2.0
            FVH2(I)=(FVH2(I-1)+FVH2(I+1))/2.0
            DVH(I)=(DVH(I-1)+DVH(I+1))/2.0
            CVH(I)=(CVH(I-1)+CVH(I+1))/2.0
            FVH1(I)=(FVH1(I-1)+FVH1(I+1))/2.0
            BVH1(I)=(BVH1(I-1)+BVH1(I+1))/2.0
            GVH1(I)=(GVH1(I-1)+GVH1(I+1))/2.0
            AVH1(I)=(AVH1(I-1)+AVH1(I+1))/2.0
         ENDIF
325   CONTINUE
C
      DO 335 I=1,ND,1
        DO 337 J=1,ND,1
           AVHDF2(I,J)=0.0
           BVHDF2(I,J)=0.0
           FVHDF2(I,J)=0.0
           GVHDF2(I,J)=0.0
           DVHDF1(I,J)=0.0
337     CONTINUE
335   CONTINUE
C
      DO 340 I=1,ND,1
        AVHDF2(I,I)=AVHDF2(I,I)-AVH2(I+1)-((AVH2(I)+AVH2(I+2)))/2.0
        BVHDF2(I,I)=BVHDF2(I,I)-BVH2(I+1)-((BVH2(I)+BVH2(I+2)))/2.0
        FVHDF2(I,I)=FVHDF2(I,I)-FVH2(I+1)-((FVH2(I)+FVH2(I+2)))/2.0
        GVHDF2(I,I)=GVHDF2(I,I)-GVH2(I+1)-((GVH2(I)+GVH2(I+2)))/2.0
340   CONTINUE
C
      DO 350 I=1,ND-1,1
        AVHDF2(I,I+1)=AVHDF2(I,I+1)+(AVH2(I+1)+AVH2(I+2))/2.0
        AVHDF2(I+1,I)=AVHDF2(I+1,I)+(AVH2(I+2)+AVH2(I+1))/2.0
        BVHDF2(I,I+1)=BVHDF2(I,I+1)+(BVH2(I+1)+BVH2(I+2))/2.0
        BVHDF2(I+1,I)=BVHDF2(I+1,I)+(BVH2(I+2)+BVH2(I+1))/2.0
        FVHDF2(I,I+1)=FVHDF2(I,I+1)+(FVH2(I+1)+FVH2(I+2))/2.0
        FVHDF2(I+1,I)=FVHDF2(I+1,I)+(FVH2(I+2)+FVH2(I+1))/2.0
        GVHDF2(I,I+1)=GVHDF2(I,I+1)+(GVH2(I+1)+GVH2(I+2))/2.0
        GVHDF2(I+1,I)=GVHDF2(I+1,I)+(GVH2(I+2)+GVH2(I+1))/2.0
        DVHDF1(I,I+1)=DVHDFI(I,I+1)+(DVH(I+2)+DVH(I+1))/2.0
        DVHDF1(I+1,I)=DVHDF1(I+1,I)-(DVH(I+2)+DVH(I+1))/2.0
350   CONTINUE
C
      DO 365 I=1,ND,1
        DO 367 J=1,ND,1
          AVHDF2(I,J)=AVHDF2(I,J)*AC1
          BVHDF2(I,J)=BVHDF2(I,J)*AC1
          FVHDF2(I,J)=FVHDF2(I,J)*AC1
          GVHDF2(I,J)=GVHDF2(I,J)*AC1
          DVHDF1(I,J)=DVHDF1(I,J)*SQRT(AC1)/2.0
```

```
367     CONTINUE
365   CONTINUE
C
      N2=ND+ND
      DO 375 I=1,NN,1
        DO 377 J=1,NN,1
          B(I,J)=0.0
377     CONTINUE
375   CONTINUE
C
      DO 380 I=1,ND,1
        DO 383 J=1,ND,1
          B(I,J)=B(I,J)+AVH1(I+1)*UNIM(I,J)-AVHDF2(I,J)
     $          +FH(I+1)*UNIM(I,J)*AC1
          B(I+ND,J+ND)=B(I+ND,J+ND)+BVH1(I+1)*UNIM(I,J)-BVHDF2(I,J)
     $          +FL(I+1)*UNIM(I,J)*AC1
          B(I+N2,J+N2)=B(I+N2,J+N2)+GVH1(I+1)*UNIM(I,J)-GVHDF2(I,J)
     $          +FSO(I+1)*UNIM(I,J)*AC1
          B(I+ND,J)=B(I+ND,J)+CVH(I+1)*UNIM(I,J)+DVHDF1(I,J)
          B(I+N2,J)=B(I+N2,J)+SQRT(2.0)*CVH(I+1)*UNIM(I,J)
     $          -DVHDF1(I,J)/SQRT(2.0)
          B(I+N2,J+ND)=B(I+N2,J+ND)+FVH1(I+1)*UNIM(I,J)+FVHDF2(I,J)
     $          -SQRT(3.0/2.0)*DVHDF1(I,J)+GDELM(I+1)*UNIM(I,J)*AC1
383     CONTINUE
380   CONTINUE
      DO 392 I=1,ND,1
        DO 394 J=1,ND,1
          B(I,J+ND)=B(J+ND,I)
          B(I,J+N2)=B(J+N2,I)
          B(I+ND,J+N2)=B(J+N2,I+ND)
394     CONTINUE
392   CONTINUE
C
      DO 404 I=1,NN,1
        DO 408 J=1,NN,1
          HUPMAT(I,J)=B(I,J)/AC1
408     CONTINUE
404   CONTINUE
      RETURN
      END
C%%%%%%%%%%%%%%%%%%%%%%%%%%%%%%%%%%%%%%
C%% EIGENVLAUE AND EIGENVECTOR FOR CB %%
C%%%%%%%%%%%%%%%%%%%%%%%%%%%%%%%%%%%%%%
      SUBROUTINE CBEIGEN(CG2,CF,DIVISI,NCB)
      PARAMETER (NB=30,NW=39,ND=NB+NW+NB,NN=ND+ND+ND,NP=ND+2,NK=50)
      REAL*8 CG2(NP),CF(NP),DIVISI
      INTEGER   NB,NW,ND,NN,NP,NK
      REAL*8 HBAR,EV,MO,TLL,RATIO,AC1
C
      REAL*8 A(ND,ND),EVALD(ND),ED(ND),NORM,CEVECD(ND,ND),ASK(ND)
      INTEGER I,J
C
      HBAR=1.0546D-34
      EV=1.60219D-19
      MO=9.1095D-31
      TLL=DIVISI*(NB+NW+NB+1)
      RATIO=MO/HEAR*TLL/HEAR*TLL
      AC1=(ND+1)**2
C
      DO 102 I=1,ND,1
        DO 104 J=1,ND,1
```

```
          A(I,J)=0.0
104     CONTINUE
102   CONTINUE
C
      CALL CBMAT(CG2,CF,ACI,A)
C
      CALL TRED2(A,ND,ND,EVALD,ED)
      CALL TQLI(EVALD,ED,ND,ND,A)
      CALL EIGSRT(EVALD,A,ND,ND)
C
      DO 222 I=ND,ND-NCB+1,-1
         NORM=0.0
         DO 225 J=1,ND,1
          NORM=NORM+A(J,I)*A(J,I)
225    CONTINUE
        DO 227 J=1,ND,1
          CEVECD(J,I)=A(J,I)/SQRT(NORM)
227     CONTINUE
222   CONTINUE
     DO 232 I=1,ND,1
       ASK(I)=EVALD(I)/RATIO*AC1/EV
232   CONTINUE
      BANDC1=ASK(ND)
C%%%%%%%% SAVING EIGENVALUES FOR CB %%%
          WRITE(*,239) (ASK(I),I=ND,ND-3,-1)
239       FORMAT(' CONDUCTION :',4F10.5)
          WRITE(10,240) (ASK(I),I=ND,ND-NCB+1,-1)
240       FORMAT(10F10.5)
C%%%%%%% SAVING EIGENVECTOR FOR CB %%%
      DO 242 I=1,ND,1
        WRITE(20,248) (CEVECD(I,J),J=ND,ND-NCB+1,-1)
248     FORMAT(10F10.5)
242   CONTINUE
C
      RETURN
      END
C%%%%%%%%%%%%%%%%%%%%%%%%%%%%%%%%%%%%%%%
C%% EIGENVLAUE ANDEIGENVECTOR FOR VB %%
C%%%%%%%%%%%%%%%%%%%%%%%%%%%%%%%%%%%%%%%
      SUBROUTINE VBEIGEN(NVB,DIVISI,
     $ AP1,AP2,AP3,AP4,AP5,AP6,GDELM,FH,FL,FSO)
      INTEGER   NB,NW,ND,NN,NP,NK
      PARAMETER (NB=30,NW=39,ND=NB+NW+NB,NN=ND+ND+ND,NP=ND+2,NK=50)
      REAL*8    DIVISI,B(NN,NN)
      REAL*8 AP1(NP),AP2(NP),AP3(NP),AP4(NP),AP5(NP),AP6(NP),GDELM(NP) !AL1
      REAL*8 FH(NP),FL(NP),FSO(NP)
      INTEGER   NVB
      INTEGER II,I,J,NLAST
      REAL*8   KPINT,KP,NORM,HBAR,AUNG,EV,MO,TLL,RATIO,AC1
      REAL*8   EVALN(NN),EN(NN),ASKN(NN),HEVEC(NN,NN)
      REAL*8   HUPMAT(NN,NN)
C
      AUNG=1.0D-10
      HBAR=1.0546D-34
      MO=9.1095D-31
      EV=1.60219D-19
      NLAST=50
      KPINT=0.01/AUNG
C
      TLL=DIVISI*(NB+NW+NB+1)   ! TOTAL LENGTH OF QW
      RATIO=MO/HBAR*TLL/HBAR*TLL
```

```
      AC1=(ND+1)**2
C
      DO 1000 II=1,NLAST   ! START
       KP=KPINT*(II-1)
C%%%%%%%%%%% DEFINE MATRIX %%%%%%%%%
      CALL VBMAT(KP,DIVISI,AP1,AP2,AP3,AP4,AP5,AP6,
     $ GDELM,FH,FL,FSO,HUPMAT)
C
      DO 402 I=1,NN,1
        DO 404 J=1,NN,1
           B(I,J)=HUPMAT(I,J)
404     CONTINUE
402   CONTINUE
C
      DO 653 I=1,NN,1
         EN(I)=0.0
         EVALN(I)=0.0
         ASKN(I)=0.0
653   CONTINUE
C
      CALL TRED2(B,NN,NN,EVALN,EN)
      CALL TQLI(EVALN,EN,NN,NN,B)
      CALL EIGSRT(EVALN,B,NN,NN)
C=== NORMALIZATION OF EIGENVECTOR =====
      DO 712 I=NN,NN-NVB+1,-1
         NORM=0.0
         DO 715 J=1,NN,1
          NORM=NORM+B(J,I)*B(J,I)
715      CONTINUE
         DO 717 J=1,NN,1
          HEVEC(J,I)=B(J,I)/SQRT(NORM)
717      CONTINUE
712   CONTINUE
C
      DO 720 I=1,NN,1
          ASKN(I)=EVALN(I)/RATIO*AC1/EV  ! EV UNIT
720   CONTINUE
C=== SAVING EIGENVALUES ======
      WRITE(*,727) KP*AUNG,(-ASKN(I),I=NN,NN-NVB+1,-1)
727   FORMAT(20F10.5)
     WRITE(50,757) (-ASKN(I),I=NN,NN-NVB+1,-1)
     WRITE(80,757) KP*AUNG,(-ASKN(I),I=NN,NN-NVB+1,-1)
757   FORMAT(20F10.5)
C
1000  CONTINUE
C
      RETURN
      END
C%%%%%%%%%%%%%%%%%%%%%%%%%%%%%%%%%%%%%%%%%%%%%%
C%% SUBROUTINES FOR EIGENVALUES     %%%
C%%%%%%%%%%%%%%%%%%%%%%%%%%%%%%%%%%%%%%%%%%%%%%
C (C) COPR. 1986-92 NUMERICAL RECIPES SOFTWARE 00.
      SUBROUTINE TQLI(D,E,N,NP,Z)
      INTEGER N,NP
      REAL*8 D(NP),E(NP),Z(NP,NP)
      INTEGER I,ITER,K,L,M
      REAL*8 B,C,DD,F,G,P,R,S,PYTHAG
      DO 11 I=2,N
       E(I-1)=E(I)
11    CONTINUE
      E(N)=0.
```

```
      DO 15 L=1,N
        ITER=0
1       DO 12 M=L,N-1
          DD=ABS(D(M))+ABS(D(M+1))
          IF (ABS(E(M))+DD.EQ.DD) GOTO 2
12      CONTINUE
        M=N
2       IF(M.NE.L)THEN
          IF(ITER.EQ.300) PAUSE 'TOO MANY ITERATIONS IN TALI'
          ITER=ITER+1
          G=(D(L+1)-D(L))/(2.*E(L))
          R=PYTHAG(G,1.0D0)
          G=D(M)-D(L)+E(L)/(G+SIGN(R,G))
          S=1.
          C=1.
          P=0.
          DO 14 I=M-1,L,-1
            F=S*E(I)
            B=C*E(I)
            R=PYTHAG(F,G)
            E(I+1)=R
            IF(R.EQ.0.)THEN
              D(I+1)=D(I+1)-P
              E(M)=0.
              GOTO 1
            ENDIF
            S=F/R
            C=G/R
            G=D(I+1)-P
            R=(D(I)-G)*S+2.*C*B
            P=S*R
            D(I+1)=G+P
            G=C*R-B
C     OMIT LINES FROM HERE ...
            DO 13 K=1,N
              F=Z(K,I+1)
              Z(K,I+1)=S*Z(K,I)+C*F
              Z(K,I)=C*Z(K,I)-S*F
13          CONTINUE
C     ... TO HERE WHEN FINDING ONLY EIGENVALUES.
14        CONTINUE
          D(L)=D(L)-P
          E(L)=G
          E(M)=0.
          GOTO 1
        ENDIF
15    CONTINUE
      RETURN
      END
C
      SUBROUTINE TRED2(A,N,NP,D,E)
      INTEGER N,NP
      REAL*8 A(NP,NP),D(NP),E(NP)
      INTEGER I,J,K,L
      REAL*8 F,G,H,HH,SCALE
      DO 18 I=N,2,-1
        L=I-1
        H=0.
        SCALE=0.
        IF(L.GT.1)THEN
          DO 11 K=1,L
```

```
            SCALE=SCALE+ABS(A(I,K))
11        CONTINUE
          IF(SCALE.EQ.O.)THEN
            E(I)=A(I,L)
          ELSE
            DO 12 K=1,L
              A(I,K)=A(I,K)/SCALE
              H=H+A(I,K)**2
12          CONTINUE
            F=A(I,L)
            G=-SIGN(SQRT(H),F)
            E(I)=SCALE*G
            H=H-F*G
            A(I,L)=F-G
            F=0.
            DO 15 J=1,L
C     OMIT FOLLOWING LINE IF FINDING ONLY EIGENVALUES
              A(J,I)=A(I,J)/H
              G=O.
              DO 13 K=1,J
                G=G+A(J,K)*A(I,K)
13            CONTINUE
              DO 14 K=J+1,L
                G=G+A(K,J)*A(I,K)
14            CONTINUE
              E(J)=G/H
              F=F+E(J)*A(I,J)
15          CONTINUE
            HH=F/(H+H)
            DO 17 J=1,L
              F=A(I,J)
              G=E(J)-HH*F
              E(J)=G
              DO 16 K=1,J
                A(J,K)=A(J,K)-F*E(K)-G*A(I,K)
16            CONTINUE
17          CONTINUE
          ENDIF
        ELSE
          E(I)=A(I,L)
        ENDIF
        D(I)=H
18    CONTINUE
C     OMIT FOLLOWING LINE IF FINDING ONLY EIGENVALUES.
      D(1)=O.
      E(1)=O.
      DO 24 I=1,N
C     DELETE LINES FROM HERE ...
        L=I-1
        IF(D(I).NE.O.)THEN
          DO 22 J=1,L
            G=O.
            DO 19 K=1,L
              G=G+A(I,K)*A(K,J)
19          CONTINUE
            DO 21 K=1,L
              A(K,J)=A(K,J)-G*A(K,I)
21          CONTINUE
22        CONTINUE
        ENDIF
C     ...TO HERE WHEN FINDING ONLY ETGENVALUES.
```

```
          D(I)=A(I,I)
C       ALSO DELETE LINES FROM HERE...
          A(I,I)=1.
          DO 23 J=1,L
            A(I,J)=0.
            A(J,I)=0.
23        CONTINUE
C       ...TO HERE WHEN FINDING ONLY EIGENVALUES.
24      CONTINUE
        RETURN
        END
C  (C) COPR. 1986-92 NUMERICAL RECIPES SOFTWARE 00.
        FUNCTION PYTHAG(A,B)
        REAL*8 A,B,PYTHAG
        REAL*8 ABSA,ABSB
        ABSA=ABS(A)
        ABSB=ABS(B)
        IF(ABSA.GT.ABSB)THEN
          PYTHAG=ABSA*SQRT(1.+(ABSB/ABSA)**2)
        ELSE
          IF(ABSB.EQ.0.)THEN
            PYTHAG=O.
          ELSE
            PYTHAG=ABSB*SQRT(1.+(ABSA/ABSB)**2)
          ENDIF
        ENDIF
        RETURN
        END
C  (C) COPR. 1986-92 NUMERICAL RECIPES SOFTWARE 00.
        SUBROUTINE EIGSRT(D,V,N,NP)
        INTEGER N,NP
        REAL*8  D(NP),V(NP,NP)
        INTEGER I,J,K
        REAL*8  P
        DO 13 I=1,N-1
          K=I
          P=D(I)
          DO 11 J=I+1,N
            IF(D(J).GE.P)THEN
              K=J
              P=D(J)
            ENDIF
11        CONTINUE
          IF(K.NE.I)THEN
            D(K)=D(I)
            D(I)=P
            DO 12 J=1,N
              P=V(J,I)
              V(J,I)=V(J,K)
              V(J,K)=P
12          CONTINUE
          ENDIF
13      CONTINUE
        RETURN
        END
C%%%%%%%%%%%%%%%%%%%%%%%%%%%%%%%%%%%%%%
C%% SUBROUTINES FOR QUADRATURE      %%%
C%%%%%%%%%%%%%%%%%%%%%%%%%%%%%%%%%%%%%%
C  (C) COPR. 1986-92 NUMERICAL RECIPES SOFTWARE 00.
        SUBROUTINE GAULEG(X1,X2,X,W,N)
        INTEGER N
```

```
      REAL*8 X1,X2,X(N),W(N)
C      REAL*8 EPS
      DOUBLE PRECISION EPS
      PARAMETER (EPS=3.OD-14)
      INTEGER I,J,M
      DOUBLE PRECISION P1,P2,P3,PP,XR,XM,Z,Z1
C
      M=(N+1)/2
         XM=0.5D0*(X2+X1)  ! ONLY HALF OF THEM BECAUSE OF THEIR SYMMETRY
         XR=0.5D0* (X2-X1)
         DO 12 I=1,M
          Z=COS(3.141592654D0*(I-0.25D0)/(N+0.5D0))
1     CONTINUE
      P1=1.ODO
      P2=0.ODO
      DO 11 J=1,N
         P3=P2
         P2=P1
         P1=((2.ODO*J-1.ODO)*Z*P2-(J-1.ODO)*P3)/J
11    CONTINUE
      PP=N*(Z*P1-P2)/(Z*Z-1.ODO)
      Z1=Z
      Z=Z1-P1/PP
      IF (ABS(Z-Z1) .GT. EPS) GOTO 1
         X(I)=XM-XR*Z
         X(N+1-I)=XM+XR*Z
         W(I)=2.ODO*XR/((1.ODO-Z*Z)*PP*PP)
         W(N+1-I)=W(I)
12    CONTINUE
      RETURN
      END
```

PROBLEMS

1. Let $\underline{V}(t)$ be the projected evolution operator of the system defined as follows:

$$\underline{V}(t)=\underline{V}\exp\left\{-i\int_0^t d\tau \underline{Q}L_s(\tau)\underline{Q}\right\} \tag{5.243}$$

then, prove that

$$\underline{V}(t)=\underline{P}+\underline{Q}\underline{U}(t)\underline{Q}. \tag{5.244}$$

2. Prove that projected operators $\underline{H}(t,\tau)$ and $\underline{S}(t,\tau)$ satisfy $\underline{H}(t,\tau)=\underline{V}(t)\underline{S}(t,\tau)\underline{V}^{-1}(\tau)$. Use the following relation with the initial condition $\underline{H}(t,\tau)=1$:

$$\frac{d}{dt}\underline{H}(t,\tau)=-i\underline{Q}L_T(t)\underline{Q}\underline{H}(t,\tau). \tag{5.245}$$

3. Prove that evolution operators $\underline{G}(t,\tau)$ and $\underline{R}(t,\tau)$ satisfy $\underline{G}(t,\tau)=\underline{U}(\tau)\underline{R}(t,\tau)\underline{U}^{-1}(t)$.

4. Let $\underline{U}_{ext}(t)$ be the evolution operator of the system in the interaction picture such that $\underline{U}(t)=\underline{U}_0(t)\underline{U}_{ext}(t)$, where $\underline{U}_0(t)=\underline{T}exp\{-i\int_0^t dsL_0(s)\}$ is the unperturbed evolution operator of the system. Then, prove that $\underline{U}_{ext}(t)=\underline{T}exp\{-i\int_0^t \underline{U}_0^{-1}(t)L_{ext}(s)\underline{U}_0(s)ds\}$.

5. Using an FDM for a Schrödinger equation given by Equation (5.127), calculate the quantized subband energies for the conduction band of a GaAs/Al$_x$Ga$_{1-x}$As ($x=0.3$) QW structure. Assume that the parameters for Al$_x$Ga$_{1-x}$As are

$$m_c^*(x)=(0.0665+0.083x), E_g(x)=(1.424+1.247x)(eV).$$

Also, use $\mathrm{L}_w=50$ and $\mathrm{L}_b=100$ Å for well and barrier widths.

6. Calculate the first three quantized subband energies in the valence band for GaAs/Al$_x$Ga$_{1-x}$As QW ($x=0.3$) using the 3×3 Hamiltonian given by Equation (5.136) and the program in the Appendix. The parameters are given in the program.

7. Derive the band edge energies Equation (5.189) of the HH, LH, and split-off bands using Hamiltonian equation (5.187).

REFERENCES

[1] M. O. Scully and M. S. Zubairy, *Quantum Optics*. Cambridge: Cambridge University Press, 1997.

[2] D. Ahn, "Theory of non-Markovian optical gain in quantum-well lasers," *Prog. Quantum Electron.*, 21, p. 249, 1997.

[3] D. Ahn, "Theory of non-Markovian gain in strained-layer quantum-well lasers with many-body effects," *IEEE J. Quantum Electron.*, 32, p. 960, 1996.

[4] W. W. Chow, S. W. Koch, and M. Sergent III, *Semiconductor-Laser Physics*. Berlin: Springer-Verlag, 1994.

[5] S. H. Park, S. L. Chuang, and D. Ahn, "Intraband relaxation time effects on non-Markovian gain with many-body effects and comparison with experiment," *Semicond. Sci. Technol.*, 15, p. 203, 2000.

[6] S. H. Park, D. Ahn, and Y. T. Lee, "Finite element analysis of valence band structures in quantum wires," *J. Appl. Phys.*, 96, p. 2055, 2004.

[7] S. C. Jain, M. Willander, and R. Van Overstraeten, *Compound Semiconductors Strained Layers and Devices*. London: Kluwer Academic Publishers, 2000.

[8] D. A. Faux, J. R. Dowries, and E. P. O'Reilly, "Analytic solutions for strain distributions in quantum-wire structures," *J. Appl. Phys.*, 82, p. 3754, 1997.

[9] J. R. Downes, D. A. Faux, and E. P. O'Reilly, "A simple method for calculating strain distributions in quantum dot structures," *J. Appl. Phys.*, 81, p. 6700, 1997.

[10] D. M. Gvozdić, N. M. Nenadović, and A. Schlachetzki, "Gain and threshold-current calculation of V-groove quantum-wire InGaAs-InP laser," *IEEE J. Quantum Electron.*, 38, p. 1565, 2002.

[11] H. Jiang and J. Singh, "Self-assembled semiconductor structures: electronic and optoelectronic properties," *IEEE J. Quantum Electron.*, 34, p. 1188, 1998.

INDEX